Discard

D0467918

State of the World 2001

Other Norton/Worldwatch Books

State of the World 1984 through *2000* (an annual report on progress toward a sustainable society)
Lester R. Brown et al.

Vital Signs 1992 through *2000* (an annual report on the environmental trends that are shaping our future)
Lester R. Brown et al.

Saving the Planet
Lester R. Brown
Christopher Flavin
Sandra Postel

How Much is Enough?
Alan Thein Durning

Last Oasis
Sandra Postel

Full House
Lester R. Brown
Hal Kane

Who Will Feed China?
Lester R. Brown

Tough Choices
Lester R. Brown

Fighting for Survival
Michael Renner

The Natural Wealth of Nations
David Malin Roodman

Life Out of Bounds
Chris Bright

Beyond Malthus
Lester R. Brown
Gary Gardner
Brian Halweil

Pillar of Sand
Sandra Postel

Vanishing Borders
Hilary French

State of the World 2001

A Worldwatch Institute Report on Progress Toward a Sustainable Society

Lester R. Brown
Christopher Flavin
Hilary French

Janet N. Abramovitz
Seth Dunn
Gary Gardner
Lisa Mastny
Ashley Mattoon
David Roodman
Payal Sampat
Molly O. Sheehan

Linda Starke, Editor

W · W · NORTON & COMPANY
NEW YORK LONDON

The text of this book is composed in Galliard, with the display set in Franklin Gothic and Gill Sans. Book design by Elizabeth Doherty; composition by Worldwatch Institute; manufacturing by the Haddon Craftsmen, Inc.

First Edition

ISBN 0-393-04866-7
ISBN 0-393-32082-0 (pbk)

W. W. Norton & Company, Inc., 500 Fifth Avenue, New York, N.Y. 10110
www.wwnorton.com

W. W. Norton & Company Ltd., 10 Coptic Street, London WC1A IPU

1 2 3 4 5 6 7 8 9 0

⊛ This book is printed on recycled paper.

Worldwatch Institute Board of Directors

Worldwatch Institute Staff

Worldwatch Institute Management Team

Acknowledgments

Every member of the talented Worldwatch staff—along with legions of friends and colleagues from outside the Institute—had a hand in producing this eighteenth edition of *State of the World*. Writers, editors, marketers, administrative support, librarians, interns, reviewers, and funders all deserve our sincere thanks.

We begin by acknowledging the foundation community, whose faithful support sustains and encourages us. The John D. and Catherine T. MacArthur Foundation and the United Nations Population Fund made grants specifically for *State of the World*, while a host of other generous donors funded the Institute's work overall: the Compton Foundation, the Geraldine R. Dodge Foundation, the Ford Foundation, the Richard and Rhoda Goldman Fund, the William and Flora Hewlett Foundation, the W. Alton Jones Foundation, the Charles Stewart Mott Foundation, the Curtis and Edith Munson Foundation, the David and Lucile Packard Foundation, the Summit Foundation, the Turner Foundation, the Wallace Genetic Foundation, the Weeden Foundation, the Winslow Foundation, and the Wallace Global Fund.

We are also grateful to the growing numbers of individual donors who support our work, including the 3,000-strong Friends of Worldwatch—our enthusiastic donors who are strongly committed to Worldwatch and its efforts to contribute to a sustainable world. And we are indebted to the Institute's Council of Sponsors, including Tom and Cathy Crain, Roger and Vicki Sant, Robert Wallace and Raisa Scriabine, and Eckart Wintzen, for their strong expression of confidence in our work through their generous annual contributions of $50,000 or more.

In this year of transition, we are particularly grateful to Board members Tom and Cathy Crain and James and Deanna Dehlsen, who provided leadership gifts of $250,000 each to support the Institute's Second Generation Fund, a general support fund that is helping the Institute in its capacity building efforts and in launching its second generation initiatives. We also want to thank the entire Board of Directors for contributing additional support for the Second Generation Fund.

The Institute was fortunate in 2000 to recruit an unusually bright and productive group of research interns who tenaciously tracked down data and other vital information from all over the world. Bryan

ACKNOWLEDGMENTS

Mignone ably compiled and analyzed statistical data for Chapter 5; Mike Montag produced much of the extensive documentation for data-rich Chapters 6 and 7; Ann Hwang applied her expertise in chemistry, and Danielle Nierenberg and Jennifer Silva worked conscientiously on Chapter 2; and David Ruppert cheerfully lent his abundant energy to the research efforts for Chapters 1, 4, and 10. Keenly intelligent and good-natured, each was a stimulating addition to our summer staff.

The interns' information-gathering process is augmented by the Institute's library staff. Research librarian Lori Brown, office assistant Jonathan Guzman, and volunteer assistant librarian Maya Johnson track down books and articles with remarkable efficiency, and help researchers to manage the information from the myriad newspapers, periodicals, reports, and books that arrive at the Institute each day. We are grateful for their patience with last-minute rush requests.

Once the research and initial writing are done, draft chapters are reviewed to ensure that findings are accurate and clearly communicated. At the 2000 staff review meeting in September, authors and interns were joined by *World Watch* magazine staffers Ed Ayres and Curtis Runyan, Worldwatch alumnus John Young, and researchers Brian Halweil and Michael Renner for a day of reflection on the state of the chapters. The authors are grateful for the excellent advice they received on how to improve the drafts. They also appreciate the inspired advice and support of their research colleagues Chris Bright and Anne Platt McGinn, who were working on other projects this year.

Experts from outside the Institute also read portions of the manuscript with a critical eye. Typically these are global authorities on our chapter topics, whose generous gifts of time and expertise we greatly appreciate. For their review comments, we are particularly grateful to Eduardo Athayde, Jesse Ausubel, Duncan Brack, Cynthia Carey, Jennifer Clapp, James P. Collins, Susan Cutter, Joseph Domagalski, Cheryl Goodman, Robert Hamilton, Gavin Hayman, Ailsa Holloway, Craig Hoover, David Hunter, Jorge Illueca, Lee Kimball, Adrian Lawrence, Karen R. Lips, Doug McKenzie-Mohr, Caroline Michellier, Larry Minear, Mary Fran Myers, Joan Nelson, Thomas Nolan, James O'Meara, Michael Penders, Deike Peters, Sue Pfiffner, Roger Pielke, Jr., Jamie K. Reaser, Forest Reinhardt, Christian Suter, Louis S. Thompson, David B. Wake, Jonathan Walter, Carol Welch, Angelika Wertz, Gilbert White, and John Williamson. As always, we benefited this year from the advice and ideas of Bill Mansfield, who is active in the international environmental policy arena.

Linda Starke and Liz Doherty shepherd the reworked chapters in the home stretch of production. Linda massaged the work of nearly a dozen authors into a harmonious whole in this, her fortieth Worldwatch Book. Liz gives our research a strong aesthetic appeal through artful formatting of chapters and design of figures and tables. Ritch Pope finalizes the production by preparing the index.

Our communications staff works on several fronts to ensure that the *State of the World* message circulates widely beyond our Washington offices. Early in the research process, Mary Caron helped researchers to think about media messages that might emerge from their work. In her work on press outreach, Mary was ably supported by the dependable efforts of Liz Hopper. When Liz moved to Germany in September, Niki Clark confidently assumed responsibility for backing up the communications team. Denise Warden also joined us this year, bringing her creativity and expertise to our marketing efforts. Dick Bell oversees these operations as well as our communications team, including the Institute's expanding Web site, which is developed by our superb Webmaster and computer systems administrator, Christine Stearn. Jason Yu, our first computer intern, was instrumental in revamping the Worldwatch bookstore on our Web site. Finally, the work of the communications team is complemented by the friendly customer service and order fulfillment provided by Millicent Johnson, Joseph Gravely, and Sharon Lapier.

All these efforts would be impossible without strong administrative backing. Reah Janise Kauffman assists board chairman Lester Brown with his time-consuming travel schedule and his many writings, and works to ensure that *State of the World* and other Institute publications are extensively translated and published outside the United States. Janet Larsen, who joined the Institute this year, provides invaluable research assistance for Lester's work. Barbara Fallin ably handles the Institute's finances and keeps the Institute well stocked and the staff well fed. Suzanne Clift cheerfully backs up the work of our president and vice president for research, and helps researchers with travel and other logistical matters. Mary Redfern, meanwhile, ably supported the Institute's fundraising activities, working closely with our foundation supporters and individual donors.

We give special thanks to our overseas colleagues for their extraordinary efforts to spread the message of environmental sustainability. *State of the World* is published in some 30 languages—in 2000, we added Estonian, Portuguese, and Ukrainian editions—thanks largely to the dedication of a host of publishers, civil society groups, and individuals who translate and publish our research. For a list of these publishers, see <www.worldwatch.org/foreign/index.html>. We would particularly like to acknowledge the help we receive from Magnar Norderhaug in Norway, Hamid Taravaty and Farsaneh Bahar in Iran, Soki Oda in Japan, Gianfranco Bologna in Italy, Eduardo Athayde in Brazil, and Jonathan Sinclair Wilson in England.

Our long-standing relationship with our publisher, W.W. Norton & Company, remains strong thanks to the production and promotion efforts of our colleagues there: Amy Cherry, Andrew Marasia, and Lucinda Bartley. As always, Norton continues to give the Institute's publications broad exposure, especially in college and university bookstores around the United States.

Finally, we welcome into the extended Worldwatch family Joseph Hugh McGinn III, the newborn son of researcher Anne Platt McGinn and her husband, Joe. Jack is a source of delight to all of us on his visits to the office—and a strong source of motivation for Anne's work this year on eliminating persistent organic pollutants. It is our fervent hope that Jack and his generation will inherit a world well on its way to sustainability.

Lester R. Brown, Christopher Flavin, and Hilary French

Contents

CONTENTS

List of Tables and Figures

Tables

1 Rich Planet, Poor Planet

2 Uncovering Groundwater Pollution

3 Eradicating Hunger: A Growing Challenge

4 Deciphering Amphibian Declines

6 Making Better Transportation Choices

8 Ending the Debt Crisis

9 Controlling International Environmental Crime

Figures

1 Rich Planet, Poor Planet

3 Eradicating Hunger: A Growing Challenge

5 Decarbonizing the Energy Economy

Foreword

Few places on Earth are as remote or pristine as Greenland. Lying just south of the North Pole between Scandinavia and northern Canada, Greenland is largely uninhabited and covered year-round by a massive ice sheet that has built up through millions of years of snowfall. With minimal human habitation and few automobiles, farms, or factories, Greenland would seem to be insulated from the kind of ecological damage that is so widespread in other parts of the world.

It came as a surprise, therefore, when in 2000 a team of U.S. scientists discovered that the vast Greenland ice sheet is melting. The ice, which covers an area greater than that of Mexico, contains enough water to fill the Mediterranean Sea two thirds full. It is estimated to be losing some 51 billion cubic meters of water a year, a rate that already approaches the annual flow of the Nile River—accounting for 7 percent of the observed annual rise in global sea levels. An article in *Science* reports that if the entire Greenland ice sheet were to melt, sea level would rise 7 meters, or 23 feet.

Although the melting of Greenland's ice sheet received far less attention last year than such events as the dramatic U.S. election in November or the Sydney Olympic Games, its long-run human significance is far greater. Almost anywhere it is found on Earth, from Himalayan glaciers to Antarctica, ice is melting away—one of the most striking signs of the rising global temperatures that are caused by the burning of billions of tons of fossil fuels each year.

If fossil fuel combustion were to continue at these levels or higher throughout the new century—as it did during the last one—virtually every natural system and human economy would be at risk from rising seas, more severe storms, or more intense droughts. A major scientific review by the Intergovernmental Panel on Climate Change in October 2000 concluded that new evidence suggests the rate of warming in this century could be even greater than estimated by earlier studies.

Climate change is beginning to exacerbate the ecological decline that is provoked by other environmental stresses. One of the most worrisome examples of these dangerous synergies is found in the worldwide decline of many species of frogs, salamanders, and other *Amphibia* described by Ashley Mattoon in Chapter 4 of this year's book—a global phenomenon that has a multiplicity of underlying causes, ranging from climate change to introduced diseases.

And it is not just other species that are suffering. The immense human toll from the rising tide of natural disasters over the last decade, described by Janet Abramovitz in Chapter 7, is a powerful reminder that human activities can dramatically alter natural processes, with unpredictable and often catastrophic results. The economies of Central America, for example, still have not recovered from the devastation of Hurricane Mitch in late 1998.

An indication of the scale of these threats came in a report from scientists in Indonesia just as this year's *State of the World* was going to press: half of the once vast coral reefs that surrounded the world's largest archipelago have been destroyed—apparently due to a combination of deforestation, marine pollution, and rising temperatures. The decline of the reefs, which account for 14 percent of the world's corals, threatens hundreds of millions of dollars of revenue from fishing and tourism that are needed to recover from Indonesia's worst financial crisis in a generation. Maritime Affairs Minister Sarwono Kusmaatmadja said the government was so distracted by recent political and economic crises that it had failed to protect a resource that is key to Indonesia's future.

Human-induced environmental change far exceeds the natural pace of evolution, so our species finds itself in a tight race to adapt quickly enough to keep pace with the environmental problems we are creating with such abandon. We live in a period of epoch change, but can *we* change in time?

Human Natures, a new book by Stanford biologist Paul Ehrlich, argues that in response to rapidly changing circumstances, human nature is now evolving much more rapidly than our underlying genetic makeup is. Technologies, institutions, and even value systems are in rapid flux due to accelerated cultural evolution. In the final chapter of this *State of the World*, Gary Gardner examines what it will take to achieve the kind of change needed to create an environmentally sustainable society. His conclusion, after reviewing past patterns of change, is that concerted efforts could lead us to one of the most dramatic periods of transformation the world has ever seen.

These efforts will not succeed, however, unless attempts to address environmental problems are paralleled by equally hard work to deal with the world's social ills. Many societies have seen inequality and poverty rise in recent years. Increased investment in health and education, expanded access to credit, and greater social and political rights for women are among the keys to economic and environmental progress in the years ahead.

Many chapters of *State of the World 2001* focus on areas in which important change is already under way. For example, in Chapter 9, Hilary French and Lisa Mastny discuss initiatives to enhance the enforcement of the array of environmental treaties agreed to in recent decades. And in Chapter 6, Molly Sheehan examines the growing interest in innovative solutions to the transportation problems that plague many parts of the world today.

One of the biggest challenges in the new century is stabilizing Earth's climate, which will require a nearly complete transformation of the world's energy systems. This monumental task now appears just a bit less out of reach due to another icebreaking development just a few hundred kilometers to the east of Greenland. The tiny nation of Iceland has launched plans to make itself a pioneer developer of a twenty-first century energy system that relies on renewable energy rather than fossil fuels.

As described by Seth Dunn in Chapter 5, Iceland's goal is to use its abundant geothermal energy and hydropower to produce hydrogen from seawater, relying on the new fuel to run its factories, motor vehicles, and fishing fleet. The scheme, which is attracting investments from scores of companies, including multinationals such as DaimlerChrysler and Royal Dutch Shell, would make the island nation of 270,000 people the first to have the kind of environmentally sustainable energy system that may be found everywhere before the century is over. Hydrogen pioneer Bragi Arnason of the University of Iceland says, "Many people ask me how soon this will happen. I tell them, We are living at the beginning of the transition. You will see the end of it. And your children, they will live in this world."

Worldwatch Institute itself has entered a period of historic change. Led by our Board of Directors, the Institute launched its second generation of environmental leadership when Senior Vice President Christopher Flavin was selected to be the Institute's second President. The Institute's founding President, Lester Brown, has become Chairman of the Board, and former Board Chairman Andrew Rice is now Chairman of the Executive Committee.

The Institute's Board and staff are developing plans to further enhance Worldwatch's reach and effectiveness, building on the organization's strong global reputation and the talents of its staff to meet the needs of a new century. The original mission of the Worldwatch Institute—to provide the information and ideas needed to create an environmentally sustainable society—will remain at the heart of its new agenda. But the rapid evolution of the issues we study, and of the information economy we rely on for dissemination, demands fresh approaches.

In the coming year, we will launch a series of initiatives that are intended to allow Worldwatch to reach key decision-makers in the public and private sectors more systematically, while continuing to build a broad public audience. We will develop and disseminate practical solutions to global problems, using the new tools of the information economy to reach millions of users, thereby expanding the already sizable market for the Institute's publications. In order to achieve these goals, the Institute will also focus on internal capacity building in the years ahead, expanding its staff with an emphasis on strengthening our marketing and development capabilities.

During the past year, we have already begun to extend our presence on the Internet, making many of our publications available for sale electronically, and even posting audio recordings of Institute events on our Web site, www.worldwatch.org. Also, Lester Brown has begun writing a series of frequent four-page Worldwatch Issue Alerts that are posted on the Web site and disseminated via e-mail. In the months ahead, we plan to create an electronic, searchable library of Worldwatch Papers that will be available on the Web site.

Meanwhile, changes are under way in the release of *State of the World* itself. This year, for the first time, the report will be simultaneously launched in January in up to five countries and in as many as three languages: the United States, India, and the United Kingdom (English editions), Brazil, and South Korea. We appreciate the exceptional efforts of the publishers—Lawrence Surendra of Earthworm Books in India, Jonathan Sinclair Wilson of Earthscan Publications in the United Kingdom, Eduardo Athayde of the Mid-Atlantic Forest Open University in Brazil, and Yul Choi of the Korean Federation of

Environmental Movement in Korea—who made this possible.

Their work, combined with the extraordinary efforts of our many other international publishers, gives us all hope that our vision of a sustainable future will be realized. We were encouraged this past year when Ukrainian was added to the 30 or so languages in which *State of the World* has been published. Information on our international publishers can be found at www.worldwatch.org/foreign/index.html.

We hope you will find *State of the World* useful in your own efforts to understand and shape our fast-changing world. Please send us any comments on this edition or suggestions for future editions by letter, fax (202-296-7365), or e-mail (worldwatch@worldwatch.org).

Lester R. Brown
Christopher Flavin
Hilary French

Worldwatch Institute
1776 Massachusetts Ave., NW
Washington, DC 20036

December 2000

State of the World 2001

Chapter 1

Rich Planet, Poor Planet

Christopher Flavin

A visit to Brazil's tropical state of Bahia provides contrasting views of the state of the world at the dawn of the new millennium. Bahia's capital, Salvador, has a population of over 3 million and a thoroughly modern veneer. Its downtown is full of large office buildings and busy construction cranes, and its highways are crammed with sport utility vehicles. The state is also rich in natural resources: the wealth provided by gold and sugarcane made Salvador the obvious location for colonial Brazil's leading port and capital for two centuries.[1]

Once a backwater—slavery was not outlawed until the end of the nineteenth century, one of the last regions to ban this practice—Bahia's economy is now booming. The state has a prospering manufacturing sector and has become popular with many leading multinationals, including automobile companies that have put some of their most advanced factories there. The information economy is in a particularly competitive frenzy. Brazilian Internet service providers are connecting customers for free, and cell phones appear to be almost as common as they are in many European cities.

Scratch the surface, however, and another Bahia is still there. The large favelas that ring Salvador's outskirts are crowded with thousands of poor people who lack more than cell phones and computers: toilets, running water, and schoolbooks are among the basic services and products that are unavailable to many of Bahia's poor. Similar gaps can be seen in the low hills that run south of Salvador along Bahia's rugged coast: the collapse of many of the country's rich cacao farms due to a devastating pathogen called witches'-broom and a sharp decline in world chocolate prices have left thousands of farm workers jobless and unable to provide for their families.

Bahia's environmental condition is just as uneven. Considered by ecologists to be one of the world's biological "hot spots," the Atlantic Rain Forest covers more than

Units of measure throughout this book are metric unless common usage dictates otherwise.

2,000 kilometers of Brazil's subtropical coast. In 1993, biologists working in an area south of Salvador identified a world record 450 tree species in a single hectare. (A hectare of forest in the northeastern United States typically contains 10 species.) In the last decade, Bahia's political and business leaders have come to recognize the extraordinary richness of their biological heritage—wildlands are being protected, ecological research facilities are being set up, and ecotourist resorts are mushrooming. A sign at the airport even warns travelers that removing endemic species from the country is a felony.[2]

And yet, signs of destruction are everywhere: cattle ranches sprawl where the world's richest forests once stood; 93 percent of the Atlantic forest is already gone, and much of the remainder is fragmented into tiny plots. Pressure on these last bits of forest is enormous—both from powerful landowners and corporations eager to sell forest and agricultural products in the global marketplace, and from poor families desperately seeking a living.[3]

This picture of Bahia in the year 2000 is replicated at scores of locations around the globe. It is the picture of a world undergoing extraordinarily rapid change amid huge and widening disparities. Unprecedented economic prosperity, the emergence of democratic institutions in many countries, and the near instantaneous flow of information and ideas throughout a newly interconnected world allow us to address challenges that have been neglected for decades: meeting the material needs of all 6 billion members of the human race, and restoring a sustainable balance between humanity and Earth's ecological systems.

This moment is historic, perhaps even evolutionary, in character. Tragically, it is not being seized. Despite a surge in economic growth in recent years and significant gains in health and education levels in many developing nations, the number of people who survive on less than $1 of income per day—the poverty threshold used by the World Bank—was 1.2 billion in 1998, almost unchanged since 1990. In some parts of the world, including sub-Saharan Africa, South Asia, and the former Soviet Union, the number living in poverty is substantially higher than the figures recorded a decade ago.[4]

The struggle to restore the planet's ecological health presents a similar picture: a number of small battles have been won, but the war itself is still being lost. Double-digit rates of growth in renewable energy markets, plus a two-year decline in global carbon emissions, for example, have failed to slow the rate of global climate change. Indeed, recent evidence, from the rapid melting of glaciers and the declining health of heat-sensitive coral reefs, suggests that climate change is accelerating. The same pattern can be seen in the increased commitment to protection of wild areas and biological diversity: new laws are being passed, consumers are demanding eco-friendly wood products, and eco-tourist resorts are sprouting almost as quickly as dot-com companies. But foresters and biologists report that this host of encouraging developments has not reversed the massive loss of forests or the greatest extinction crisis the world has seen in 65 million years.[5]

Long considered distinct issues, consigned to separate government agencies, ecological and social problems are in fact tightly interconnected and mutually reinforcing. The burden of dirty air and water and of decimated natural resources invariably falls on the disadvantaged. And the poor, in turn, are often compelled to tear town down the last nearby tree or pollute the local stream in order to survive. Solving one problem without addressing the other is simply not feasible. In fact, poverty and environmental decline are both embedded deeply in today's economic systems. Neither is a peripheral problem that can be considered in isolation. What is needed is what Eduardo Athayde, General Director of Bahia's Atlantic Forest Open University, calls "econology," a synthesis of ecology, sociology, and economics that can be used as the basis for creating an economy that is both socially and ecologically sustainable—the central challenge facing humanity as the new millennium begins.[6]

The challenge is made larger by the fact that it must be met simultaneously at national and global levels, requiring not only cooperation but partnership between North and South. Responsibility for the current health of the planet and its human inhabitants is shared unequally between rich and poor countries, but if these problems are to be resolved, the two groups of nations will need to bring their respective strengths and capabilities to bear. This will require a new form of globalization—one that goes beyond trade links and capital flows to strengthened political and social ties between governments and civil society.

A select group of large industrial and developing countries—a collection that can be called the E–9, given that they are key environmental as well as economic players—could have a central role in closing the North-South gap. Together, this group of countries accounts for 57 percent of the world's population and 80 percent of total economic output. (See Table 1–1.) This chapter uses data on these nine diverse countries and areas to illuminate key economic, social, and ecological trends. But this grouping has more than just analytical value. As argued at the end of the chapter, E–9 cooperation could be a key to achieving accelerated economic and environmental progress in the new century.[7]

Table 1–1. The E–9: A Population and Economic Profile

Country or Grouping	Population, 2000 (million)	Gross National Product, 1998 (billion dollars)
China	1,265	924
India	1,002	427
European Union[1]	375	8,312
United States	276	7,903
Indonesia	212	131
Brazil	170	768
Russia	145	332
Japan	127	4,089
South Africa	43	137

[1]Data for European Union do not include Luxembourg.

SOURCE: World Bank, *World Development Indicators 2000* (Washington, DC: 2000), 10–12; Population Reference Bureau, "2000 World Population Data Sheet," wall chart (Washington, DC: June 2000).

A Tale of Two Worlds

Halfway through the year 2000, two stories from the Philippines made headlines around the world. In June, a computer virus dubbed the "love bug" appeared almost simultaneously on every continent, crashing the computer systems of scores of multinational corporations and government offices, ranging from the U.S. Pentagon to the British Parliament. The estimated total cost of the resulting disruptions: $10 billion. Computer security experts and FBI agents quickly traced the diabolical love bug to a small Manila technical college and a 24-year-old student named Onel de Guzman. For computer experts, this may have been an indication of the vulnerability of the global Internet, but in the Philippines it quickly became a source of national pride. People took the love bug debacle as an encouraging sign that their developing nation was leapfrogging into the top ranks of the global economy's hottest sector.[8]

Economic successes and social failures are found side by side around the world in this supposed time of plenty.

Across town, a Manila neighborhood called the Promised Land was hit by a different kind of news a month later: more than 200 people were killed in a massive landslide and subsequent fire. Although this tragedy was precipitated by Typhoon Kai-Tak, it was anything but a natural disaster. The Promised Land, it turns out, is a combination garbage dump/shantytown that is home to 50,000 people, most of whom make their living by scavenging the food and materials discarded by Manila's growing middle class. When two days of heavy rain loosened the mountain of garbage, it came crashing down on hundreds of homes as well as the dump's electrical lines, starting a massive fire. Scores of Promised Land residents were buried, others were burned alive, and still more were poisoned by toxic chemicals released by the fire.[9]

Economic successes and social failures are now found side by side, not just in the Philippines, but around the world in this supposed time of plenty. The annual output of the world economy has grown from $31 trillion in 1990 to $42 trillion in 2000; by comparison, the total output of the world economy in 1950 was just $6.3 trillion. And in 2000, the growth of the world economy surged to a 4.7-percent annual rate, the highest in the last decade. This increase in economic activity has allowed billions of people to buy new refrigerators, televisions, and computers, and has created millions of jobs. Global telephone connections grew from 520 million in 1990 to 844 million in 1998 (an increase of 62 percent), and mobile phone subscribers went from 11 million to 319 million in that time (up 2,800 percent). The number of "host" computers, a measure of the Internet's expansion, grew from 376,000 in 1990 to 72,398,000 in 1999—an increase of 19,100 percent.[10]

The economic boom of the last decade has not been confined to the rich countries of the North. Much of the growth is occurring in the developing nations of Asia and Latin America, where economic reforms, lowered trade barriers, and a surge in foreign capital have fueled investment and consumption. Between 1990 and 1998, Brazil's economy grew 30 percent, India's expanded 60 percent, and China's mushroomed by a remarkable 130 percent. China now has the world's third largest economy (second if measured in terms of purchasing power parity), and a booming middle class who work in offices, eat fast food, watch color television, and surf the Internet. China alone now has 420 million radios, 344 million television sets, 24 million mobile phones, and 15 million computers.[11]

Still, the global economy remains tarnished by vast disparities. (See Table 1–2.) Gross national product (GNP) per person ranges from $32,350 in Japan to $4,630 in Brazil, $2,260 in Russia, and just $440 in India. Even when measured in purchasing power terms, GNP per person among these countries varies by a factor of 10. Per capita income has increased 3 percent annually in 40 countries since 1990, but more than 80 nations have per capita incomes that are lower than they were a decade ago. Within countries, the disparities are even more

Table 1–2. Economic Trends in E–9 Nations

Country	GNP per Person, 1998	Purchasing Power per Person, 1998	Population Earning Below $2 per Day, 1993–99[1]	Share of Income or Consumption: Lowest 20 percent, 1993–98[1]	Share of Income or Consumption: Highest 10 percent, 1993–98[1]
	(dollars)		(percent)	(percent)	
Japan	32,350	23,592	–	10.6	21.7
United States	29,240	29,240	–	5.2	30.5
Germany[2]	26,570	22,026	–	8.2	23.7
Brazil	4,630	6,460	17.4	2.5	47.6
South Africa	3,310	8,296	35.8	2.9	45.9
Russia	2,260	6,180	25.1	4.4	38.7
China	750	3,051	53.7	5.9	30.4
Indonesia	640	2,407	66.1	8.0	30.3
India	440	2,060	86.2	8.1	33.5

[1]Data are from a single year within the time frame. [2]Comparable data for European Union not available; Germany is most populous EU member.
SOURCE: World Bank, *World Development Indicators 2000* (Washington, DC: 2000), 10–12, 62–64, 66–68.

striking. In the United States, the top 10 percent of the population has six times the income of the lowest 20 percent; in Brazil, the ratio is 19 to 1. More than 10 percent of the people living in "rich" countries are still below the poverty line, and in many, inequality has grown over the last two decades.[12]

The boom in global consumption over the past decade has been accompanied by improvements in living standards in many countries and declines in others. The U.N. Development Programme estimates that the share of the world's population suffering from what it calls "low human development" fell from 20 percent in 1975 to 10 percent in 1997. Still, World Bank figures show that 2.8 billion people, nearly half the world's population, survive on an income of less than $2 per day, while a fifth of humanity, 1.2 billion people, live on less than $1 per day. An estimated 291 million sub-Saharan Africans—46 percent of the region's population—now live on less than $1 a day, while in South Asia, the figure is 522 million. This is a staggering number of people to enter the new century without the income needed to purchase basic necessities such as food, clean water, and health care.[13]

Worldwide, some 1.1 billion people are currently estimated to be malnourished. Most of these are poor people in rural areas who have insufficient land to grow the food they need, and not enough income to buy it from others. Many of these people live in countries with food surpluses, but while well-off farmers sell their products to middle-class consumers in distant nations, the proceeds have no benefit for millions of starving children. In some African countries, such as Kenya, Zambia, and Zimbabwe, as much as 40 percent of the population is malnourished.[14]

Roughly 1.2 billion people do not have access to clean water. In China, the portion that fall in this category is 10 percent (125 million people), in India it is 19 percent, and in South Africa, 30 percent. Toilets are even rarer in many countries: 33 percent of Brazil's population does not have one, nor does 49 percent of Indonesia's or 84 percent of India's.[15]

Polluted water is a major contributor to one of the largest disparities today's world faces: the health gap. Although infant mortality rates have dropped 25–50 percent in many countries in the past decade, they still stand at 43 per thousand live births in China and 70 per thousand in India. (See Table 1–3.) Much of the wide difference in this number around the world results from undernutrition and common infectious diseases that remain rampant in many poor countries. More intractable diseases such as cholera and tuberculosis are also becoming epidemic in many areas.

More alarming still is the fact that AIDS, which has been brought under control in some rich countries, is spreading rapidly in many developing nations. The crisis is particularly acute in southern Africa, which a decade ago had relatively low rates of infection. By 2000, HIV infection rates had reached a stunning 20 percent in South Africa, 25 percent in Zimbabwe, and 36 percent in Botswana. Decades of rising life expectancy are being reversed in a matter of years, as hundreds of thousands of young adults and children succumb to the disease. Health care budgets are being overwhelmed, and education undermined by the early deaths of many teachers. It is no accident that the countries most ravaged by AIDS are those with high rates of social disruption and limited government health services. In China, poor people who sell their blood in order to make ends meet are paying a high price in the form of HIV infection from contaminated needles. Ironically, in parts of Africa, it is those who are just emerging from poverty that are being hit the hardest—devastating a generation of

Table 1–3. Health Indicators in E–9 Nations

Country	Health Expenditures per Person, 1990–98[1]	Infant Mortality 1980	Infant Mortality 1998	Tuberculosis Incidence, 1997	HIV Prevalence Among Adults, 1997
	(dollars of purchasing power)	(per thousand live births)		(per 100,000)	(percent)
United States	4,121	8	4	7	0.76
Germany[2]	2,364	12	5	15	0.08
Japan	1,757	13	7	29	0.01
South Africa	571	42	31	394	12.91
Brazil	503	70	33	78	0.63
Russia	404	22	17	106	0.05
China	142	90	43	113	0.06
India	73	115	70	187	0.82
Indonesia	38	67	51	285	0.05

[1]Data are from the most recent year available. [2]Comparable data for European Union not available; Germany is most populous EU member.
SOURCE: World Bank, *World Development Indicators 2000* (Washington, DC: 2000), 90–92, 102–04, 106–08.

educated young workers, a cataclysm that may forestall the growth of an economically secure middle class.[16]

One of the key ingredients of economic progress is education, and on this front, the world is doing better than it was two decades ago. (See Table 1–4.) In India, the share of children in secondary school has risen from 41 percent to 60 percent; in China, it has gone from 63 to 70 percent; and in South Africa, from 62 to 95 percent. But even with these improvements, many countries are failing to invest adequately in their young people, who are unlikely to be able to participate in or benefit from today's most vibrant economic sectors, which demand not only basic literacy but often specialized training. Girls in particular are receiving inadequate education in many countries. Adult female illiteracy rates remain as high as 25 percent in China and 57 percent in India, levels that virtually guarantee a host of social and economic problems—and that make environmental threats more difficult to address.

Testing the Limits

When the Russian icebreaker *Yamal* reached the North Pole in July 2000, the scientists aboard were confronted with a strange sight: an expanse of open, calm water in place of the two or three meters of pack ice that is common to the region even at the height of summer. In the 91 years since Robert Peary and Matthew Henson reached the North Pole by dogsled in 1909, nothing like this had been reported. But human memory is the wrong scale on which to measure this development: scientists estimate that the last time the polar region was completely ice-free was 50 million years ago.[17]

The dynamic, shifting character of the Arctic ice pack suggests that the open water over the pole itself was, for now, a fleeting phenomenon. But recent scientific evidence confirms the underlying trend: Earth's frozen top is melting at an extraordinary rate. Submarine sonar measurements

Table 1–4. Education in E–9 Nations

	Adult Illiteracy Rate				Share of Children in Secondary School	
	Female		Male			
Country	1980	1998	1980	1998	1980	1997
	(percent)				(percent)	
Germany[1]	–	–	–	–	82	95
Japan	–	–	–	–	93	100
United States	–	–	–	–	94	96
Russia	2	1	1	0	98	88
Brazil	27	16	23	16	46	66
South Africa	25	16	22	15	62	95
Indonesia	40	20	21	9	42	56
China	48	25	22	9	63	70
India	74	57	45	33	41	60

[1]Comparable data for European Union not available; Germany is most populous EU member.
SOURCE: World Bank, World *Development Indicators 2000* (Washington, DC: 2000), 74–76, 82–84.

indicate a 40-percent decline in the average thickness of summer polar ice since the 1950s, far exceeding the rate of melting previously estimated. Based on these observations, scientists now estimate that by the middle of this century the Arctic could be ice-free in summer.[18]

Among the myriad signs of human-induced global climate change—fossil fuel combustion was recently estimated to have raised atmospheric concentrations of carbon dioxide to their highest levels in 20 million years—this one may be the most dramatic. In late 2000, the Intergovernmental Panel on Climate Change (IPCC), the scientific body that advises government negotiators, produced its latest report. It included the strongest consensus statement yet that societies' release of carbon dioxide and other greenhouse gases "contributed substantially to the observed warming over the last 50 years." By the end of the century, the IPCC concluded, temperatures could be 5 degrees Celsius higher than in 1990—an increase greater than the change in temperature between the last Ice Age and today.[19]

While the shipping industry is already beginning to view the Arctic meltdown as a potential short-term opportunity—perhaps cutting the transit distance between Europe and the Far East by as much as 5,000 kilometers—the full economic and ecological consequences would be far more extensive and hard to predict. Scientists have recently learned that Arctic ice is a key part of the "engine" that drives the powerful oceanic conveyor belt—the warm Gulf Stream—that provides northern Europe with the relatively temperate and stable climate that allowed European societies to flourish. Shutting it down could change the climate of Europe more than at any time since the last Ice Age. And because the Gulf Stream is a dominant feature in the oceanic circulation system, any major change in its course would have ripple effects globally. Moreover, with less ice to reflect the sun's rays, the warming of Earth that caused the ice to melt in the first place would accelerate.[20]

Some 10,000 kilometers south of the North Pole lies a very different environment—the world's tropical oceans and their abundant coral reefs, a biologically rich ecosystem that has been described as the rainforest of the ocean (65 percent of fish species are reef dwellers). One of the richest is the Belize Barrier Reef on the Yucatan Peninsula in the Caribbean, the site of a recent diving expedition by marine biologist Jonathan Kelsey and journalist Colin Woodard. What was intended to be an exciting exploration of the region's spectacular, multihued marine life turned out to be a disturbing disappointment: "Bright white boulders dotted the seascape in all directions, a sign of severe coral distress," Woodard reported. "A centuries-old stand of elkhorn coral as big as an elephant was now dead and smothered in a thick two-year growth of brown algae....Across the plane, the corals appeared to be dying."[21]

Around the world, from the Caribbean to the Indian Ocean and Australia's Great Barrier Reef, similar observations have been reported in the past two years. Coral polyps are temperature-sensitive, and often sicken or die when ocean surface temperatures rise even slightly. The temporary warming of ocean waters that accompanies El Niño anomalies in the Pacific is generally hard on coral reefs, but the 1998 El Niño was something different: reports of sick coral were soon being filed by marine biologists around the world, who estimated that more than one quarter of the coral reefs were sick or dying. In some areas of the Pacific, the figure is as high as 90 percent. For many small island nations, the loss in income from fishing and tourism, as well as increased storm damage from the loss of coral reefs, may be enough to trigger the collapse of their economies.[22]

Following another serious episode of coral bleaching just a decade earlier, this recent epidemic of coral disease is another strong indication that the world is warming. But it is more than that: coral reefs are sort of a marine version of the famous canary in a coalmine—vulnerable to many environmental stresses that now run rampant, including urban sewage, agricultural runoff, and the sedimentation that comes from deforestation. The recent decimation of coral reefs and the growing frequency of such events suggest that the world's ecological balance has been profoundly disturbed.

Whether it is Arctic ice, tropical corals, oceanic fisheries, or old-growth forests, the forces driving ecological destruction are varied, complex, and often dangerously synergistic. Population is one factor. The nearly fourfold expansion in human numbers over the past century has drastically increased demands on natural resources. The combination of population growth and deforestation, for example, has cut the number of hectares of forest per person in half since 1960—increasing pressures on remaining forests and encouraging a rapid expansion in plantation forestry. Demand for water, energy, food, and materials have all been driven up by the unprecedented expansion in human numbers. And increasingly, it is in the world's developing countries that natural systems are declining the fastest and people face the most serious environmentally related stresses. (See Table 1–5.)[23]

Population growth alone could not have tested environmental limits this severely, however. The pressures it imposes have been magnified by rising consumption levels as each individual demands more from

Table 1–5. Ecological Health of E–9 Nations

Country	Share of Land Area That is Forested, 1995[1]	Change of Average Annual Deforestation, 1990–95	Share of Mammals Threatened, 1996	Share of Flowering Plants Threatened, 1997	Share of Land Area Nationally Protected, 1996
			(percent)		
Russia	22	0	11.5	–	3.1
Brazil	16	0.5	18.0	2.4	4.2
United States	6	– 0.3	8.2	4.0	13.4
China	4	0.1	19.0	1.0	6.4
Germany[2]	3	0	10.5	0.5	27.0
Indonesia	3	1	29.4	0.9	10.6
India	2	0	23.7	7.7	4.8
Japan	0.7	0.1	22.0	12.7	6.8
South Africa	0.2	0.2	13.4	9.5	5.4

[1]Data may refer to earlier years. [2]Comparable data for European Union not available; Germany is most populous EU member.

SOURCE: World Bank, *World Development Indicators 2000* (Washington, DC: 2000), 126–28.

nature. Meat-based diets and automobile-centered transportation systems are among the highly consumptive practices first adopted by the billion or so people living in rich countries, and now proliferating quickly in many parts of the developing world. Meanwhile, government regulations and emission control technology have lagged well behind the pace of adoption in richer countries. As a consequence, the most serious air pollution is now found in cities such as Jakarta and São Paulo. (See Table 1–6.)

The combination of population growth and increased consumption is projected to cause the number of people living in water-deficit counties to jump from 505 million to over 2.4 billion in the next 25 years. In countries that already face severe water shortages, such as Egypt, India, and Iran, water scarcity is likely to require large-scale food imports. In northern China, the water table under Beijing fell 2.5 meters in 1999, bringing the total decline since 1965 to 59 meters. Similarly, surging demand for oil—particularly in North America and East Asia—contributed in the year 2000 to the highest sustained oil prices the world has seen since the early 1980s. Beyond the proximate political reasons for higher oil prices, the underlying cause is clear: world oil production is nearing its eventual all-time peak, and producers are struggling to meet the combined demands of first-time car owners in China and those who are buying the large SUVs now found in nearly half of U.S. garages.[24]

While the last decade's growth in affluence contributed to many environmental problems, keeping people poor is not the answer—either morally or practically. In impoverished areas around the world, the rural poor are pushed onto marginal, often hilly lands, from which they must hunt bushmeat, harvest trees, or clear land for pasture or crops in order to survive. A 2000 study on the root causes of biodiversity loss, sponsored by the World Wide Fund for Nature (WWF), concluded that together with other forces, poverty often plays a major role.[25]

Table 1–6. Air Pollution in E–9 Nations

Country	Sulfur Dioxide, 1995	Suspended Particulates, 1995	Nitrogen Dioxide, 1995
		(micrograms per cubic meter)	
Germany (Frankfurt)[1]	11	36	45
Japan (Tokyo)	18	49	68
South Africa (Cape Town)	21	–	72
United States (New York)	26	–	79
India (Mumbai)	33	240	39
Brazil (São Paulo)	43	86	83
China (Shanghai)	53	246	73
Russia (Moscow)	109	100	–
Indonesia (Jakarta)	–	271	–

[1]Comparable data for European Union not available; Germany is most populous EU member.
SOURCE: World Bank, *World Development Indicators 2000* (Washington, DC: 2000), 162–64.

In the Philippines, for example, the country's rich array of coral reefs, forests, and mangroves—home to an estimated 40,000 species—are shrinking rapidly in area, while the remaining pockets lose much of their original diversity. According to the WWF study, rural poverty and the unequal distribution of land in the Philippines are among the major causes of biodiversity loss that must be remedied if the country's natural wealth is to be preserved for future generations. Similarly, a study in the southern Mexican state of Campeche found that much of the pressure on the Calakmul Biosphere Reserve is coming from the efforts of local indigenous people to meet their material needs. Meeting those needs sustainably is a key component of any effective program to reverse environmental decline.[26]

Seizing the Moment

The last year has been marked by a roiling worldwide debate on the merits of economic globalization and how best to assure accelerated human and ecological progress in the decades ahead. Virtually every important meeting of international financial institutions has been met by thousands of protesters seeking to influence or shut down the discussions. While the demonstrations have been colorful, the endless argument over whether market liberalization and globalization are good or bad for people and the planet is not a particularly productive note on which to start a new century. Each side tends to simplify and demonize the position of the other, resulting in a rhetorical standoff.

There can be little doubt that opening markets in countries with weak governments, inadequate legal systems, and rampant corruption can exacerbate both social and environmental problems. However, more-open markets are at the same time potentially powerful tools for building economic and social opportunities for the poor and contributing to the development of civil society. In many parts of the developing world, capital is now more available to start small businesses, new ideas are flowing more freely, and the number of nongovernmental organizations (NGOs) is burgeoning. People are excited and energized as they consider future possibilities, in ways they were not a decade ago.

The multifaceted debate that has flowed from the protests about globalization has been echoed within the World Bank, where preparation of the annual *World Development Report*—focused in 2000 on the theme of poverty—created a vigorous internal and external debate that led its main author, Cornell economist Ravi Kanbur, to resign in protest. At the center of the World Bank debate was the view of Kanbur and other experts and NGO representatives that market liberalization and economic growth are insufficient by themselves to reduce poverty.[27]

To the surprise of many, the published version of the Bank's report, which appeared a few months later, acknowledged the failure of economic growth to reduce the numbers in poverty or close the equality gap in many nations. The report urged a broader, more comprehensive strategy to fight poverty, noting that, "facilitating the empowerment of poor people—by making state and social institutions more responsive to them—is also key to reducing poverty." Around the world, a strengthened consensus is emerging that reducing poverty is a complex undertaking that requires extensive but delicate government interventions, including investments in education and health, strong legal and financial systems, land reform, and strong anti-corruption policies. The experience of Russia, where market liberalization has been accompanied by an increase in poverty rates of 50 percent, is an important reminder that a healthy political system in which all of a society's interests are adequately represented and a strong legal and regulatory structure are key ingredients to meeting the needs of the poor.[28]

In southern India, a group called Myrada is a broker between banks and groups of poor people who use the money they borrow to start small businesses.

Social and ecological progress will also require a shared commitment to an agreed set of goals—goals that go beyond the expansion-of-wealth model that predominates in many political systems today. Expanding human options, eliminating poverty, and bringing the human economy into balance with Earth's natural systems are challenges large enough to frame a new millennium, but pressing enough that they must be met within the current century. Together, these changes would represent a revolution as fundamental as any in human history—an "Econological" Revolution that will test our technological abilities, our economic capacities, and even our humanity.

Transitions of this magnitude do not occur without strong pressures to change, since people generally resist disruptions to their existing patterns unless there is a clear need to do so. (See also Chapter 10.) Anthropologists believe that the Agricultural Revolution occurred in regions where environmental stress or population pressures were making the traditional hunter-gatherer way of life less viable. Similarly, the Industrial Revolution was precipitated in part by the social and economic limits of the prevailing eighteenth-century economy. An Econological Revolution must be accompanied by wide recognition that change is necessary—that without it, human progress will slow and then reverse. But it is also essential that people understand the opportunities that lie ahead if the revolution succeeds. As a Greenpeace representative told a conference in Oslo in June 2000, "If you want people to build boats, you must first create a longing for the islands."[29]

This should not be difficult. The world at the dawn of the millennium is extraordinarily dynamic, and despite the continued deterioration reflected in many ecological and social indicators, hundreds of success stories can be pointed to—seeds of change that will grow and spread if properly nurtured.

One of the most encouraging success stories of recent years is the growing attention to directly meeting the needs of the poor in many countries. Government investments in education and health care have increased substantially in some nations, spurred in part by increased commitments by international financial institutions. Latin America, in particular, which has historically been marked by enormous inequalities, has been closing the gap by investing more heavily in people. Since 1980, for example, the share of Brazilian children in secondary school has risen from 46 to 66 percent, while the proportion of women who are illiterate has fallen from 27 to 16 percent. And in the poor northeastern state of Ceara, a preventive health program that relied on 7,300 community health agents, 235 trained nurses, and a media campaign contributed to a decline in infant mortality from 102 per thousand live births to 65.[30]

Closing the large gender gaps that still exist in many countries is one of the keys to social progress. In many parts of Asia, Africa, the Middle East, and Latin America, women still lack many of the legal rights that men enjoy, and they are denied equal access to education, credit, and other ingredients to economic progress. This not only disadvantages half the human population, it impedes the advance of small business and agriculture, which are female-dominated in many countries. But this situation, too, is beginning to change, as women organize NGOs such as the Self-Employed Women's Association in India, which has provided day-to-day support for women as well as giving them a voice within the established political system. Around the world, many of societies' impediments to women's progress are slowly being erased.[31]

One of the recent social innovations that has proved particularly helpful to women is micro-credit, a concept pioneered by the Grameen Bank in Bangladesh and Bolivia's BancoSol. Over the last decade, this approach has been adapted to scores of countries, reaching over 10 million borrowers with tiny loans that turn them into small entrepreneurs, able to own and operate their own businesses. In southern India, a group called Myrada is serving as a broker between banks and groups of poor people who use the money they borrow to start small businesses. Such efforts have helped educate many policymakers to the fact that lack of access to affordable capital is limiting economic progress in many poor communities. International financial institutions and industrial countries are now providing funds to support many micro-credit programs.[32]

Social progress also requires a healthy environment, particularly in rural areas where the poor generally depend on local resources to provide food, water, shelter, and energy—a factor that is left out of the development equations used by most economists. In many cases, deforestation, soil erosion, and groundwater depletion have left villagers unable to meet basic needs, and without the financial resources needed to invest in social progress. Experience in India, however, has shown that empowering communities and providing them with assistance in managing their local forests and watersheds can lead to rapid improvement in living standards.[33]

Another innovation that has taken root in recent years is organic farming. More than 7 million hectares of farmland are now devoted to organic agriculture, up roughly 10-fold over the last decade. Recent food scares, particularly in Europe, have spurred consumer demand for food that is free of artificial pesticides and fertilizer, as has growing recognition of the ecological benefits of these new methods of farming. Government agencies have contributed to the growth in organics by certifying organic foods, and in some cases providing subsidies. Private farmers have taken advantage of the higher prices for organic crops by planting more land using the new techniques. (See Figure 1–1.) And with nation-

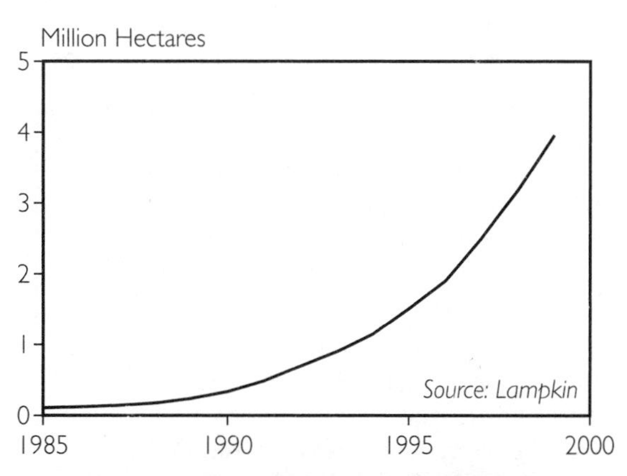

Figure 1–1. Certified Organic Agricultural Area in European Union, 1985–99

al trade barriers falling, farmers in countries such as Argentina, India, and Uganda are growing organic foods for export to industrial countries. At an estimated $22 billion annually, the global organic food market is still a tiny fraction of the total, but recent growth rates suggest a critical mass is being reached that may soon make it possible for most food to be grown in this way.[34]

Matching the boom in organic agriculture is the recent surge in interest in environmentally certified wood products. What started slowly with a small number of consumers who were upset about wood products that came from old-growth forests has recently mushroomed, attracting large buyers such as Home Depot and two of the largest U.S. home building companies. The key to this shift in the wood products market is the Forest Stewardship Council, which set up the first certification system in the early 1990s. Working with the World Wide Fund for Nature in the United Kingdom, it set up the first buyers' groups; more than 600 different companies in 18 countries now belong to the Global Forest and Trade Network, and some 25 countries are developing sustainable forestry standards. Currently, 20 million hectares of forest are under independently certified sustainable management, a number that is projected to grow to 200 million hectares by 2005.[35]

The greening of the wood products market is being followed by the more recent emergence of a market for "green" power. Electricity, which is mainly generated from coal and nuclear power in most countries, has in the past been sold as a single, undifferentiated commodity. But some governments are now requiring "labeling" of electricity on power bills, and allowing both electric utilities and independent power producers to market electricity from different sources—renewable energy from biomass, wind, and solar energy generally being the most popular. Among the countries where sizable numbers of power customers have signed up for green power are the United States (particularly in California, Colorado, and Pennsylvania), Australia, Germany, Japan, and the Netherlands. Green power is proving popular both with businesses and with consumers, who are sending a strong signal to the market about the energy sources they prefer. Over the next few years, this is likely to lead to substantial additional investment in renewable energy.[36]

One manifestation of the growing interest in green energy is the booming market in wind power in the last two years (see Figure 1–2); in 1999, wind turbine sales grew by 65 percent—almost as fast as mobile phone sales. Although the more than 18,000 megawatts of wind power that was projected to be in place by the end of 2000 produces less than 1 percent of the world's electricity, the share has already surpassed 2 percent in Germany, and exceeds 10 percent in Denmark.[37]

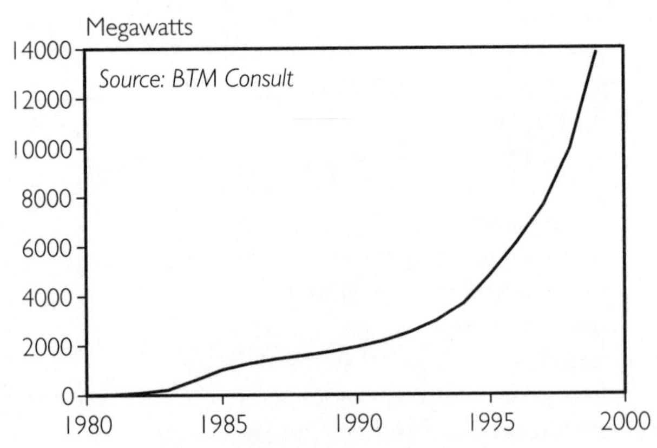

Figure 1–2. World Wind Energy Generating Capacity, 1980–99

The concentration of wind energy development in just a few countries is a reflection of policies that provide market access at favorable terms, motivated by the local jobs and tax revenues generated by wind energy investments. However, because wind power technology is based on standard components and manufacturing techniques, it has been disseminated rapidly from one nation to others as electricity policies are changed. In 2000, sizable wind energy projects started up in China, Japan, India, and the United States, suggesting that some of the world's largest power markets are becoming more favorable for wind development. Another sign of the times is the announcement by the power equipment giant ABB that it is shifting from its historical focus on multibillion-dollar thermal power plants to smaller scale generators, including wind power and other renewable energy technologies.[38]

From micro-credit to micro-power, the lesson of the past few years seems to be that rapid change is possible—when the conditions are right. Even as the sudden melting of the North Pole reminds us that environmental problems do not proceed in a neat, linear progression, we can take some comfort in knowing that environmental and social solutions can also unfold at exponential rates. This process of change—and how to accelerate it—is explored at length in Chapter 10.

North Meets South

Scientists studying the early evolution of technologies have noted that some of the most successful new devices and practices emerged when one human society took an idea developed elsewhere, and then adapted and improved upon it. Indeed, even before the age of exploration opened much of the globe to rapid diffusion of ideas, the slow spread of innovation from one village to the next allowed cultures as distant as those of Western Europe and China to learn from each other. But such diffusion has always occurred more rapidly in an East-West direction than between North and South, where variations in climate and difficult-to-cross deserts made travel and communication more difficult.

In today's world, these traditional barriers are largely gone. A person can travel physically to most points on the globe within a day, and the Internet has provided nearly instantaneous connections between diverse cultures in distant places. The question now is whether this potential for communication, and the growing level of commerce between nations, can be translated into a common effort to tackle shared problems.

The greatest benefits would come from a shared North-South commitment to a sustainable world, since differences in perspective between rich and poor countries have plagued efforts to deal effectively with issues ranging from population growth to biological diversity and climate change. In many international negotiations, finger-pointing has delayed action and slowed the adoption of effective policies. It is time for industrial countries to accept their historical responsibility for the current state of the planet—and time for developing countries to recognize that they are at great risk from environmental problems but will also benefit from the economic opportunities unleashed by a new development path. Shared burdens and leadership are now critically important.

Take one example: Climate change is an unequally distributed problem if ever there was one. Industrial countries produced most of the global warming gases that cause climate change, and yet it is developing countries that are likely to feel the most severe effects. The densely populated nations of South Asia, East Asia, and West Africa, where millions of people live on vast deltas at or below sea level, are most vulnerable to rising sea levels. In Bangladesh, for example, a 1-meter sea level rise would inundate 3 million hectares and displace 15–20 million people. In Viet Nam's Mekong Delta, the figure is 2 million hectares and 10 million people. In Nigeria, up to 70 percent of the coast could be covered and close to 4 million people displaced, including many residents of Lagos, the capital. Another North-South discontinuity is seen in the fact that the technologies to combat climate change, such as more benign industrial chemicals, fuel cells, and solar photovoltaics, are coming primarily from the R&D centers of northern countries, and yet the booming industrial and energy markets where they are most needed are in the South.[39]

Bridging these gaps between North and South will require a combination of innovative market reforms and a common commitment by governments to fill the gaps left by the private sector. Most of the recent emphasis has been on the market, pointing to developments such as the certified forest products market and booming consumer interest in ecotourism. And even government-negotiated treaties such as the Kyoto Protocol on climate change now rely on market mechanisms as primary tools for achieving their goals. Greenhouse gas trading schemes are being viewed as a way of not only trimming emissions as efficiently as possible, but also distributing the burden of addressing the problem among various countries.

Market mechanisms are often effective, and private innovation is key to solving many problems, but North-South cooperation will have to be based on something more than commercial relationships if the world's current problems are to be surmounted. Cooperation among NGOs, for example, allows innovative social programs and political techniques to be transferred rapidly from one country to another, dramatically speeding the rate of progress. The recent surge in the number of these groups in the developing world is being spurred by the support of foundations in industrial countries, as well as by the spread of democracy in many poor nations. And the Internet is proving a boon to the spread of civil society in countries where it has been weak in the past. The ability of citizens to communicate easily among themselves—and with people in distant lands with similar concerns—is rapidly transforming the political equation in many countries, and is creating more favorable conditions for addressing social and ecological problems.

Government leadership is also key: governments need to forge strong partnerships and provide sufficient funding to invest in the public infrastructure needed to support a sustainable economy. The failure of many industrial countries to meet the financial commitments they have agreed to under various international agreements and the failure of some developing countries to carry through on political and economic reforms have left a residue of distrust that must be overcome. Although it is unlikely that foreign aid levels will ever return to the figures that were typical in the 1960s and 1970s, a steady flow of well-targeted grants is essential to sustain progress. And with private capital taking up much of the burden of industrial growth and large-scale infrastructure, government aid can be targeted at pressing needs, with multiplier effects on human progress and environmental protection: areas such as education, health care, the status of women, micro-credit, and broad Internet access. One essential step is reducing the developing-country debt burden, which has reached onerous levels in recent years. (See Chapter 8.)

The economic and political weakness of many developing countries has prevented them from taking the more central position on the world stage that is now logically theirs. With 80 percent of the world's population, the bulk of its natural resources, and an opportunity to learn from the historical mistakes of today's industrial countries, it seems clear that the South will increasingly dominate the twenty-first century. Today's industrial powers will likely resist this shift, but they will soon find that they cannot achieve their own goals without the cooperation of the South. The summer of 2000 saw an intriguing sign of the changing balance of power when Mexico elected its first president from outside the traditional ruling party. Vicente Fox, a charismatic modern leader, traveled to Washington and called for allowing workers to travel as freely across the Mexico–U.S. border as capital now does.[40]

The existing structure of international institutions such as the World Bank and the World Trade Organization will have to be reformed to allow developing countries to take the more central role that is now essential to solving the world's most difficult problems. With shared power will come shared responsibility—a role that seems far more achievable today than it did two decades ago, when participatory political systems were still rare in the developing world.

One new organizing principle for countries that is particularly appropriate is the E–9 group described earlier—a coalition of northern and southern countries that between them have far greater impact on global social and ecological trends than do the Group of Eight (G–8) industrial countries. Between them, the E–9 have 60 percent of the world's population, 73 percent of the carbon emissions, and 66 percent of higher plant species. (See Table 1–7.). They have both the ability and the responsibility to lead the world in addressing the main challenges of the twenty-first century.

North-South cooperation will have to be based on something more than commercial relationships if the world's current problems are to be surmounted.

It is time for the E–9 to be organized as a semi-official group of nations that meets regularly to consider the range of economic, social, and environmental issues facing the world. Although such meetings may be less harmonious and more freewheeling than the current G–8 meetings, they would also be far more consequential, as they would involve a group of countries with the ability to shape global trends and to help forge a worldwide, North-South consensus on key issues. Under this model, the E–9 would not be a substitute for wider international bodies that represent all nations large and small, but rather it would spur those broader institutions, as well as the private sector, to action.

One example of the potential impact is climate change, where the E–9 accounts for nearly three quarters of the world market for oil, coal, and natural gas, whose combustion is the principal cause of climate change. A commitment by these nine to shift rapidly to energy efficiency, renewable energy, and zero-emission cars would put the global climate on a new trajectory. Similarly, a strong commitment by the E–9 to address the underlying causes of poverty—implementing economic and legal reforms and providing resources—would go far to close the equity gap.[41]

Table 1–7. The E–9: Leaders for the Twenty-first Century

	Share of				
Country	World Population, 1999	PPP Gross Domestic Product, 1998	World Carbon Emissions, 1999	World Forest Area, 1995	World Vascular Plant Species, 1997
			(percent)		
China	21.0	10.2	13.5	4	11.9
India	16.5	5.4	4.5	2	5.9
European Union	6.3	20.5	14.5	3	–
United States	4.6	21.3	25.5	6	6
Indonesia	3.5	1.3	.9	3	10.9
Brazil	2.8	2.9	1.5	16	20.8
Russia	2.4	2.4	4.6	22	–
Japan	2.1	8.0	6.0	0.7	2.1
South Africa	0.7	0.9	2.0	0.2	8.7
E–9 Total	**59.9**	**72.9**	**73**	**56.9**	**66.3**

SOURCE: Worldwatch calculations based on Population Reference Bureau, "1999 World Population Data Sheet," wall chart (Washington, DC: June 1999); World Bank, *World Development Indicators 2000* (Washington, DC: 2000), 10–12; BP Amoco, *BP Amoco: Statistical Review of World Energy* (London: June 2000), 38; U.N. Food and Agriculture Organization, *State of the World's Forests 1999* (New York: 1999), 125–30; World Conservation Union–IUCN, *1997 IUCN Red List of Threatened Plants* (Cambridge, U.K.: 1998), xvii, xxvii–xxxiii.

In the end, the greatest challenge will not be technological or even economic. As University of Maryland economist Herman Daly has written, a sustainable economy "would make fewer demands on our environmental resources, but much greater demands on our moral resources." One of those demands will be to reorganize international institutions so that power is based not on who has the biggest GNP but on a human sense of fairness, balance—and what is ultimately needed to ensure a healthy future for humanity and the planet. This may seem like a big leap in the first few years of the new century. But as we leave a century that began with women prohibited from voting in most countries, and with war viewed as the accepted means of settling differences among major powers, we should set a high standard for the decades ahead.[42]

Chapter 2

Uncovering Groundwater Pollution

Payal Sampat

A young man in London turns on his tap. As he watches a tea kettle fill, chances are he isn't giving much thought to the water's history. After all, some of the water is already boiling away into steam or has disappeared down the drain; the rest will be consumed within minutes. How could something so seemingly short-lived have a history? He'd be surprised to learn, then, that some of the water flowing out of his tap might have fallen as rain many millennia ago, around the time when woolly mammoths were tramping over today's Trafalgar Square. London draws much of its water from the Chalk aquifer, a huge underground reservoir of fresh water that lies hundreds of feet beneath the city; some of the water it holds trickled underground as long ago as the last Ice Age.[1]

It's natural to think of water as something that flows or evaporates. We see it coming down as rain, or running in streams or rivers. But most of our fresh water is not so easily observed because it lies deep underground in aquifers—geological formations made of porous materials such as sand and gravel, or spaces between subterranean rocks. These formations can retain enormous quantities of water. Aquifers are recharged—if they are recharged—by rainfall, runoff from rivers, or glacial melts. So most of our water is not visible on the surface: some 97 percent of the planet's liquid fresh water is stored in aquifers.[2]

In the last half-century, as global population and food demand have more than doubled, and rivers and streams have become more polluted, we have increasingly turned to aquifers for drinking and irrigation water—and in the process, we have made a sobering discovery. Despite the popular impression that groundwater is shielded from contaminants, scientists are uncovering cases of pollution in aquifers near farms, factories, and cities on every continent. We are now learning that the water buried

---

An expanded version of this chapter appeared as Worldwatch Paper 154, *Deep Trouble: The Hidden Threat of Groundwater Pollution.*

beneath our feet is not only susceptible to pollution, it is in many ways more vulnerable than water above ground.

This is a distinction of enormous consequence. Because water moves through the earth with glacial slowness, aquifers become sinks for pollutants, decade after decade. Some aquifers recharge fairly quickly, while others, like the Chalk, store their water for millennia. But the average residence time for groundwater is 1,400 years, as opposed to just 16 days for river water. So instead of being flushed out to the sea or becoming diluted with constant additions of fresh water, its pollutants accumulate. And in these sources, unlike rivers, the pollution is generally irreversible.[3]

Just as the onset of climate change has awakened us to the fact that the air over our heads is an arena of titanic forces, the water crisis has revealed that, slow-moving though it may be, groundwater is part of a system of powerful hydrological interactions—between earth, surface water, sky, and sea—that we ignore at our peril. A few years ago, reflecting on how human activity is beginning to affect climate, Columbia University scientist Wallace Broecker warned, "The climate system is an angry beast and we are poking it with sticks." A similar statement might now be made about the system under our feet. If we continue to drill holes into it—expecting it to swallow our waste and yield fresh water in return—we may unpredictably jeopardize supplies of the world's most important natural resource.[4]

Valuing Groundwater

For most of human history, groundwater was tapped mainly in arid regions where surface water was in short supply. Over the centuries, as populations and cropland expanded, water became such a valuable resource that some cultures developed elaborate mythologies imbuing underground water and its seekers with special powers. In medieval Europe, people called water witches or dowsers claimed the ability to detect groundwater using a forked stick and mystical insight.[5]

In the second half of the twentieth century, the soaring demand for water turned the dowsers' modern-day counterparts into a major industry. Today, massive aquifers are tapped on every continent, and groundwater is the primary source of drinking water for 1.5–2 billion people worldwide. (See Table 2–1.) The aquifer that lies beneath the Huang-Huai-Hai plain in eastern China alone supplies drinking water to nearly 160 million people. Some of the largest cities in the developing world—including Dhaka, Jakarta, Lima, and Mexico City—depend on aquifers for almost all their water. And in rural areas, where centralized supply systems are undeveloped, groundwater is typically the sole source of water. Almost 99 percent of the rural U.S. population and 80 percent of rural Indians depend on groundwater for drinking.[6]

A principal reason for the explosive rise in groundwater use since 1950 has been a dramatic expansion in irrigated agriculture. In fact, irrigation accounts for about two thirds of the fresh water drawn from rivers and wells each year. In India, the leading country in total irrigated area and the world's third largest grain producer, the number of shallow tubewells used to draw groundwater surged from 3,000 in 1950 to 6 million in 1990. Today aquifers supply water to more than half of India's irrigated land. About 40 percent of India's agricultural output comes from areas irrigated with groundwater, bringing groundwater's contribution to the gross domestic product

Table 2–1. Groundwater as a Share of Drinking Water Use, by Region, Late 1990s[1]

Region	Share of Drinking Water from Groundwater	People Served
	(percent)	(million)
Asia-Pacific	32	1,000 to 1,200
Europe	75	200 to 500
Latin America	29	150
United States	51	135
Australia	15	3
World		1,500 to 2,000

[1]Data on Africa not available.
SOURCE: Asia-Pacific and Latin America from United Nations Environment Programme, *Groundwater: A Threatened Resource* (Nairobi: 1996), 10–11; Europe from Organisation for Economic Co-operation and Development, *Water Resources Management: Integrated Policies* (Paris: 1989); U.S. Environmental Protection Agency, Office of Water, *The Quality of Our Nation's Water* (Washington, DC: 1998); Environment Australia, *State of the Environment Report 1996* (Canberra: 1996).

to about 9 percent. The United States, with the third highest irrigated area in the world, uses groundwater for 43 percent of its irrigated farmland.[7]

While agriculture is the largest groundwater consumer, other sectors of the economy have been expanding their water use even faster—and generating much higher profits in the process. On average, a ton of water used in industry generates roughly $14,000 worth of output—about 70 times as much profit as the same amount of water used to grow grain. Thus as the world has industrialized, substantial amounts of water have been shifted from farms to more lucrative factories. Industry's share of total consumption has reached 22 percent and is likely to continue rising rapidly. The amount of water available for drinking is thus constrained not only by a limited resource base, but by competition with other, more powerful users.[8]

As rivers and lakes are stretched to their limits—many of them dammed, dried up, or polluted—people are growing more and more dependent on groundwater for all their needs. In Taiwan, for example, the share of water supplied by groundwater almost doubled in just eight years—from 21 percent in 1983 to over 40 percent in 1991. Bangladesh, which was once almost entirely river- and stream-dependent, dug over a million wells in the 1970s to substitute for its badly polluted surface-water supply. Today, 95 percent of its people use only groundwater for drinking. In wealthier countries, sales of bottled spring water (supposedly from underground sources) are soaring: in the United States, bottled water use grew ninefold between 1978 and 1998.[9]

Even as our dependence on groundwater increases, the resource is becoming less available. On almost every continent, many major aquifers are being drained much more rapidly than their natural rates of recharge. Groundwater depletion is most severe in parts of India, China, the United States, North Africa, and the Middle East, resulting in a worldwide water deficit of an estimated 200 billion cubic meters a year. Removing large amounts of water from an aquifer can magnify the concentration of pollutants in the groundwater that remains. And in some cases, polluted surface flow or salty ocean water may pour into the aquifer to replace the depleted groundwater—thus further shrinking supplies.[10]

To compound the problem, groundwater overdraft can cause aquifer sediments to compact under certain geological conditions, permanently shrinking the aquifer's storage capacity. This loss can be quite considerable and is irreversible. The amount of water storage capacity lost because of aquifer compaction in California's Central Valley, for example, is equal to more than 40 percent of the combined storage capacity of all human-made reservoirs across the state. Compacted aquifer sediments can also cause the land above to sink. Such "land subsidence" has occurred in some of the world's most populous places, including Mexico City, Beijing, and some 45 other Chinese cities.[11]

As the competition among factories, farms, and households intensifies, it is easy to overlook the extent to which groundwater is also required for essential ecological services. It is an important component of the planet's hydrological cycle. When it rains, some water trickles into soil and soaks underground into an aquifer. Over centuries, the aquifer gradually releases the water to the surface, and eventually to the sea. Therefore it is not just rainfall but also groundwater welling up from beneath that replenishes rivers, lakes, and streams. In a study of 54 streams in different parts of the United States, the U.S. Geological Survey (USGS) found that groundwater is the source for more than half the total flow, on average. The 492 billion gallons (1.86 cubic kilometers) of water that aquifers add to U.S. surface water bodies each day is nearly equal to the daily flow of the Mississippi. Groundwater provides the base contribution for the Mississippi, the Niger, the Yangtze, and many more of the world's great rivers—some of which would otherwise not flow year-round. Wetlands, important habitat for birds, fish, and other wildlife, are often entirely groundwater-fed, created in places where the water table overflows to the surface on a constant basis. Where too much groundwater has been depleted, the result is often dried up riverbeds and desiccated wetlands.[12]

In addition to providing enough water to keep surface bodies stable, aquifers also help prevent them from flooding: when it rains heavily, aquifers beneath rivers soak up the excess water, preventing the surface flow from rising too rapidly and overflowing onto neighboring fields and villages. In tropical Asia, where the hot season can last as long as nine months and where monsoon rains can be very intense, this dual hydrological service is of critical value. Aquifers also provide a way to store fresh water without losing much liquid to evaporation—another service that is especially valuable in hot, drought-prone regions, where such losses can be quite high. In Africa, for instance, on average a third of the water removed from reservoirs each year is lost via evaporation.[13]

Tracking the Hidden Crisis

In 1940, during World War II, the U.S. Department of the Army acquired 70 square kilometers of land around Weldon Spring and its neighboring towns near St. Louis, Missouri. Where farmhouses and barns had been, the Army established the world's largest TNT-producing facility. In this sprawling warren of plants, toluene (a component of gasoline) was treated with nitric acid to produce more than a million tons of the explosive compound each day at the peak of production.[14]

Part of the manufacturing process involved purifying the TNT—washing off unwanted "nitroaromatic" compounds left behind by the chemical reaction between the toluene and nitric acid. Over the years, millions of gallons of this red-colored muck were generated. Some of it was treated at wastewater plants, but much of it ran off from the leaky treatment facilities into ditches and ravines, and soaked into the ground. In 1945, when the Army left the site, soldiers burned down the contaminated buildings but left the red-tinged soil and the rest of the site as they were. For decades, the site was abandoned and unused.[15]

In 1980, the U.S. Congress passed "Superfund" legislation that required the cleanup of several sites in the country that were contaminated with hazardous waste. Weldon Spring made it to the list of high priority sites. The Army Corps of Engineers was assigned the task, and what the workers found baffled them. While they expected the soil and vegetation around the site to be contaminated with discarded nitroaromatic wastes, they found that the chemicals were also showing up in people's wells, in towns several miles from the site—a possibility that no one had anticipated because the original pollution had been completely localized. Eventually, geologists determined that there was an enormous plume of contamination in the water below the TNT factory—a plume that over 35 years had flowed through fissures in the limestone rock to other parts of the aquifer.[16]

The Weldon Spring story may sound like an exceptional case of clumsy planning combined with a particularly vulnerable geological structure. But in fact there is nothing exceptional about it at all. Across the United States, as well as in parts of Europe, Asia, and Latin America, human activities are still unwittingly sending dangerous pollutants into groundwater. This is not entirely new, of course; the subterranean world has always been a sink for whatever we need to dispose of—whether it be our sewage, our garbage, or our dead. But prior to the twentieth century, these practices did not usually result in serious damage to groundwater. This has changed as the sheer volume of materials used has escalated—and as scientists have introduced thousands of chemicals that did not exist a century ago. And many of these new substances not only endure far longer in the environment, they are often more toxic than their predecessors. Pesticide formulations available today, for instance, are between 10 and 100 times more potent than those sold in 1975.[17]

In many parts of the world, we are only just beginning to discover contamination caused by practices of 30 or 40 years ago.

What happened in Weldon Spring shows that we cannot always anticipate where the pollution is going to turn up in our water, or how long after it is deposited it will reappear. Because it can often take months or years for a chemical to make its way from the surface into groundwater, damage done to aquifers may not show up for decades. In many parts of the world, we are only just beginning to discover contamination caused by practices of 30 or 40 years ago. Some of the most egregious cases of contamination now being unearthed date back to cold war–era nuclear testing and weapons-making, for example. And once it gets into groundwater, the pollution usually persists. Aquifers usually contain less in the way of dissolved oxygen, minerals, microbes, and organic matter than soils—conditions that do not encourage chemical breakdown. And many aquifers are very large—the Ogallala in the United States, for instance, spans portions of eight midwestern states. These giant formations can contain enormous volumes of water, and may be very hard to access—and thus virtually impossible to purify.[18]

As this covert crisis unfolds, we are only beginning to understand its dimensions. Even hydrogeologists and health officials have only a hazy impression of the likely extent of groundwater damage in different parts of the world. Few countries regularly track the health of their aquifers, as the enormous size and remoteness make them extremely expensive to monitor. Nonetheless, given the data we now have, it is possible to sketch a rough map of the regions affected and the principal threats they face. (See Table 2–2.)

Jack Barbash, an environmental chemist at the U.S. Geological Survey, points out that we may not need to wait for expensive tests to alert us to what to expect in our groundwater. "If you want to know what you're likely to find in aquifers near Shanghai or Calcutta, just look at what's used above ground," he says. "If you've been applying DDT to a field for 20 years, for example, that's one of the chemicals you're likely to find in the underlying groundwater." While the full consequences of today's chemical-dependent and waste-producing economies may not become apparent for another generation, Barbash and other scientists are beginning to get a sense of just how serious those consequences are likely to be if present consumption and disposal practices continue.[19]

The Slow Creep of Nitrogen

Fertilizers and pesticides applied to cropland have seeped into groundwater beneath farming regions in many parts of the world. Since the early 1950s, farmers have stepped up their use of nitrogen fertilizers 20-fold in an attempt to boost yields. But the larger doses of nutrients often cannot be fully used by plants. A study of a 140,000-square-kilometer region of northern China, for example, found that crops used on average only 40 percent of the nitrogen that was applied. The U.S. National Research Council estimates that in the United States, between a third and half of nitrogen fertilizer applied to plants is wasted. In an aerobic (oxygen-containing) environment, nitrogen is converted to nitrate—a form more readily used by plants. Much of the unused nitrate dissolves in rain and irrigation water, eventually trickling through the soil into underlying aquifers.[20]

Joining the excess chemical fertilizer from farm crops is the organic waste generated by farm animals and the sewage produced by cities—both of which have a high nitrate content. Because of its enormous volume, livestock waste forms a particularly potent tributary to the stream of excess nutrients flowing into the environment. In the United States, farm animals produce 130 times as much waste as people do—with the result that millions of tons of cow and pig feces are washed into streams and rivers or seep into groundwater. To this burden can be added the innumerable leaks and overflows from urban sewage systems; the fertilizer runoff from suburban lawns, golf courses, and landscaping; and the nitrates leaking (along with other pollutants) from landfills.[21]

Nitrate pollution of groundwater has become particularly severe in places where human population—and the demand for high food productivity—is most concentrated. In the northern Chinese counties of Beijing, Tianjin, Hebei, and Shandong, nitrate concentrations in groundwater exceeded 50 milligrams per liter (mg/liter) in more than half of the locations studied. (The World Health Organization (WHO) drinking water guideline is 45 mg/liter of nitrate.) In some places, the concentration had risen as high as 300 mg/liter. It is likely that these

Table 2–2. Some Major Threats to Groundwater

Threat	Sources	Health and Ecosystem Effects at High Concentrations	Principal Regions at Risk
Nitrates	Fertilizer runoff; manure from livestock operations; septic systems	Restricts amount of oxygen reaching brain, which can cause death in infants ("blue-baby syndrome"); linked to digestive tract and other cancers; causes algal blooms and eutrophication in surface waters	Parts of midwestern and mid-Atlantic United States, north China plain, northern India, parts of Eastern Europe
Pesticides	Runoff from farms, backyards, golf courses; landfill leaks	Organochlorines linked to reproductive and endocrine disorders in wildlife; organophosphates and carbamates linked to nervous system damage and cancers	Parts of United States, China, India
Petrochemicals	Underground petroleum storage tanks	Benzene and other petrochemicals can be cancer-causing even at low exposure	United States, United Kingdom, parts of former Soviet Union
Chlorinated solvents	Metals and plastics degreasing; fabric cleaning; electronics and aircraft manufacture	Linked to reproductive disorders and some cancers	California, industrial zones in East Asia
Arsenic	Naturally occurring	Nervous system and liver damage; skin cancers	Bangladesh, West Bengal, India, Nepal, Taiwan
Other heavy metals	Mining waste and tailings; landfills; hazardous waste dumps	Nervous system and kidney damage; metabolic disruption	United States, Central America, Eastern Europe
Radioactive materials	Nuclear testing and medical waste	Increased risk of certain cancers	Western United States, parts of former Soviet Union
Fluoride	Naturally occurring	Dental problems; crippling spinal and bone damage	Northern China, northwestern India; parts of Sri Lanka, Thailand, and East Africa
Salts	Seawater intrusion	Freshwater unusable for drinking or irrigation	Coastal China and India, Gulf coasts of Mexico and Florida, Australia, Thailand

SOURCE: Compiled from various sources cited throughout the chapter.

levels have increased, as fertilizer applications have escalated since the tests were carried out in 1995. They may increase even more as China's population (and demand for food) swells and as more farmland is lost to urbanization, industrial development, nutrient depletion, and erosion.[22]

Reports from other regions show similar results. (See Table 2–3.) In India's breadbasket states of Punjab and Haryana, where nitrogen fertilizer is applied intensively, wells tested in the early 1990s contained nitrate at levels 5 to 15 times higher than the safe limit. The USGS found that about 15 percent of shallow groundwater sampled below agricultural and urban areas in the United States in the mid-1990s had nitrate concentrations above the drinking water guideline; in some states, such as Nebraska, a third of the wells exceeded this limit.[23]

Although there is little historical information available about trends in the pollution of aquifers, several studies indicate that nitrate concentrations have increased as fertilizer applications and population size have grown. In California's Central Valley, for instance, nitrate levels in groundwater increased 2.5 times between the 1950s and 1980s—a period in which fertilizer inputs grew sixfold. Levels in Danish groundwater have nearly tripled since the 1940s.[24]

What happens when nitrates get into drinking water? Consumed in high concentrations—at levels above 45 mg/liter—they can cause infant methemoglobinemia, or so-called blue-baby syndrome. Because of

Table 2–3. High Nitrate Levels in Groundwater, Selected Regions, 1990s[1]

Region	Nitrate Levels	Source
Northern China	Above 50 mg/liter in more than half the locations tested	Fertilizer runoff from farms
Yogyakarta, Indonesia	Above 50 mg/liter in half the wells tested	Septic tanks
Canary Islands	Ranged between 70 and 265 mg/liter in tested wells beneath banana plantations	High nitrogen fertilizer use on banana plantations
Central Nigeria	Ranged between 50 and 500 mg/liter in wells tested near small towns	Human and animal waste disposal
Romania	Above 50 mg/liter in 35 percent of groundwater tested	Unsewered wastewater
East Anglia, United Kingdom	Above 50 mg/liter in 142 locations tested	Fertilizer leaching from fields
Yucatán Peninsula, Mexico	Shallow groundwater had levels above 45 mg/liter at more than half the locations tested	Domestic animal and human waste; agricultural runoff
Nebraska and Kansas, United States	Above 45 mg/liter in 35 percent of samples tested	Fertilizer runoff from farms

[1]Levels found exceeded the WHO drinking water limit of 45 mg of nitrate per liter.

SOURCE: See endnote 23.

low gastric acidity, infant digestive systems convert nitrate to nitrite, which blocks the oxygen-carrying capacity of a baby's blood, causing suffocation and death. Since 1945, about 3,000 cases have been reported worldwide—nearly half of them in Hungary, where private wells have particularly high concentrations of nitrates. Ruminant livestock such as goats, sheep, and cows are vulnerable to methemoglobinemia in much the same way infants are, because their digestive systems also quickly convert nitrate to nitrite. Nitrates have been linked to miscarriages in women and to an increased risk of non-Hodgkin's lymphoma. They have also been implicated in digestive tract cancers, although the epidemiological link is still uncertain.[25]

In cropland, nitrate pollution of groundwater can have a paradoxical effect. When nitrate-laden water is used to irrigate crops, the net result may be to reduce rather than increase production. In Libya, for example, grape vines irrigated with water containing 50 mg/liter of nitrogen bore almost no fruit. Too much nitrate can also weaken plants' immune systems, making them more vulnerable to pests and disease. Over-fertilizing makes wheat more susceptible to wheat rust, for example, and it makes pear trees more vulnerable to fire blight. At concentrations above 30 mg/liter, but sometimes as low as 5 mg/liter, nitrate applications can delay crop maturity, can severely damage plant roots, and can thin stems and branches, making it difficult for plants to bear their own weight. The U.N. Food and Agriculture Organization (FAO) reports that, in general, too much nitrogen in irrigation water has the same effect on crops as too much fertilizer.[26]

Nitrates in groundwater can also damage surface ecosystems. Consider the Chesapeake Bay in the mid-Atlantic United States. Once a thriving ecosystem and vibrant fishery, much of the bay is now suffering the consequences of too many nutrients. Enormous volumes of nitrogen and phosphorus are washed into its waters each day from the region's chicken farms, cropland, and septic systems; these nutrients spur the growth of algae, which now cover the water's surface. These massive algal blooms slowly atrophy the ecosystem by blocking sunlight from sea grasses—important habitat for fish and shellfish, and food for waterfowl. And when the algae die, their decomposition sucks up dissolved oxygen, killing off other aquatic species.[27]

Nitrates have been linked to miscarriages in women and to an increased risk of non-Hodgkin's lymphoma.

The bay's plight has alarmed many of the region's residents, but they might not be aware of groundwater's role in the ecosystem's collapse. Almost half of the nutrients that pour into the bay are carried there by aquifers either directly or, more typically, via the region's streams: groundwater contributes more than half of the 190 billion liters of water that rivers and streams empty into the Chesapeake Bay every day.[28]

The amount of a chemical that reaches groundwater depends on a number of factors: the amount used above ground, the geology of the region, climate, cropping practices, and the characteristics of the chemical itself, such as how mobile and soluble it is in water. Aquifers that are fractured in many places and that lie below coarse-textured and porous soils can be very vulnerable to pollution. This is true for the basalt and the sand and gravel deposits that lie beneath southeastern Washington state's potato and corn fields. These fields are heavily irrigated, thus expediting the flow of water and chemicals into the underlying aquifer. Other aquifers may be less susceptible: the relatively impermeable clay soils in some parts of the U.S. Midwest, for example, make it difficult for water and chemicals to seep underground. Because the soils drain water so poorly, the region's farmers have constructed tile drains and ditches to divert the excess irrigation water. As a result, farm chemicals run off over land into streams and lakes, which is why the region's surface water has among the highest nitrate levels in the country.[29]

In places where farms are adjacent to woods and forests, groundwater nitrate levels are often significantly lower. This is because forested areas create conditions that prevent the biological transformation of nitrogen into nitrate. Vegetation can also act as a filter, absorbing some of the nutrients before they enter groundwater. In the nutrient-laden Chesapeake Bay watershed, for instance, the lowest levels of nitrate in groundwater were reported in areas where farms were interspersed with woodlands. On the other hand, groundwater beneath homogeneous farmland that is heavily fertilized is more likely to carry high levels of nitrate.[30]

Pesticides on Tap

Pesticides are designed to kill. Yet it took several years after the first synthetic pesticides were introduced in the 1940s before it became apparent that these chemicals were also injuring non-target organisms—including humans. Even when the health concerns about some pesticides were widely recognized in the 1960s, it was assumed that the real dangers lay in the dispersal of these chemicals among animals and plants—not deep underground. It was generally assumed that very little pesticide would leach below the upper layers of soil, and that if it did, it would be degraded before it could get any deeper. Soil, after all, is known to be a natural filter that purifies any water that trickles through. It was thought that industrial or agricultural chemicals, like such natural contaminants as bacteria or leaf mold, would be filtered out as the water percolated through the soil.[31]

But over the past 35 years, this assumption of safety has proved mistaken. Cases of extensive pesticide contamination of groundwater have come to light in farming regions of the United States, Western Europe, and South Asia. In the United States, for instance, nearly 60 percent of wells sampled in agricultural areas in the mid-1990s contained these chemical compounds. And because they are also used to get rid of weeds on front lawns and golf courses, and to kill mosquitoes and other disease-carrying insects, pesticides also lurk in aquifers below cities and suburbs.[32]

We now know that pesticides not only leach into aquifers, but sometimes remain there long after the chemical is no longer used. The organochlorine pesticide DDT, for instance, is still found in U.S. groundwater even though its use was banned 30 years ago. In the San Joaquin Valley of California, the soil fumigant DBCP (dibromochloropropane), which was used intensively in fruit orchards before it was banned in 1977, still lingers in the region's water supplies. Of 4,507 wells sampled by the USGS between 1971 and 1988, nearly a third had DBCP levels that were at least 10 times higher than the maximum allowed by drinking water standards. And dieldrin, an organochlorine that was used for termite control around Atlanta until its use was prohibited in 1987, showed up in the city's well water in tests conducted in the mid-1990s.[33]

In places where organochlorines are still widely used, the risks continue to mount. After half a century of spraying in the eastern Indian states of West Bengal and Bihar, for example, the Central Pollution Control Board found DDT in groundwater at levels as high as 4,500 micrograms per liter—several thousand times higher than what is considered acceptable. Organochlorines are especially dangerous because they accumulate in body fat and tissue, and because their concentration magnifies as they move up the food chain.[34]

In recent decades, chemical companies have developed hundreds of compounds that are highly toxic but considered less ecologically damaging because they are short-lived. A pesticide's persistence is measured in terms of its half-life—the time it takes for 50 percent of a chemical's mass to decay—in soil. What scientists are learning, however, is that pesticides are far more persistent in groundwater than they are in soil. The herbicide alachlor, for instance, has a half-life of 20 days in soil, but of nearly four years in groundwater.[35]

Many of the pesticides that have replaced the organochlorines are known to be acutely toxic to humans and wildlife. The organophosphate and carbamate insecticides, for instance, are neurotoxins, or nerve poisons. A number of herbicides frequently detected in groundwater—including alachlor, atrazine, and triazine—are thought to interfere with the body's reproductive systems. And several pesticides are known to cause cancers, suppress the body's immune systems, or interfere with childhood development.[36]

Pesticides are often found in combination, because most farms use a range of toxins to destroy different kinds of insects, weeds, and plant diseases. The USGS detected two or more pesticides in groundwater at a quarter of the sites sampled across the United States. (See Table 2–4.) In the Central Columbia Plateau aquifer, under the states of Washington and Idaho, two thirds of water samples contained multiple pesticides. And even when the original pesticide does not appear to be in groundwater, its breakdown components, or degradates, often show up. When USGS researchers tested groundwater for this phenomenon, degradates of herbicides turned up more frequently than the original, or parent, compounds. For example, although just 1 percent of wells sampled in Iowa contained alachlor at levels above 0.2 micrograms per liter, more than half contained its degradates. These compounds can be as persistent and toxic as the original pesticide, or more so.[37]

In the United States, nearly 60 percent of wells sampled in agricultural areas in the mid-1990s contained pesticides.

Scientists are not entirely sure what happens when these chemicals come together. Water quality standards do not exist for the many hundred individual pesticides in use—the U.S. Environmental Protection Agency (EPA) has drinking water standards for just 33 of these compounds—to say nothing of the infinite variety of toxic blends now trickling into the groundwater. But there is some indication of possible additive or synergistic surprises we can expect. When researchers at the University of Wisconsin examined the effects of aldicarb, atrazine, and nitrate blends in groundwater—a mixture typically found beneath U.S. farms—they found that "more biological responses occur in the presence of mixtures of common groundwater contaminants than if contaminants occur singly." Fluctuation in concentrations

Table 2–4. Groundwater Contamination in the United States, Selected Chemicals, 1990s

	Share of Groundwater		
Chemical Group	Containing at Least Chemical Tested For	Containing Two or More Chemicals in Group	Above Drinking Water Guidelines for a Single Chemical
		(percent)	
Nitrates	71	not applicable[1]	15
Pesticides	50	25	not significant
Volatile Organic Compounds[2]	47[3]	29	6

[1]Nitrates are typically found in aquifers where pesticides are detected, however. [2]A small share of these VOCs are used as pesticides. [3]Samples from urban areas only.

SOURCES: Bernard T. Nolan and Jeffrey D. Stoner, "Nutrients in Groundwaters of the Conterminous United States, 1992–1995," Environmental Science & Technology, vol. 34, no. 7 (2000), 1156; U.S. Geological Survey, The Quality of Our Nation's Waters—Nutrients and Pesticides (Reston, VA: 1999) 57–58, 76; Paul J. Squillace et al., "Volatile Organic Compounds in Untreated Ambient Groundwater of the United States," Environmental Science & Technology, vol. 33, no. 23 (1999), 4176.

of the thyroid hormone, for example, is a typical response to mixtures, but not usually to individual chemicals. Other research found that combinations of pesticides increased the incidence of fetal abnormalities in the children of pesticide sprayers.[38]

While the most direct impacts may be on drinking water, there is also concern about what occurs when the pesticide-laden water below farmland is pumped back up for irrigation. One apparent consequence is a reduction in crop yields. In 1990, the congressional Office of Technology Assessment reported that herbicides in shallow groundwater had the effect of "pruning" crop roots, thereby retarding plant growth.[39]

Although most studies on pesticide contamination have been conducted in temperate regions, these chemicals pose serious risks to groundwater in the tropics as well. Researchers found such extensive atrazine contamination beneath sugarcane plantations in Barbados that they concluded the chemical was "more or less ubiquitous" in the island's coral limestone aquifer. (Barbadians get almost all their water from this aquifer.) And high levels of the soil insecticide carbofuran and its more persistent degradate, carbofuran-phenol, were detected in groundwater under vegetable farms on Sri Lanka's northwest coast, where soils are sandy and permeable.[40]

The Pervasiveness of Volatile Organic Compounds

San Jose, California, is the capital of the world's high-tech industry. As you drive into the city, it may be hard to imagine that the squeaky-clean veneer of the computer industry conceals a dirty underbelly: Silicon Valley has more Superfund sites than any other area its size in the United States. Its pollution is reflected not in smokestacks but in the contaminated aquifers beneath the valley. The source? Thousands of underground tanks that leak chlorinated solvents and other chemicals that are stored or discarded by electronics firms. In industrial countries, waste that is too hazardous to landfill is routinely buried in underground tanks. But as these caskets age, they eventually spring leaks. As of February 2000, there were about 386,000 confirmed leaks from underground storage tanks in the United States. In Silicon Valley, local groundwater authorities found that 85 percent of the tanks they inspected had creaks.[41]

Pull into any gas station in the United States and you will likely park over a second pervasive groundwater threat: an underground storage tank for petroleum. Like the tanks that store solvents, many of these were installed two or three decades ago. Left in place long past their expected lifetimes, many have rusted through—allowing a steady trickle of chemicals into the ground. EPA estimates that in the United States, about 100,000 of these tanks are leaking. In 1993, petroleum giant Shell reported that a third of its 1,100 gas stations in the United Kingdom were known to have contaminated soil and groundwater. Because the tanks are underground, they are expensive to dig up and repair, so the leakage in some cases continues for years. Petroleum and its associated chemicals—benzene, toluene, and gasoline additives such as MTBE, a fuel oxygenate added to reduce carbon monoxide emissions—is the most common category of groundwater contaminant found in aquifers in the United States.[42]

Both petrochemicals and chlorinated solvents are types of synthetic chemicals known as volatile organic compounds (VOCs), whose chemical and physical properties allow them to move freely between water and air. VOCs often turn up in groundwater beneath industrial areas and cities: they were detected in almost half the wells sampled near U.S. cities between 1985 and 1995. Between 35 million and 50 million people in these urban areas may be drinking water containing varying levels of these compounds.[43]

In many ways, solvents are ideal candidates as groundwater pollutants. One reason is that they are produced and used so widely. Synthetic organic chemical production expanded from less than 150,000 tons in 1935 to 150 million tons in 1995; these compounds are now used in paints, adhesives, gasoline, plastics, and hundreds of other everyday products. They are also used for cleaning and degreasing in the electronics and aerospace industries, as well as by small firms such as photo developers and dry cleaners.[44]

A second reason for this pollution is chlorinated solvents' physical and chemical properties. They do not stick to soils, meaning that almost none of the chemical is absorbed by sediments on its underground journey. Because many solvents are denser than water, they can sink deep into an aquifer. And since they do not degrade easily, they remain there for a long time, dissolving slowly and circulating to other parts of the aquifer—as happened in Weldon Spring.[45]

VOCs can be dangerous to human and animal health when they are consumed even in tiny concentrations. Petrochemicals such as benzene, for example, at extremely low levels can cause cancer. Women exposed to chlorinated solvents have a two- to fourfold higher incidence of miscarriage. These compounds have also been linked to kidney and liver damage and childhood cancers. An early case came from the town of Woburn, Massachusetts, in the 1970s, where a cluster of childhood leukemia cases was traced back to high levels of the chlorinated solvents perchloroethylene (PERC) and trichloroethylene in the city's wells.[46]

Ironically, a major factor in such contamination is that in most places people have learned to dispose of waste—to remove it from sight and smell—so effectively that it is easy to forget that Earth is a closed ecological system in which nothing permanently disappears. The methods normally used to conceal garbage and other waste—landfills, septic tanks, and sewers—become the major conduits of chemical pollution of groundwater. In the United States, businesses drain almost 2 million kilograms of assorted chemicals into septic systems each year, contaminating the drinking water of 1.3 million people. In many parts of the developing world, factories still dump their liquid effluent onto the ground and wait for it to disappear.[47]

Today, arsenic in drinking water could threaten the health of 20–75 million Bangladeshis—more than half the country's population.

In part, the "volatility" of these compounds has made it seem deceptively easy to get rid of them. When poured over the earth, as they were in Woburn, VOCs seem to disappear completely, to evaporate into the atmosphere. But in fact, some amount seeps underground and gets into groundwater. Even protected landfills can be a potent source of aquifer pollution: EPA found that a quarter of the landfills in Maine, for example, had contaminated groundwater.[48]

Sometimes waste is funneled directly into aquifers. Sixty percent of the most hazardous liquid waste in the United States—34 billion liters of solvents, heavy metals, and radioactive materials—is injected straight into deep groundwater via thousands of "injection wells" across the country. Although EPA requires that these effluents be injected below the deepest source of drinking water, some of these wastes have entered aquifers used for water supplies in parts of Florida, Texas, Ohio, and Oklahoma. And in India, a study across five industrializing states—Gujarat, Haryana, Punjab, Andhra Pradesh, and Karnataka—found that scores of factories were illegally injecting their wastes into tubewells that were used to pump out drinking and irrigation water.[49]

Like pesticides, VOCs are typically found in combination. In the United States, 29 percent of wells tested near urban areas contained multiple VOCs; overall, a total of 46 different kinds of these compounds turned up in groundwater. But Paul Squillace, the lead researcher in the study, notes that "because current health criteria are based on exposure to a single contaminant, the health implications of these mixtures are not known."[50]

VOCs have been detected in groundwater in other industrial countries as well. In the Netherlands, one study found that 28 percent of groundwater used for drinking contained PERC, a solvent used widely in dry cleaning, at levels greater than 10 micrograms per liter. Chlorinated solvents were found in close to half the groundwater used for drinking in England in 1985. And a survey of 15 Japanese cities found that 30 percent of all groundwater supplies contained varying levels of chlorinated solvents, although just 3 percent were above prescribed limits. The source was leaky storage tanks from electronics companies. Few data are available from other rapidly industrializing countries of East Asia, but this is a region where VOC use is accelerating. For instance, the production of semiconductor chips, which involves chlorinated solvents, is expected to nearly triple in the Asia-Pacific region (excluding Japan) between 1999 and 2003.[51]

Some of the greatest shocks may be felt in places where chemical use and disposal have climbed in the last few decades, but where the most basic measures to shield groundwater have not been taken. In India, for example, the Central Pollution Control Board surveyed 22 major industrial zones and found that groundwater in every one of them was unfit for drinking. When asked about these findings, the Board's chairman D.K. Biswas remarked, "The result is frightening, and it is my belief that we will get more shocks in the future."[52]

The Threat of Natural Contaminants

In the early 1990s, several villagers living near India's West Bengal border with Bangladesh began to complain of skin sores that wouldn't heal. A researcher at Calcutta's Jadavpur University, Dipanker Chakraborti, recognized the lesions immediately as early symptoms of chronic arsenic poisoning. In later stages, the disease can lead to gangrene, skin cancer, damage to vital organs, and eventually death. In the months that followed, Chakraborti learned that patients with similar symptoms were streaming into hospitals in Bangladesh. By 1995, it was clear that the region faced a crisis of untold proportions, and that the source of the poisoning was water from tubewells, from which 95 percent of Bangladesh gets its drinking water.[53]

Experts estimate that today, arsenic in drinking water could threaten the health of 20–75 million Bangladeshis—more than half the country's population—and another 6–30 million people in West Bengal, India. As many as 1 million wells in the region may be contaminated with the heavy metal at levels between 5 and 100 times the WHO drinking water guideline of 0.01 mg/liter. Arsenic poisoning has already caused at least 7,000 deaths, say local officials in Bangladesh. WHO predicts that within a few years, 1 in 10 deaths in southern Bangladesh may be from arsenic-related cancers.[54]

How did the arsenic get into groundwater? Until the early 1970s, rivers and ponds supplied most of Bangladesh's drinking water. Concerned about the risks of waterborne disease, international aid agencies launched a well-drilling program to tap groundwater instead. The agencies, unaware that soils of the Ganges aquifers are naturally rich in arsenic, did not test the sediment before drilling tubewells. Scientists are still debating what chemical reactions released the arsenic from the mineral matrix in which it is naturally bound up. But since the effects of chronic arsenic poisoning can take up to 15 years to appear, the epidemic was not recognized until it was well under way.[55]

Salt is another naturally occurring groundwater pollutant that is often introduced by human activity. Normally, water in coastal aquifers empties into the sea. But when too much water is pumped out of these, the process is reversed: seawater moves inland and enters the aquifer. Because of its high salt content, just 2 percent of seawater mixed with fresh water makes the water unusable for drinking or irrigation. And once salinized, a freshwater aquifer can remain contaminated for a very long time. Brackish aquifers frequently simply have to be abandoned because desalinization is very expensive. This has been the case in Bangkok, Thailand, and in parts of Gujarat state and the city of Madras in India.[56]

In Manila, where groundwater levels have fallen 50–80 meters because of overdraft, seawater has flowed as far as 5 kilometers into the Guadalupe aquifer that lies below the city. Similarly, salt water has traveled several kilometers inland into aquifers beneath Jakarta, parts of Florida in the United States, and along parts of coastal Turkey and China. Saltwater intrusion is also a serious problem on islands such as the Maldives and Cyprus, which are so dependent on aquifers for water supply.[57]

Fluoride is another natural contaminant that threatens millions in parts of Asia. Aquifers in the drier regions of northwestern India, northern China, and parts of Thailand and Sri Lanka are naturally rich in fluoride deposits. Fluoride is an essential nutrient for bone and dental health, but when consumed in high concentrations it can lead to crippling damage to the neck and back and to a range of dental problems. WHO estimates that 70 million people in northern China and 30 million in northwestern India are drinking water with excessive fluoride levels.[58]

Changing Course

The various incidents of aquifer pollution described thus far may seem isolated. A group of wells in northern China have nitrate problems; another cluster in the United Kingdom are laced with solvents. In each place it might seem that the problem is local and can be contained. But put them together and the bigger picture starts to emerge. Some of the world's most populous and rapidly expanding regions are in essence unintentionally poisoning their own wells—thus giving up their supplies of a vital source of fresh water. Perhaps most worrisome is that despite only limited monitoring and testing of underground water, we have discovered as much damage as we have. And because of the time lags involved—and given our high levels of chemical use and waste generation in recent decades—the future may bring even more unpleasant surprises.

In most cases, responses to groundwater pollution have been largely "end-of-pipe" solutions: utilities have installed filters or have abandoned groundwater supplies altogether. Several cities around the world have had to seek out alternate supplies of water because their groundwater became unusable. (See Table 2–5.) In recent years, half of all wells in Santa Monica, California, for instance, have been shut down because of dangerously high MTBE levels. In places where alternate supplies are not easily available, utilities will have to resort to increasingly elaborate filtration set-ups to make the water safe for drinking. In heavily contaminated areas, hundreds of different filters may be necessary. By one estimate, utilities in the U.S. Midwest spend an added $400 million each year to treat water for just one chemical—atrazine, a commonly detected pesticide in U.S. groundwater. When chemicals are found in unpredictable mixtures rather than discretely, providing safe water may become even more expensive.[59]

Where engineers have actually tried to "clean" contaminated aquifers, difficult though that might be, the favored technology has been "pump-and-treat." Three quarters of the highly contaminated Superfund sites in the United States where cleanup is under way use this technology. Groundwater is sucked out of the aquifer, its contaminants are flushed out or chemically treated above ground, and the water is injected back into the aquifer. This technology works on the principle that decades of such treatment will ultimately dilute the underground contamination. The amount of water needed to

Table 2–5. Selected Examples of Aquifers Abandoned due to Chemical Pollution

Region	Chemical(s)	Comments
Bangkok, Thailand	Salt	Excessive pumping of groundwater caused seawater to enter the aquifer. Chloride levels increased 60-fold; many wells have been abandoned.
Santa Monica, California, United States	MTBE, a gasoline additive	A petroleum spill produced MTBE levels 30 times higher than the guideline. Wells supplying half the city's water had to be closed down.
Shenyang, China	Nitrate, ammonium, oils, phenol and other industrial pollutants	Overpumping and pollution have forced authorities to replace groundwater with more expensive surface water supplies.

SOURCE: See endnote 59.

purge an aquifer is "unimaginably large," say scientists at the U.S. National Research Council. And they add that "simple calculations...show that predicted cleanup times range from a few years to tens, hundreds, and even thousands of years."[60]

The National Research Council estimates that in the United States, the costs of cleaning up the 300,000–400,000 heavily contaminated sites where groundwater is polluted will be as high as $1 trillion over the next 30 years. (So far, cleanup work has begun on just 4,000 of these sites since Superfund laws were passed in 1980.) Experts concur that in most cases, complete cleanup is nearly impossible. This is in large part due to the enormous size of aquifers and the persistent nature of many synthetic chemicals now found underground. Some of the radioactive waste that has leaked into Washington State's Central Columbia Plateau Aquifer has a half-life of 250,000 years. (Since 1943, hundreds of billions of gallons of radioactive wastes have been dumped into the aquifer and soils by the U.S. Department of Energy's Hanford Nuclear Reservation.) When such long-lived waste gets into aquifers—as it has in Washington—cleanup is not even an option.[61]

In many places, various authorities and industries have tried to fight contamination leak by leak, or chemical by chemical, only to find that individual fixes simply do not add up. As landfills are lined to reduce leakage, for instance, tons of pesticide may be running off nearby farms and into aquifers. As holes in underground gas tanks are mended, acid from mines may be seeping into groundwater. Clearly, it is essential to control damage already inflicted, and to protect communities and ecosystems from the poisoned fallout. But given what is already known—that damage done to aquifers is mostly irreversible, that it can take years before groundwater pollution reveals itself, that chemicals react synergistically and often in unanticipated ways—it is equally clear that a patchwork response will not be effective. (See Table 2–6.) Given how much

Table 2–6. Evaluating Responses to Groundwater Pollution

Strategy	Evaluation
End-of-pipe filters	Necessary if groundwater is the only available drinking water source. Expensive; does not protect people against multiple chemicals. Does not prevent pollution of aquifers.
Cleaning aquifers using "pump-and-treat" and other remediation technologies	Very expensive; can take decades or centuries before water quality improves significantly. Technically impossible to completely clean an aquifer. Does not prevent pollution of aquifers.
Restricting chemical use and disposal above vulnerable aquifers	Transfers polluting activity away from most vulnerable aquifers; does not prevent pollution from entering the environment altogether.
Changing polluting systems by moving toward closed-loop agricultural, urban, and industrial systems	Reduces the amount of pollution in the entire system, lightening the load on aquifers as well as other ecological systems.

SOURCES: National Research Council, *Alternatives for Ground Water Cleanup* (Washington, DC: National Academy Press, 1994); U.N. Environment Programme, *Groundwater: A Threatened Resource* (Nairobi: 1996).

damage this pollution inflicts on public health, the environment, and the economy once it gets into the water, it is critical that emphasis shift from filtering out toxins to not using them in the first place. Andrew Skinner, who heads the International Association of Hydrogeologists, puts it this way: "Prevention is the only credible strategy."[62]

To do this requires looking not just at individual factories, gas stations, cornfields, and dry cleaning plants, but at the social, industrial, and agricultural systems of which these businesses are a part. The ecological untenability of these systems is what is poisoning the world's water. It is the predominant system of high-input agriculture, for example, that floods the land—and the underlying water—with massive applications of agricultural chemicals. It is the system of car-dominated, geographically expanding cities that flood aquifers and soils with petrochemicals, heavy metals, and sewage. An adequate response will require a thoughtful overhaul of each of these systems.

Perhaps the most dramatic gains will come from reorienting agriculture so that it is far less dependent on chemicals. Farm runoff is a leading cause of groundwater pollution in many parts of Europe, the United States, China, and India—and it is impossible to control this pollution in a piecemeal fashion. Lessening its impact thus calls for adopting practices that sharply reduce this runoff—or, better still, that require far smaller amounts of pesticides and fertilizers to begin with. Even without radical shifts in the way we farm, there is plenty of room to improve the efficiency of chemical use. On average, roughly 85–90 percent of pesticides used for agriculture never reach target organisms, but instead spread through the environment. In Brazil, for example, farmers spray orchards with almost 10,000 liters of pesticide per hectare each week. FAO experts say that with modified application techniques, these chemicals could be applied at one tenth that amount and still be effective.[63]

In the Netherlands, some 550 farmers have reduced their levels of inputs by between 30 and 50 percent—and have completely eliminated their insecticide use—by monitoring and altering their farming practices: testing soil fertility to estimate how much additional nutrients are needed, for example, and planting a diverse array of crops. Since they have been able to maintain crop yields while using less input, their profit margins have increased. Meanwhile, their farms pollute far less: nitrogen and phosphorus levels in drainage water have declined 40–80 percent.[64]

But while greater efficiency constitutes a major improvement, there is also the possibility of replacing chemicals with nonpolluting methods of improving soil fertility and controlling pests. Recent studies suggest that farms can still maintain high yields using this approach. One decade-long investigation by the Rodale Institute in Pennsylvania, for example, compared a high-intensity system with traditional manure and legume-based cropping systems that used no synthetic fertilizer or pesticides. The researchers found that the two traditional systems retained more soil organic matter and nitrogen—indicators of soil fertility—and leached 60 percent less nitrate than the high-intensity system. Yields for maize and soybean crops differed by less than 1 percent between the three cropping systems over 10 years.[65]

Nearly 150 years ago, Charles Darwin observed in *On the Origin of Species* that wheat fields planted in diverse varieties of the grain were more productive than those planted in single varieties. This is because mixtures check the spread of pathogens and, therefore, of disease. For centuries, farmers have practiced such "polycropping" to protect their fields against disease. And in China, all the farmers in the Yunnan Province recently reconfirmed Darwin's observations. Until 1998, they planted monocultures of just two kinds of hybrid rice—and constantly battled the fungal disease, rice blast. But by growing multiple varieties of rice in the same paddies, they were able to double yields and at the same time to completely eliminate their use of fungicide.[66]

Even without radical shifts in the way we farm, there is plenty of room to improve the efficiency of chemical use.

Since 1986, Indonesian farmers have practiced ecologically based pest control—a set of practices known as integrated pest management (IPM). In addition to planting diverse species, strategies include introducing a pest's natural predators and interspersing crops with plants that repel pests. In the first four years of Indonesia's nationwide program, pesticide use on rice fell by half and yields increased by 15 percent. Similar programs have helped rice farmers in seven other Asian countries to cut pesticide use by nearly half while raising yields on average 10 percent. IPM has also taken root in places as diverse as Kenya, Cuba, Peru, and Iowa in the United States.[67]

Reining in agriculture's excessive dependence on chemicals will call for innovative responses not just from farmers, but from policymakers and businesses too. For instance, Indonesia's success was due in large part to parallel changes in policies: 57 pesticides were banned for use on rice, and pesticide subsidies were removed—saving the government $120 million. Several European countries, including Denmark, Finland, Norway, and Sweden, now tax pesticide sales to encourage farmers to use fewer inputs. Sweden's 7.5-percent tax (per kilogram of active ingredient) has had impressive results: it helped slash the country's pesticide use by 65 percent between 1986 and 1993. The program is still under way, with the goal of reducing pesticide use by another half. The French government is considering a steep tax on fertilizers and pesticides, spurred largely by water pollution problems in the country.[68]

In Germany, private water supply companies have discovered the economic benefits of organic farming. Firms in Munich, Osnabrück, and Leipzig pay local farmers as much as 550 Deutsche marks ($250) per hectare for three years to convert to organic operations. The firms are responsible for supplying clean water to their customers, and have found that it costs less to invest in sustainable farming than to strip pesticides out of polluted water.[69]

New York City has also worked with farmers to protect its water sources. The city's water supply depends on 5,000 square kilometers of watershed located 250 kilometers to its north. The watershed is also home to many of New York's small dairy farmers, who grow corn and hay for feed. Policymakers in the 1980s knew that farm runoff would affect water quality, but believed that the only way to protect New York's water supply was by installing elaborate filtration systems. The filters would have doubled the costs of New York's water, and mandatory farm regulations would have put many small farmers out of business.[70]

Concerned about the expense and shortsightedness of this solution, in 1991 Al Appleton, the Commissioner of the city's Department of Environmental Protection, began to develop an alternate plan. Today, the farmer-led Watershed Agricultural Council helps New York's farmers better manage their operations to protect water quality—for example, to time their fertilizer applications better and to be more efficient with their use of inputs. Although it is too early to quantify the water quality benefits, the program's financial savings have been considerable. While the filtration system would have cost upwards of $4 billion, New York City's watershed protection will cost an estimated $1.5 billion. Nationwide, EPA estimates that reducing agricultural pollution could save at least $15 billion in avoided costs of constructing advanced water treatment facilities.[71]

In industrial settings, building "closed-loop" production and consumption systems can help slash the amounts of waste that factories and cities send to landfills, sewers, and dumps—thus protecting aquifers from leaking pollutants. In places as far-ranging as Tennessee in the United States, Fiji, Namibia, and Denmark, environmentally conscious investors have begun to build "industrial symbiosis" parks in which the unusable wastes from one firm become the input for another. Such waste exchanges help an industrial park in Kalundborg, Denmark, to keep more than 1.3 million tons of effluent out of landfills and septic systems each year, while preventing some 135,000 tons of carbon and sulfur from leaking into the atmosphere.[72]

By reusing spent materials and chemicals, individuals and firms can help reduce the risk of groundwater pollution from heavy metals, insulation chemicals, cleaning solvents, and other toxic substances that leak out of landfills. The Xerox Corporation, for example, remanufactures more than a third of its photocopiers by using components from older machines—a strategy that saved 143,000 tons of materials from bring dumped in landfills in 1999 alone. Each remanufactured machine must meet the same standards as a newly minted one and comes with the same warranty.[73]

Some regions have set waste reduction as a collective goal. The Netherlands, for instance, has a national goal of cutting its wastes 70–90 percent. Pollution taxes have already helped the nation slash discharges of heavy metals into waterways by 72–99 percent between 1976 and the mid-1990s. The city of Canberra in Australia aims for a "No-Waste-by-2010" goal. As part of the campaign, authorities in the city have set up an online resource exchange—an information center that helps match suppliers of waste material with buyers.[74]

In some cases, the costs of using a particular chemical may be so great that the only way to protect human health and the environment adequately is to eliminate its use completely. In 1987, for instance, the global community signed a treaty phasing out the use of substances that were found to deplete the ozone layer. Since that time, their use has fallen substantially—by 88 percent in the case of chlorofluorocarbons, chemicals that were commonplace in refrigerators and air conditioners just a few years earlier. Currently under negotiation is an international treaty on a class of dangerous synthetic materials known as persistent organic pollutants. Under discussion is a proposed phaseout of 12 of these chemicals nicknamed The Dirty Dozen—9 of which are pesticides, including DDT and dieldrin. Many participants in the treaty process argue that the list of chemicals to be phased out should be expanded to include dozens, even hundreds, of other persistent chemicals whose presence in the environment poses an unacceptable risk to humanity.[75]

In Germany, private water supply companies have discovered that it costs less to invest in sustainable farming than to strip pesticides out of polluted water.

As it becomes clearer to decisionmakers that the most serious threats to human security are no longer those of military attack but instead pervasive environmental and social decline, experts worry about the difficulty of mustering sufficient political will to bring about the kinds of systemic—and therefore revolutionary—changes in human life necessary to turn the tide in time. In confronting the now heavily documented assaults of climate change and biodiversity loss, leaders seem on one hand to be paralyzed by how bleak the big picture appears to be, and on the other hand too easily drawn into inaction by the seeming lack of immediate consequences of delay.

But the need to protect aquifers may provide a more immediate incentive for change. It simply may not be possible to live with contaminated groundwater for as long as we could make do with a gradually more irritable climate or polluted air or impoverished wildlife. Although we have damaged portions of some aquifers to the point of no return, scientists believe that a large part of the resource still remains pure—for the moment. That's not likely to remain the case if we continue to depend on simply stepping up the present reactive tactics of cleaning up more chemical spills, replacing more leaking gasoline tanks, placing more plastic liners under landfills, or issuing more fines to careless hog farms and copper mines.

Protecting our water in time requires the same fundamental restructuring of the global economy that stabilization of the climate and biosphere as a whole rests on—the rapid transition from a resource-depleting, oil- and coal-fueled, high-input industrial and agricultural economy to one that is based on renewable energy, compact cities, and a very light human footprint. We have been slow to come to grips with this, but it may be our thirst that finally makes us act.

Chapter 3

Eradicating Hunger: A Growing Challenge

Lester R. Brown

In 1974, U.S. Secretary of State Henry Kissinger made a pledge at the World Food Conference in Rome: "By 1984, no man, woman, or child would go to bed hungry." Those attending the conference, including many political leaders and ministers of agriculture, came away inspired by this commitment to end hunger.[1]

More than 26 years later, hunger is still very much part of the social landscape. Today, 1.1 billion of the world's 6 billion people are undernourished and underweight. Hunger and the fear of starvation quite literally shape their lives. A report from the U.N. Food and Agriculture Organization (FAO) describes hunger: "It is not a transitory condition. It is chronic. It is debilitating. Sometimes, it is deadly. It blights the lives of all who are affected and undermines national economies and development processes... across much of the developing world."[2]

Kissinger's boldly stated goal gave the impression that there was a plan to eradicate hunger. In fact, there was none. Kissinger himself had little understanding of the difficult steps needed to realize his goal. Unfortunately, this is still true of most political leaders today.

In 1996, governments again met in Rome at the World Food Summit to review the food prospect. This time delegates from 186 countries adopted a new goal of reducing the number who were hungry by half by 2015. But as in 1974, there was no plan for how to do this, nor little evidence that delegates understood the scale of the effort needed. FAO projections released in late 1999, just three years after the new, modest goal was set, acknowledge that the objective for 2015 is not likely to be reached because "the momentum is too slow and the progress too uneven."[3]

Assertions such as Kissinger's and those of other political leaders may make people feel good, but if they are not grounded in a carefully thought out plan of action and supported by the relevant governments, they ultimately undermine confidence in the public process. This in turn can itself undermine progress.

In its most basic form, hunger is a productivity problem. Typically people are hungry because they do not produce enough food to meet their needs or because they do not earn enough money to buy it. The only lasting solution is to raise the productivity of the hungry—a task complicated by the ongoing shrinkage in cropland per person in developing countries.

The war against hunger cannot be won with business as usual. Given the forces at work, it will no longer be possible to stand still. If societies do not take decisive steps, they face the possibility of being forced into involuntary retreat by continuing population growth, spreading land hunger, deepening hydrological poverty, increasing climate instability, and a shrinking backlog of unused agricultural technology. Eradicating hunger—never easy—will now take a superhuman effort.

A Hunger Report: Status and Prospects

As noted, 1.1 billion people are undernourished and underweight as the new century begins. The meshing of this number with a World Bank estimate of 1.3 billion living in poverty, defined as those living on $1 a day or less, comes as no surprise. Poverty and hunger go hand in hand.[4]

The alarming extent of hunger in the world today comes after a half-century during which world food output nearly tripled. The good news is that the share of population that is hungry is diminishing in all regions except Africa. Since 1980, both East Asia and Latin America have substantially reduced the number and the share of their populations that are hungry. In the Indian subcontinent, results have been mixed, with the number of hungry continuing to increase but the share declining slightly. In Africa, however, both the number and the share of hungry people have increased since 1980.[5]

The decline in East Asia was led by China, which brought the share of its people who are hungry down from 30 percent in 1980 to 11 percent in 1997. China's economic reforms initiated in 1978 led to a remarkable surge in agricultural output, one that boosted the grain harvest from roughly 200 kilograms per person a year to nearly 300 kilograms. This record jump in production in less than a decade led to the largest reduction in hunger on record. For most countries, the best nutrition data available are for children, who are also the segment of society most vulnerable to food scarcity. In Latin America, the share of children who are undernourished dropped from 14 percent in 1980 to 6 percent in 2000.[6]

These gains in eradicating hunger in East Asia and Latin America leave most of the world's hungry concentrated in two regions: the Indian subcontinent and sub-Saharan Africa. In India, with more than a billion people, 53 percent of all children are undernourished. In Bangladesh, the share is 56 percent. And in Pakistan, it is 38 percent. In Africa, the share of children who are undernourished has increased from 26 percent in 1980 to 28 percent today. In Ethiopia, 48 percent of all children are underweight. In Nigeria, the most populous country in Africa, the figure is 39 percent.[7]

Within the Indian subcontinent and sub-Saharan Africa, most of the hungry live in the countryside. The World Bank reports that 72 percent of the world's 1.3 billion poor live in rural areas. Most of them are undernourished; many are sentenced to a short life. These rural poor usually do not live on the productive irrigated plains, but on the semiarid/arid fringes of agriculture or in the upper reaches of watersheds on steeply sloping land that is highly erodible. Eradicating hunger depends on stabilizing these fragile ecosystems.[8]

Recognizing that malnutrition is largely the result of rural poverty, the World Bank is replacing its long-standing agricultural development strategies, which were centered around crop production, with rural development strategies that use a much broader approach. The Bank planners believe that a more systemic approach to eradicating poverty in rural areas—one that embraces agriculture but that also integrates human capital development, the development of infrastructure, and social development into a strategy for rural development—is needed to shrink the number living in poverty. One advantage of encouraging investment in the countryside in both agribusiness and other industries is that it encourages breadwinners to stay in the countryside, keeping families and communities intact. In the absence of such a strategy, rural poverty simply feeds urban poverty.[9]

Demographically, most of the world's poor live in countries where populations continue to grow rapidly, countries where poverty and population growth are reinforcing each other. The Indian subcontinent, for example, is adding 21 million people a year, the equivalent of another Australia. India's population has nearly tripled over the last half-century, growing from 350 million in 1950 to 1 billion in 2000. According to the U.N. projections, India will add 515 million more people by 2050, in effect adding roughly twice the current U.S. population. Pakistan's numbers, which tripled over the last half-century, are now expected to more than double over the next 50 years, going from 156 million to 345 million in 2050. And Bangladesh is projected to add 83 million people during this time, going from 129 million to 212 million. The subcontinent, already the hungriest region on Earth, is thus expected to add another 787 million people by mid-century.[10]

The war against hunger cannot be won with business as usual.

No single factor bears so directly on the prospect of eradicating hunger in this region as population growth. When a farm passes from one generation to the next, it is typically subdivided among the children. With the second generation of rapid population growth and associated land fragmentation now unfolding, farms are shrinking to the point where they can no longer adequately support the people living on them.

Between 1970 and 1990, the number of farms in India with less than 2 hectares (5 acres) of land increased from 49 million to 82 million. Assuming that this trend has continued since then, India may now have 90 million or more families with farms of less than 2 hectares. If each family has six members, then 540 million people—over half India's population—are trapped in a precarious balance with their land.[11]

Whether measuring changes in farm size or in grainland per person, the results of continuing rapid population growth are the same. Pakistan's projected growth from 156 million today to 345 million by 2050 will shrink its grainland per person from 0.08 hectares at present to 0.03 hectares, an area scarcely the size of a tennis court. African countries, with the world's fastest population growth, are facing a similar reduction. For example, as Nigeria's population increases from 111 million today to a projected 244 million in 2050, its per capita grainland, most of it semiarid and unirrigated, will shrink from 0.15 hectares to 0.07 hectares. Nigeria's food prospect, if it stays on this population trajectory, is not promising.[12]

In Bangladesh, average farm size has already fallen below 1 hectare. According to one study, Bangladesh's "strong tradition of bequeathing land in fixed proportions to all male and female heirs has led to increasing landlessness and extreme fragmentation of agricultural holdings." In addition to the millions who are now landless, millions more have plots so small that they are effectively landless.[13]

Further complicating efforts to expand food production are water shortages. Of the nearly 3 billion people to be added to world population in the next 50 years, almost all will be added in countries already facing water shortages, such as India, Pakistan, and many countries in Africa. In India, water tables are falling in large areas as the demands of the current population exceed the sustainable yield of aquifers. For many countries facing water scarcity, trying to eradicate hunger while population continues to grow rapidly is like trying to walk up a down escalator.[14]

Even as the world faces the prospect of adding 80 million people a year over the next two decades, expanding food production is becoming more difficult. In each of the three food systems—croplands, rangelands, and oceanic fisheries—output expanded dramatically during the last half of the twentieth century. Now all that has changed.

Between 1950 and 2000, the world production of grain, the principal product of croplands, expanded from 631 million tons to 1,860 million tons, nearly tripling. In per capita terms, production went from 247 kilograms per person in 1950 to an all-time high of 342 kilograms in 1984, a gain of nearly 40 percent as growth in the grain harvest outstripped that of population. After 1984, production slowed, falling behind population. Production per person declined to 308 kilograms in 2000, a drop of 10 percent from the peak in 1984. (See Figure 3–1.) This decline is concentrated in the former Soviet Union, where the economy has shrunk by half since 1990, and in Africa where rapid population growth has simply outrun grain production.[15]

Roughly 1.2 billion tons of the world grain harvest are consumed directly as food, with most of the remaining 660 million tons being consumed indirectly in livestock, poultry, and aquacultural products. The share of total grain used for feed varies widely among the "big three" food producers—ranging from a low of 4 percent in India to 27 percent in China and 68 percent in the United States.[16]

Over the last half-century, the world's demand for animal protein has soared. Expanded output of meat from rangelands

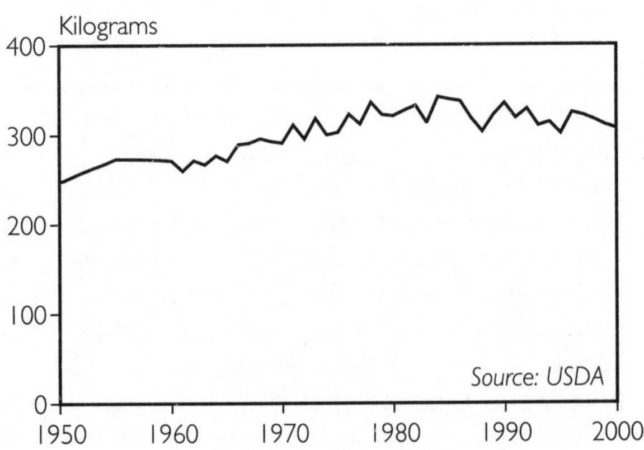

Figure 3–1. World Grain Production Per Person, 1950–2000

and of seafood from oceanic fisheries has satisfied most of this demand. World production of beef and mutton increased from 24 million tons in 1950 to 67 million tons in 1999, a near tripling. Most of the growth, however, occurred between 1950 and 1990, when output went up 2.5 percent a year. Since then, beef and mutton production has expanded by only 0.6 percent a year. (See Figure 3–2.)[17]

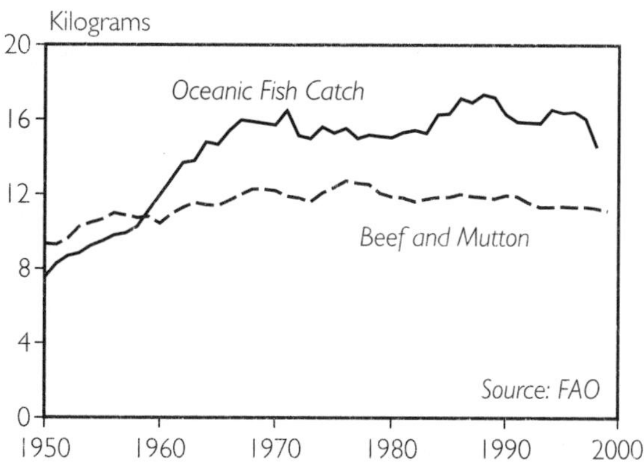

Figure 3–2. Oceanic Fish Catch and Beef and Mutton Production Per Person, 1950–99

An estimated four fifths of the 67 million tons of beef and mutton produced worldwide in 1999, roughly 54 million tons, comes from animals that forage on rangelands. If the grain equivalent of the forage-based output is set at seven kilograms of grain per kilogram of beef or mutton, which is the conversion rate in feedlots, the beef and mutton produced on rangeland are the equivalent of 378 million tons of grain.[18]

The growth in the oceanic fish catch exceeded even that of beef and mutton, increasing from 19 million tons in 1950 to 86 million tons in 1998, the last year for which data are available. This fourfold growth, too, was concentrated in the period from 1950 to 1990, a time during which the annual growth in the oceanic catch—at 3.8 percent—was easily double that of world population. As a result, seafood consumption per person worldwide roughly doubled, climbing from 8 kilograms in 1950 to 16 kilograms in 1990. Since then, it has fallen by some 10 percent. Assuming the fish farm conversion of less than two kilograms of grain for each kilogram of live weight added, then the grain equivalent of the 86-million-ton fish catch in 1998 was 172 million tons of grain.[19]

The new reality is that fishers and ranchers can no longer contribute much to the growth of the world's food supply. For the first time since civilization began, farmers must carry the burden alone.

For a sense of the relative importance of rangelands and oceanic fisheries in the world food economy, compare the grain equivalent of their output with the world grain harvest. With rangelands accounting for the equivalent of 378 million tons of grain and with fisheries at 172 million tons, rangelands contributed 16 percent of the world grain supply and oceanic fisheries 7 percent. (See Table 3–1.)

Thus rangelands and oceanic fisheries provide the equivalent of nearly one fourth of the world grain supply. With their output no longer expected to expand, all future growth in the food supply must come from the 77 percent of total grain equivalent that is represented by croplands. With little new land to plow, the world's ability to eradicate hunger in the years ahead will depend heavily on how much cropland productivity can be raised. This is also one of the keys to eliminating poverty.

Eradicating hunger in the Indian subcontinent and Africa will not be easy. It is difficult to eradicate for the same reasons it exists in the first place—rapid population

Table 3–1. Cropland, Rangeland, and Oceanic Fishery Contributions to World Food Supply, Measured in Grain Equivalent, 1999

Source	Quantity of Grain (million tons)	Share of Total (percent)
Grain production from cropland	1,855	77
Grain equivalent of rangeland beef and mutton	378	16
Grain equivalent of oceanic fish catch[1]	172	7
Total	2,405	100

[1]Fish production data from 1998.

SOURCE: USDA, *Production, Supply, and Distribution*, electronic database, Washington, DC, updated September 2000; FAO, *Yearbook of Fishery Statistics: Capture Production* (Rome: various years).

growth, land hunger, and water scarcity. But there are also new forces that could complicate efforts to eliminate hunger. For example, Bangladesh, a country of 129 million people, has less than one tenth of a hectare of grainland per person—one of the smallest allotments in the world—and is threatened by rising sea level. The World Bank projects that a 1-meter rise in sea level during this century, the upper range of the recent projections by the Intergovernmental Panel on Climate Change, would cost Bangladesh half its riceland. This, combined with the prospect of adding another 83 million people over the next half-century, shows just how difficult it will be for Bangladesh, one of the world's hungriest countries, to feed its people.[20]

Water shortages are forcing grain imports upward in many countries. North Africa and the Middle East is now the world's fastest growing grain import market. In 1999, Iran eclipsed Japan, which until recently was the world's leading importer of wheat. And Egypt, another water-short country, has also edged ahead of Japan.[21]

Many developing countries that are facing acute land and water scarcity will rely on industrialization and labor-intensive industrial exports to finance needed food imports. This brings a need to expand production in exporting countries so they can cover the import needs of the growing number of grain-deficit countries. Over the last half-century, grain-importing countries, now the overwhelming majority, have become dangerously dependent on the United States for nearly half of their grain imports.

This concentration of dependence applies to each of the big three grains—wheat, rice, and corn. Just five countries—the United States, Canada, France, Australia, and Argentina—account for 88 percent of the world's wheat exports. Thailand, Viet Nam, the United States, and China account for 68 percent of all rice exports. For corn, the concentration is even greater, with the United States alone accounting for 78 percent and Argentina for 12 percent.[22]

With more extreme climate events in prospect if temperatures continue rising, this dependence on a few exporting countries leaves importers vulnerable to the vagaries of weather. If the United States were to experience a summer of severe heat and drought in its agricultural heartland like the summer of 1988, when grain production dropped below domestic consumption for the first time in history, chaos would reign in world grain markets simply because the near-record reserves that cushioned the huge U.S. crop shortfall that year no longer exist.[23]

The risk for the scores of low-income, grain-importing countries is that prices could rise dramatically, impoverishing more people in a shorter period of time than any event in history. The resulting rise in hunger would be concentrated in the cities of the Third World.

Raising Cropland Productivity

In a world where there is little new land to plow, raising the productivity of existing cropland is the key to feeding the 80 million people added each year. It is also essential for protecting the environment. If farmers had not been able to raise grain yield per hectare since 1950, it would have been necessary to clear the equivalent of half of the world's remaining forestland for food production. In addition, raising cropland productivity is the key to eradicating rural poverty and hunger in the Indian subcontinent and sub-Saharan Africa.

There are at least three ways of raising cropland productivity. One, the most conventional, is raising the yield per crop per hectare. The second is increasing the number of crops per hectare through multiple cropping. And the third is to get more out of the existing harvest. In the last case, there is a huge unrealized potential for using crop residues, such as wheat or rice straw or corn stalks, to produce meat and milk by "processing" them through ruminants.

Raising grain yields in the two regions where the world's hungry are concentrated will not be easy. India's wheat yield, for example, has tripled since 1960. The rise in rice yield, which went from just under 1 ton per hectare in 1965 to 1.9 tons in 1993, has slowed, failing to reach 2 tons per hectare during the seven years since then. Lifting land productivity in India is constrained by the country's proximity to the equator. Day length during the summer is relatively short, and since rice is typically grown during the summer monsoon season, when cloud cover is heavy, solar intensity is low.[24]

The area that is multiple cropped can be expanded by harvesting and storing water during the monsoon season so that more crops can be produced during the dry season. In some situations in India, particularly where there is an abundance of labor, polycropping—which is growing more than one crop in a field—may provide a higher yield.

With the rise in land productivity slowing, continuing rapid population growth makes eradicating rural hunger much more difficult, if not impossible. Perhaps the single most important thing India can do to enhance its future food security is to accelerate the shift to smaller families. This would enable it to move to the low-level U.N. population projection instead of the medium-level one, thereby adding only 216 million people instead of 515 million in the next 50 years.[25]

Eradicating hunger in Africa, which has the world's lowest crop yields, is an even more daunting challenge. Africa has the fastest population growth of any region and a largely semiarid climate, which limits the potential for irrigation and fertilizer use. It has not had a Green Revolution for the same reason that Australia has not: it is too dry to use much fertilizer, the key to raising yields. And now a new variable—the HIV epidemic—is decimating the adult population in sub-Saharan Africa, reducing the number of able-bodied people who can work in the fields.

Africa can substantially increase its production of food, but this depends on strengthening its agricultural infrastructure, including farm-to-market roads and grain and fertilizer storage facilities, as well as strengthening institutions in agricultural research, agricultural extension, and farm credit. Minimum government price supports for food staples can encourage investment in agriculture. With women responsible for so much of the food production in Africa, small loans to those who are working the land can pay large dividends. Even while waiting for infrastructure to develop, African farmers can do much more with nitrogen-fixing green manure crops, which will help improve the structure of their soils while providing nitrogen, one of the key nutrients (along with phosphate and potash) needed to restore soil fertility.[26]

The world is now having difficulty sustaining rises in land productivity. Over the last century or so, plant breeders greatly boosted the genetic yield potential of grains, particularly the three major ones—wheat, rice, and corn. At the heart of this effort was an increase in the share of photosynthate, the product of photosynthesis, going to the seed. While the originally domesticated wheats did not use much more than 20 percent of their photosynthate to produce seed, today's highly productive varieties devote half or more to seed formation. The theoretical upper limit for the share that can go toward seed while still maintaining the rest of the plant is estimated at 60 percent.[27]

Realizing the genetic potential of the new seeds depends on removing any nutrient or moisture constraints of soils. Fertilizers are designed to remove the limits imposed by nutrient deficiencies. As cities have grown over the past century, there has been a massive disruption of the nutrient cycle, making the world ever more dependent on fertilizer. In earlier times, when food was produced and consumed locally, nutrients were automatically recycled back onto the land in the form of livestock and human waste. But as cities developed and as international trade expanded, the net loss of nutrients from food-producing areas was offset with the use of fertilizer.

Where fertilizer use is excessive, nutrient runoff into rivers and oceans can lead to algal blooms that then use up all available oxygen in the water as the algae decompose, creating dead zones with no sea life. Nutrient runoff from the U.S. Mississippi basin creates a dead zone each year about the size of New Jersey in the Gulf of Mexico. Food output on land is expanding in part at the expense of that from the oceans.[28]

As world fertilizer use climbed from 14 million tons in 1950 to 134 million tons in 2000, it began in some countries to press against the physiological limits of plants to absorb nutrients. In response, the use of fertilizer has leveled off in the United States, Western Europe, and Japan, and now possibly China as well. In these countries, applying additional nutrients has little effect on production. Some parts of the world, such as the Indian subcontinent and Latin America, can still profitably use additional fertilizer. But for the world as a whole, the ninefold growth in the use of fertilizer—the engine that helped triple the world grain harvest during the last 50 years—is now history.[29]

Just as fertilizer removes nutrient constraints on production, irrigation can remove moisture constraints, enabling plants to realize their full genetic potential.

In some cases, irrigation simply boosts land productivity, but in others it permits an expansion of cropping onto land that is otherwise too dry to support agriculture.

While the world as a whole has nearly tripled land productivity since 1950, some countries have done even better. Over the last half-century, China, France, the United Kingdom, and Mexico have quadrupled wheat yield per hectare. The United States has quadrupled its corn yield per hectare. India has nearly done the same with its wheat yield.[30]

For several decades scientists generated a steady flow of new technologies designed to raise land productivity, but that is now slowing down. As it does, the backlog of unused technology is shrinking. In some countries, farmers are now literally looking over the shoulder of scientists at agricultural experiment stations, seeking new technologies. In countries where yields have already tripled or quadrupled, it is becoming difficult for farmers to continue raising yields. For example, wheat yields in the United States have increased little since 1983. Rice yields in Japan have gone up hardly at all since 1984.[31]

Even some developing countries are now experiencing a plateauing of grain yields. Between 1961 and 1977, rice yields in South Korea increased from 3.1 to 4.9 tons per hectare, a gain of nearly 60 percent. But they rose by only 1 percent between 1977 and 2000. Similarly, wheat yields in Mexico climbed from 1.7 tons per hectare in 1961 to 4.4 tons in 1982, a rise of 160 percent. Since then there has been little change. (See Figure 3–3.) As yields level off in more and more countries, both industrial and developing, expanding global grain output will become progressively more difficult.[32]

For the world as a whole, the rise in world grain yield per hectare since 1990 has slowed markedly. From 1950 until 1990, world grain yield per hectare rose 2.1 percent a year. Between 1990 and 2000, however, the annual gain was only 1.2 percent. (See Table 3–2.) Most of the rise in cropland productivity since agriculture began was compressed into the four-decade span between 1950 and 1990, when average yield per hectare climbed from 1.1 to 2.5 tons per hectare. The challenge during this decade will be to sustain a steady climb in grain yield per hectare, preventing it from dropping below the 1.2 percent of the 1990s.

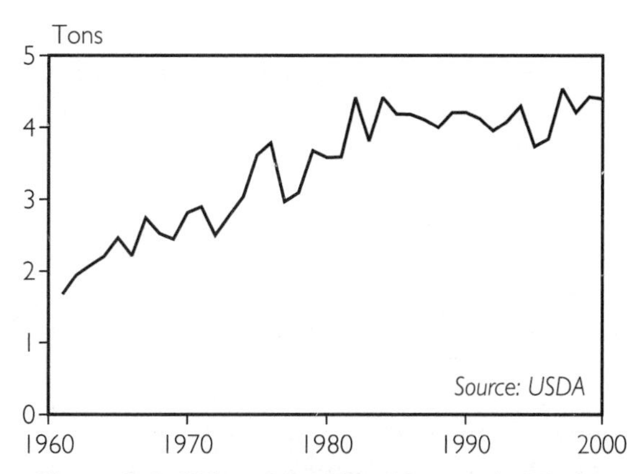

Figure 3–3. Wheat Yield Per Hectare in Mexico, 1961–2000

Table 3–2. Gains in World Grain Yield Per Hectare, 1950–2000

	Yield Per Hectare[1]	Annual Increase
	(tons)	(percent)
1950	1.06	
1990	2.47	2.1
2000	2.78	1.2

[1]Yield for 1990 is three-year average.

SOURCE: USDA, *Production, Supply, and Distribution*, electronic database, Washington, DC, updated September 2000.

Biotechnology is often cited as a potential source of higher yields, but although biotechnologists have been engineering new plant varieties for two decades, they have yet to produce a single variety of wheat, rice, or corn that can dramatically raise yields. The reason is that conventional plant breeders have already done most of the things they could think of to raise grain yields. One area where biotechnology might prove useful is in the development of more drought-tolerant crop varieties. Perhaps the largest question hanging over the future of biotechnology is the lack of knowledge about the possible environmental and human health effects of using genetically modified crops on a large scale over the long term.

Beyond increasing the yield per crop, land productivity can also be raised by increasing the crops per year where temperature and soil moisture permit. In China, for instance, double cropping of winter wheat and corn is widespread, enabling farmers in the north China plain to harvest two high-yielding grain crops each year. In northern India, the double cropping of winter wheat and summer rice is now commonplace, a key to sustaining India's population of 1 billion. Argentina and the United States both double crop winter wheat with soybeans as a summer crop.[33]

Although the United States and China occupy similar latitudes, double cropping is not as common in the former, partly because until recently, farmers' eligibility for government support prices depended on their willingness to restrict the area planted, thus discouraging multiple cropping. While there was surplus land, there was little reason to seriously consider double cropping.

Future increases in double cropping in the United States could provide a strategic assist in expanding the world food supply, while at the same time disrupting pest cycles and—because the land is covered with vegetation throughout the year—reducing soil erosion. At present, 2.5 million hectares (6.2 million acres) of the U.S. soybean harvested area of 29.1 million hectares (72.8 million acres) are double cropped, mostly with wheat, making this by far the most common double cropping combination.[34]

The expanding market for U.S. farm products is stimulating interest in the potential for double cropping in midwestern state agricultural experiment stations. In Missouri, for example, double cropping sunflowers, which are more drought-resistant, with wheat is an attractive alternative to double cropping with soybeans in dry years. In Nebraska, the double cropping of wheat and soybeans can be expanded if one of the crops is harvested early for forage. Harvesting the wheat as hay before the grain is ripe means it can be harvested two or three weeks earlier, thus making it possible to double crop with soybeans further north, where the growing season is somewhat shorter. And similarly, soybeans can be grown for forage, either as hay or ensilage, and harvested while the beans are still green in the pod. These are but a few examples, but if world demand for U.S. farm products increases as projected, many new double cropping combinations are likely to emerge in order to get the highest possible output from each hectare.[35]

Raising Water Productivity

Over the last half-century, world irrigated area tripled, climbing from 90 million hectares in 1950 to an estimated 270 million in 2000. Most of the growth occurred from 1950 to 1978, when irrigated area expanded faster than population and boosted irrigated land per person from 0.037

hectares to 0.047 hectares, an increase of one fourth. Since 1978, however, the growth in irrigated area has slowed, falling behind that of population and shrinking irrigated area per person by 8 percent in 2000. (See Figure 3–4.)[36]

During the next half-century, the combination of aquifer depletion and the diversion of irrigation water to nonfarm uses may bring the historical growth in irrigation area to an end. If so, the world will be facing a steady shrinkage in the irrigated area per person as long as population continues to grow. This in turn could translate into falling food production per person.

In many countries, the competition for water between the countryside and cities is intensifying. As countries industrialize, water use in industry climbs, often at the expense of agriculture. Although projections of the future diversion of irrigation water to residential and industrial uses do not exist for most countries, a World Bank forecast for South Korea—a relatively well-watered country—gives some sense of what may lie ahead. Like many countries, Korea today is using virtually all available water. The Bank calculates that if the Korean economy grows 5.5 percent annually until 2025, growth in water withdrawals for residential and industrial use will reduce the yearly supply remaining for irrigation from 13 billion to 7 billion tons, cutting it by nearly half. The study notes that higher water prices and specific steps to use water more efficiently will ameliorate the reduction in water for irrigation, but this analysis nonetheless shows how difficult it may be in some countries even to maintain existing irrigated area.[37]

Farmers everywhere face an uphill battle in the competition for water since the economics of water use do not favor agriculture. In China, for example, a thousand tons of water can be used in agriculture to produce one ton of wheat worth perhaps $200 or it can be used to expand industrial output by $14,000. Wherever economic growth and the creation of jobs are a central preoccupation of political leaders, scarce water will likely go to industry. China, the world's largest grain producer, now gives cities and industry top priority in the competition for scarce water, leaving agriculture as the residual claimant.[38]

In addition, key food-producing countries that are overpumping, such as China, India, the United States, and Mexico, will lose irrigation water from aquifer depletion. Once the rising demand for water surpasses the sustainable yield of an aquifer, the gap between demand and sustainable yield widens each year. As it does so, the drop in the water table also increases each year. In this situation, food production becomes progressively more dependent on the aquifer's depletion. When it is depleted, pumping is necessarily reduced to the amount of annual recharge. The resulting precipitous drop in food production could lead to rising food prices and political instability.[39]

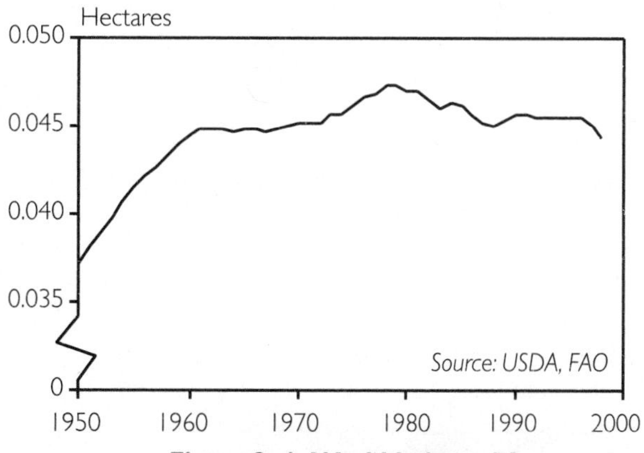

Figure 3–4. World Irrigated Area Per Person, 1950–98

The need for water in the Indian subcontinent as a whole is already outrunning the supply. Water tables are falling in much of India, including the Punjab, the country's breadbasket. The excessive use of water is encouraged by electricity subsidies to farmers, who use mostly electric pumps for irrigation. The low cost of pumping encourages farmers to use more water rather than invest in water efficiency. Phasing out water subsidies may be the only way to avoid potentially precipitous drops in food output when aquifers are depleted. The adverse economic effect on farmers often can be offset with programs to train them in water use efficiency.[40]

The key to raising water productivity is pricing water at its market value.

In sub-Saharan Africa, the potential for irrigation is limited simply because so much of the continent is arid and semiarid. The greater promise here may lie in water harvesting and systematically building soil organic matter so that soils can absorb and retain more of the low rainfall. The construction of earthen terraces supported by rocks retains water and reduces erosion of soil by water. Leguminous trees planted as windbreaks reduce wind erosion and add nitrogen to the soil.

For the world as a whole, the situation with water today is similar to that with cropland at the middle of the last century: the opportunities for developing new supplies are fast disappearing. By 1950, the frontiers of agricultural settlement had largely vanished, leaving little productive new land to plow. In response, governments launched a broad-based effort to raise land productivity, one that included government price supports for farm commodities that encouraged farmers to invest in yield-raising inputs and land improvements, heavy public investment in agricultural research to raise crop yields, and the building of public institutions to support this effort—from agricultural extension services to farm credit banks. Societies mobilized a wide array of resources to raise land productivity.

As the new century begins, a similar broad-based effort is needed to raise water productivity, but we lack even the vocabulary and indicators to talk about this goal, not to mention the data-gathering systems to measure progress toward it. Lacking precise data, we use the rule of thumb that it takes 1,000 tons of water to produce one ton of grain, and then use this ratio to roughly calculate future water needs in agriculture. In reality, we need a common measure of water productivity, such as kilograms of each grain produced per ton of irrigation water. To do this, we need more precise data on water use by individual crops in specific situations if we are to systematically boost water efficiency in agriculture. For instance, farmers will want to ask seed salesmen about the grain yield per ton of water used, much as they have traditionally asked for the yield per hectare of land used.[41]

There are several avenues to raising water productivity, but the key is pricing water at its market value, a step that leads to systemic advances in efficiency. The attraction of market pricing is that it promotes rational water use throughout the economy, affecting decisions by all water users.

With 70 percent of the water that is diverted from rivers or pumped from underground being used for irrigation, any gains in irrigation water efficiency have benefits that extend far beyond agriculture. Indeed, getting enough water for cities and industry while maintaining food production may be possible only if irrigation productivity is systematically raised worldwide.[42]

Among the specific steps needed to boost water productivity is the use of more water-efficient irrigation practices. There are many ways to irrigate crops, including furrow, flood, overhead sprinkler, and drip irrigation. Furrow irrigation, probably the earliest form, is used with row crops, with a small trench being cut near each row of plants. Flood irrigation, traditionally used on rice, is now being reconsidered since recent research indicates that in at least some situations periodic flooding will produce the same yield as continuous flooding, but use much less water.[43]

Overhead sprinkler irrigation, which is widely used in the U.S. Great Plains, is often coupled with the use of underground water. The circles of green crops that can be seen when flying over the central and southern Great Plains of the United States during the summer are created with water from center-pivot overhead sprinklers that use well water to irrigate. (In this region, most of the water is drawn from the Ogallala aquifer—essentially a fossil aquifer since its recharge is limited.) Shifting from a high-pressure to a low-pressure overhead sprinkler system can increase irrigation efficiency from, say, 65 percent to 80 percent. Shifting to a low-energy precision application sprinkler system can raise efficiency of water use to 90 percent or better.[44]

Drip irrigation, a technology pioneered in Israel, is extraordinarily efficient; it uses a pipe or a plastic hose with holes that is either laid on the soil surface or installed several inches below the surface. This high-cost, labor-intensive form of irrigation is generally profitable only on high-value crops such as fruits and vegetables. Where profitable, however, it can easily cut water use by half.[45]

Another way to raise irrigation water productivity is to shift to more water-efficient crops. For example, wheat typically produces half again as much grain per unit of water as rice does. Shaping policies to take advantage of the varying water efficiencies of grains has not progressed far simply because data on the water productivity of individual grains are not widely available.[46]

As a general matter, the higher the yield of a crop, the more productive the water use. For example, a rice crop that yields four tons per hectare uses little more water than one that yields two tons per hectare simply because so much of the water used to produce rice by flooding is lost through evaporation from the water surface. Since this evaporation is essentially a constant, producing more tons per hectare substantially reduces the water use per ton. Simply put, raising land productivity also raises water productivity.

Restructuring the Protein Economy

The demand for meat—beef, pork, poultry, and mutton—is everywhere on the rise. Perhaps driven by the taste for meat acquired during our 4 million years as hunter-gatherers, this innate hunger for animal protein, which manifests itself in every society when incomes begin to rise, has lifted the world demand for meat each year for 40 consecutive years. One of the most predictable trends in the global economy, world meat consumption climbed from 44 million tons in 1950 to 217 million tons in 1999, an increase of nearly fivefold. (See Figure 3–5.) This growth, roughly double that of population, raised meat intake per person worldwide from 17 kilograms in 1950 to 36 kilograms in 1999.[47]

Once the limits of rangelands and fish-

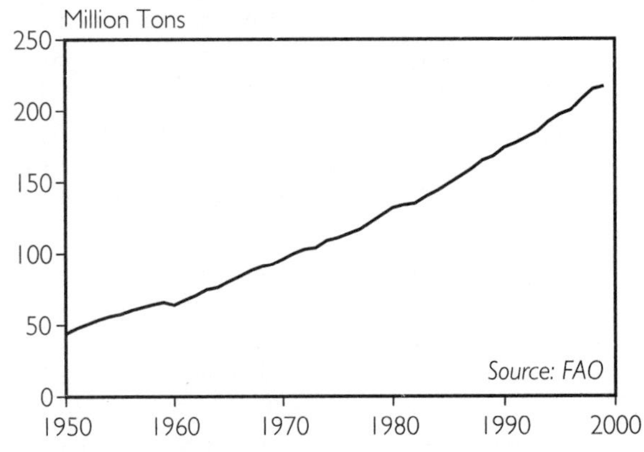

Figure 3–5. World Meat Production, 1950–99

eries are reached, then the growing demand for animal protein can be satisfied by feeding cattle in feedlots or fish in ponds; by expanding the production of pork, poultry, and eggs, all largely dependent on feed concentrates; or by producing more milk.

In this new situation, the varying efficiency with which grain is converted into protein—beef, pork, poultry, and fish—begins to shape production trends. Cattle in feedlots require roughly seven kilograms of feed concentrate per additional kilogram of live weight. For pigs, the ratio is less than four to one. Chickens are much more efficient, using scarcely two kilograms of grain concentrate for one of weight gain. Fish, including both herbivorous and omnivorous species, require less than two kilograms of grain concentrate to gain an additional kilogram.[48]

There are three ways to increase animal protein supply without consuming more grain. One is to improve the efficiency of grain conversion into animal protein through breeding and better management. A second is to shift from the less efficient forms of conversion, such as beef in the feedlot or pork, to the more efficient ones, such as poultry or farmed fish. A third approach is to rely on ruminants to convert more roughage into either meat or milk.

Not surprisingly, the economics of the varying conversion rates is accelerating growth in output among the more efficient converters. The world's existing feedlots are being maintained, but there is little new investment in feedlots simply because of the higher cost of fed beef. From 1990 to 1999, world beef production increased very little—only 0.5 percent a year compared with 2.5 percent a year for pork. Not surprisingly, the most rapidly growing source of meat during this period was poultry, expanding at 4.9 percent annually. (See Table 3–3.)

The oceanic fish catch has increased hardly at all since 1990, thus falling far behind the soaring growth in demand for seafood. In response, aquacultural output expanded from 13 million tons of fish produced in 1990 to 31 million tons in 1998, growing by more than 11 percent a year. Even if the growth of aquacultural output slows somewhat during the current decade, world aquacultural output is still on track to overtake the production of beef before 2010.[49]

Public attention has focused on the forms of aquacultural production that are particularly disruptive environmentally, such as salmon, a carnivorous species, and shrimp. Yet world aquaculture is dominated by herbivorous species, importantly carp in China and India, but also catfish in the United States and tilapia in several countries.[50]

China is the world's leading aquacultural producer, accounting for 21 million tons of the global output of 31 million tons in 1998. Its output is rather evenly divided between coastal and inland areas. Coastal output is dominated by shellfish—mostly oysters, clams, and mussels. It also includes small amounts of shrimp or prawns and some fish. Coastal aquaculture is sometimes environmentally damaging because it depends on converting wetlands into fish farms or because it concentrates waste, leading to damaging plankton blooms.[51]

Most of China's aquacultural output is produced inland in ponds, lakes, reservoirs, and rice paddies. Some 5 million hectares of land are devoted exclusively to fish farming, much of it in carp polyculture. In addition, 1.7 million hectares of rice land are used to produce rice and fish together.[52]

Over time, China has evolved a fish polyculture using four types of carp that feed at different levels of the food chain, in effect emulating natural aquatic ecosystems. Silver carp and bighead carp are filter feeders, eating phytoplankton and zooplankton, respectively. The grass carp, as its name implies, feeds largely on vegetation, while the common carp is a bottom feeder, living on detritus that settles to the bottom. Most of China's aquaculture is integrated with agriculture, enabling farmers to use agricultural wastes, such as pig manure, to fertilize ponds, thus stimulating the growth of plankton. Fish polyculture, which typically boosts the fish yield per hectare over that of monocultures by at least half, also dominates fish farming in India.[53]

As land and water become scarce, China's fish farmers are intensifying production by feeding more grain concentrates to raise pond productivity. Between 1990 and 1996, China's farmers raised the annual pond yield per hectare from 2.4 tons of fish to 4.1 tons.[54]

In the United States, catfish, which require only 1.8 kilograms of feed to gain 1 kilogram of live weight, are the leading aquacultural product. U.S. catfish production of 600 million pounds (270,000 tons) is concentrated in four states: Mississippi, Louisiana, Alabama, and Arkansas. Mississippi, with some 174 square miles (45,000 hectares) of catfish ponds and easily 60 percent of U.S. output, is the catfish capital of the world.[55]

Although rangelands are being grazed to their full capacity and beyond, there is a large unrealized potential for feeding agricultural residues—rice straw, wheat straw, or corn stalks—to ruminants, such as cattle, sheep, and goats. Ruminants have a highly sophisticated digestive system, one that can convert straw and corn stalks into meat and milk without using the grain that can be consumed by humans. At present, most food consumption comes from the photosynthate going into seed, but by feeding animals straw and corn stalks, the photosynthate that goes into stems and leaves also can be converted into food.[56]

In India, both water buffalo, which are particularly good at converting coarse

Table 3–3. World Production of Animal Protein by Source, 1990–2000

Source	Annual Rate of Growth
	(percent)
Aquaculture[1]	11.4
Poultry	4.9
Pork	2.5
Beef	0.5
Oceanic fish catch[1]	0.1

[1] 1990–98 only.

SOURCE: FAO, *Yearbook of Fishery Statistics: Capture Production and Aquaculture Production* (Rome: various years); FAO, *FAOSTAT Statistics Database*, <apps.fao.org>, updated 5 April 2000; USDA, FAS, *Livestock and Poultry: World Markets and Trade* (Washington, DC: March 2000).

roughage into milk, and cattle figure prominently in the dairy industry. India has been uniquely successful in converting crop residues into milk, expanding production from 19 million tons in 1966, the first year for which data are available, to 79 million tons in 2000—more than a fourfold increase in 34 years. Following a path of steady growth, milk became India's leading farm product in value in 1994. In 1997, India overtook the United States to become the world's leading milk producer. (See Figure 3–6.) Remarkably, it did so by using farm byproducts and crop residues rather than grain for feed. India was able to expand the protein supply without diverting grain from human consumption to cattle.[57]

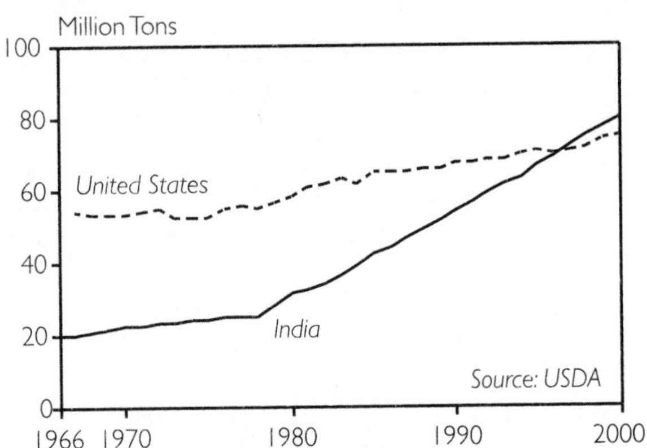

Figure 3–6. Milk Production in India and the United States, 1966–2000

Between 1966 and 2000, India's milk production per person increased from 0.7 liters per week to 1.5 liters, or roughly 1 cup of milk per day. Although this is not a high level of consumption by industrial-country standards, it is a welcome expansion of the animal protein supply in a protein-hungry country.[58]

In 1997, India overtook the United States to become the world's leading milk producer.

The dairy industry structure in India is unique in that the milk is produced almost entirely by small farmers, who typically have only one to three cows. Milk production in India is integrated with agriculture, involving an estimated 70 million farmers. For them it is a welcome source of supplemental income. Dairying, even on a small scale, is a labor-intensive process, including gathering the roughage to stall-feed the cows, milking them, and transporting the milk to market. Ownership of only a few cows or buffalo, which is typical in India, also means a supply of manure for cooking fuel and for fertilizer. One of India's challenges is to introduce new energy sources for cooking in order to reduce the cow manure used as fuel, thereby increasing that available for fertilizer.[59]

China also has a large potential in the grain-producing regions to feed corn stalks and wheat and rice straw to cattle or sheep. As the world's largest producer of both rice and wheat and the second ranked producer of corn, China produces some 500 million tons of straw, corn stalks, and other crop residues. At present, much of this either is burned, simply to dispose of it, or it is used in villages as fuel. Fortunately, China has vast wind resources that can be harnessed to produce electricity for cooking, thus freeing up roughage for feeding additional cattle or sheep.[60]

In a scientific analysis of the potential for feeding straw and corn stalks to cattle, Gao Tengyun of the Veterinary College at Hunan Agricultural University discusses the value of ammoniating crop residues (that is, incorporating nitrogen) to increase both the palatability and the digestibility of roughage. The addition of nitrogen helps the microbial flora in the rumen of the cattle and sheep to digest the roughage more completely. The use of this technology in the north China plain, particularly in the crop-producing provinces of Heibei, Shandong, Henan, and Anhui, has created what U.S. Department of Agriculture analysts call the "Beef Belt." Beef output in these four provinces now dwarfs that of the grazing provinces of Inner Mongolia, Qinghai, and Xinjiang.[61]

Even though the Chinese, like other East Asians, have traditionally not used dairy products and are often lactose-intolerant, milk consumption is now catching on. Since the economic reforms in 1978, milk production has climbed from 2.8 million to 10.4 million tons in 1999 as an embryonic dairy industry has begun to develop. Milk output is scarcely one eighth that of India, but it is expanding rapidly.[62]

The feeding of crop residues to ruminants means a given crop can yield two harvests—the grain that is initially harvested and then the meat or the milk that is produced with the straw and corn stalks. With human demands for food now pressing against the photosynthetic capacity of croplands, this processing of corn stalks and straw by ruminants into food enables farmers to make fuller use of existing photosynthetic products.

Ruminants also produce soil-enriching manure. This valuable fertilizer not only returns nutrients to the soil, it also adds organic matter, enhancing its productivity by improving both soil aeration and water retention capacity. Roughage-based livestock systems are almost necessarily local in nature because roughage is so bulky, and not easily transported.

Satisfying the demand for protein in a protein-hungry world where water scarcity is likely to translate into grain scarcity is a challenge to agricultural policymakers everywhere. As pressures mount on the world's food supply and grain becomes scarce, as now seems likely, other countries, such as the United States, Canada, and France, may follow India's example of using ruminants to systematically convert crop residues into food.

Eradicating Hunger: The Key Steps

This chapter began by noting that eradicating hunger will now take a superhuman effort, one that goes far beyond agriculture. It may depend as much on the efforts of family planners as on farmers and as much on the decisions made in ministries of energy that shape future climate trends as on decisions made in ministries of agriculture.

Within agriculture, raising land productivity deserves even greater priority today than in the past. This means raising crop yields in biological terms wherever possible, including minor crops. It also means more multiple cropping, a potential not yet fully realized in all countries.

Given the constraints on crop yields imposed by inadequate soil moisture, raising water productivity is a key to further gains in land productivity. Governments running the risk of an abrupt drop in food production as a result of aquifer depletion may be able to avoid such a situation only by simultaneously slowing population growth and raising water productivity in order to stabilize water tables.

Eradicating hunger also means getting more out of an existing harvest. It means increasing the efficiency with which the 35 percent of the world grain harvest that is fed to livestock, poultry, and fish is converted into animal protein. Simply put, it means eating less feedlot beef and less pork and more poultry and farmed fish or plant sources or protein. Within fish farming, it may also mean replacing fish monocultures with polycultures following the highly successful Chinese model. For the world's affluent, it means moving down the food chain by consuming less fat-rich livestock products, something they should be doing for health reasons anyway.[63]

For sub-Saharan Africa, eradicating hunger depends on curbing the HIV/AIDS epidemic.

Ruminants can be used to convert crop byproducts into animal protein. In effect, this offers a second harvest from a given crop while converting the roughage into manure that goes back on the land to maintain fertility. India, with its enormously successful effort to mobilize small farmers to use cattle to convert roughage into milk, is the model in this case.

Now that the backlog of agricultural technology available to raise land productivity is shrinking, providing enough food for children to realize their full physical and mental potential may depend on strengthening international agricultural research assistance. Appropriations for agricultural research are lagging far behind needs. For some farmers, the technology pipeline is running dry. It is not just wheat farmers in the United States or rice farmers in Japan who are exhausting the supply of new technologies to raise land productivity, but also wheat farmers in Mexico and rice farmers in India and South Korea. More locally oriented investment in agricultural research that will help expand multiple cropping and intercropping could pay large dividends.

Eliminating hunger in this new century requires an effort that goes far beyond agriculture. Stabilizing population is as essential as it is difficult. If rapid population growth continues in many developing countries, it will lead to further land fragmentation of holdings, as well as to hydrological poverty on a scale that is now difficult to imagine. Literally hundreds of millions of people will not have enough water to meet their most basic needs, including food production.

Addressing this challenge requires an educational effort on two fronts. The first links various family sizes, say two-child and four-child families, to population projections and then relates these to the availability of land, water, and other basic resources. Without this information, individuals may not understand the urgency of shifting to smaller families. In addition, if policymakers lack this information, they will not be able to make responsible decisions on population and related policies, such as investment in family planning services.

The second front is the educational level of women. As female educational levels rise, fertility falls. As women's education rises, the nutrition of their children improves, even if incomes do not rise, apparently because more education brings a better understanding of nutrition. We should be educating young women in developing countries as though future progress depends on it, because indeed it may.[64]

The numbers used in discussing future population growth in this chapter are the U.N. medium projections, those that have world population going from 6 billion at present to nearly 9 billion by 2050. There is also a high projection, which has human numbers approaching 11 billion by 2050, and a low projection, which has population peaking at 7.5 billion in 2040 and then declining. (See Figure 3–7.)[65]

This low number assumes that the entire world will quickly move to replacement-level fertility of two children per couple.

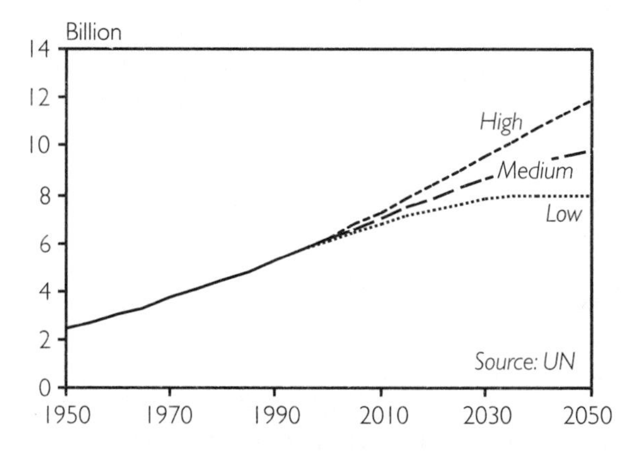

Figure 3–7. Total World Population, 1950–98, with Projections to 2050

This is not only achievable, but it may be the only humane population option, simply because if the world does not shift quickly to smaller families, the spreading land and water scarcity that is already translating into increasing hunger and mortality in some countries could reach many more.

Many of the key decisions to achieve this lower figure have to be made by national governments, but unless world leaders—the Secretary-General of the United Nations, the President of the World Bank, and the President of the United States—urge governments and couples everywhere to adopt a goal of two surviving children per couple, hunger will not likely be eradicated, or even greatly reduced. The issue today is not whether individual couples can afford more than two children, but whether Earth can afford couples having more than two children.

For sub-Saharan Africa, eradicating hunger depends on curbing the HIV/AIDS epidemic, which is depriving societies of healthy adults to work in the fields. If Africa continues to lose its adult breadwinners, it will further complicate efforts to eradicate hunger. Fortunately, this epidemic can be curbed. Uganda, one of the earliest countries hit by the HIV epidemic, has reduced the share of the adult population infected with the virus from 14 percent to 8 percent. The key to this admirable success has been the active personal leadership of President Yoweri Museveni at every step of the way. He has regularly addressed the issue in public, describing the mechanisms for transmitting the virus and the changes in behavior needed to curb its spread. Uganda is the model for curbing the spread of HIV/AIDS.[66]

A half-century ago, no one was concerned about climate change. But if we cannot now accelerate the phaseout of fossil fuels, more extreme climate events may disrupt food production, threatening food security. Of particular concern is the rise in sea level that could inundate the river floodplains in Asia that produce much of the region's rice. The rise in sea level over the last century of 20 centimeters (8 inches) or more is already affecting some low-lying coastal regions. If sea level rises by 1 meter during this century, the upper level projected, it will take a heavy toll on food production, especially in Asia. Here the principal responsibility lies with the United States, the leading source of carbon emissions. If the United States does not assume a leadership role in phasing out fossil fuels, the global effort to stabilize climate is almost certain to fail.[67]

In countries such as India, where the average farm size is continuing to shrink, it becomes more difficult to raise land productivity enough to provide adequate nutrition. The challenge in these areas is to mobilize capital both through domestic savings and by attracting investment from abroad to build the factories needed to provide employment and income in rural areas. This will keep rural families and communities intact. In this case, the model is China, which has achieved high savings rates and attracted record amounts of foreign capital.[68]

In a report outlining steps to eradicate hunger, the World Bank talks about the "indifference of many governments" to these issues. This indifference is often evident in the priorities in the use of public resources. In some ways, India today is paying the price for its earlier indiscretions when, despite its impoverished state, it invested in a costly effort to design and produce nuclear weapons to become a member of the exclusive nuclear club. Spending three times as much for military purposes as for health and family planning, India now has a nuclear arsenal capable of protecting the largest concentration of hungry people on Earth.[69]

Unless political leaders are willing to commit themselves to taking the difficult steps just described, bland assertions that we will eradicate hunger are meaningless. The time for such statements is past. We need to be honest with ourselves and either acknowledge that with business as usual there will be even more hungry people in the future, or decide to take the demanding actions needed to reverse the trends that cause hunger.

If we do not act decisively, there is a real possibility that the food situation could deteriorate rapidly in some developing countries. The world could begin to slide backward, with hunger affecting ever more people. Spreading food insecurity could lead to political instability on a scale that would disrupt global economic progress. Everyone has a stake in ensuring that the difficult steps needed to eradicate hunger are taken, lest spreading hunger and its consequences spiral out of control.

Chapter 4

Deciphering Amphibian Declines

Ashley Mattoon

In 1973, an Indonesian scientist visiting Australia discovered a nondescript, muddy-brown frog hidden in the streams of the remote, densely forested mountains of southeastern Queensland. The frog attracted a great deal of interest, since it was the first aquatic stream frog known to inhabit the continent. It was unusual in other ways as well. Michael Tyler, an Australian herpetologist, noticed that the frogs would occasionally collect in pools of water in groups of three or four—forming sort of a ring of frogs, touching by the tips of their toes.[1]

But the strangest characteristic was discovered in November 1973, when two researchers in Brisbane were cleaning an aquarium that held one of the frogs. They saw it swim up to the surface of the water and spit out six live tadpoles. They couldn't quite believe what they saw: where could the tadpoles have come from? Perhaps the species was related to a South American frog whose fertilized eggs develop in the father's vocal sacs. But they also noticed a peculiar mass of wriggling bodies under the skin of the frog's stomach. Soon another tadpole was found in the aquarium. Then two more appeared. The researchers finally decided to remove the frog and dissect it. As they reached into the aquarium and grabbed the animal, it flexed backward and opened its mouth wide. As the researchers described in *Science*, "eight juvenile frogs were then propulsively ejected in groups of two or three in no more than 2 seconds. Two further juveniles were ejected after a few minutes and, 30 seconds after the frog had been transferred to preservative, three more were disgorged similarly."[2]

Scientists found that the frog incubated its eggs in the mother's stomach—a form of parental care that was completely unprecedented among vertebrates. That gave the species its common name: the gastric-brooding frog. Normally, whatever a frog swallows would be digested by the stomach's natural chemicals. But somehow this species was able to turn off its digestive enzymes for the six or so weeks it took for the eggs to develop. This was particularly intriguing as it suggested the frog's physiology could hold clues to treating people with gastric ulcers and other stomach ailments. But in 1981, as research was getting under way, the frog suddenly disappeared from the mountain streams where it had once been so abundant. It has not been found again since.[3]

By the end of the 1980s, reports of declines and disappearances emerged from most regions where amphibians were reasonably well monitored.

It was just pure luck that scientists happened to stumble on the gastric-brooding frog in the twilight of its existence. Here was a species humans had been living beside for millennia, yet somehow it had escaped notice. And this creature had a unique adaptation for brooding its young—if we had not actually seen it, we would never have thought it possible. The frog had a lot to teach us, but for some unknown reason, our time with it was cut short.

The loss of the gastric-brooding frog attracted a great deal of attention in the scientific community, largely because it was such an unusual creature. But it turns out that its mysteriously sudden disappearance was not exactly unique to the species. By the end of the 1980s, at least 13 other amphibian species had gone into serious decline or disappeared entirely from mountainous regions of eastern Australia. The southern dayfrog, which had commonly been found in the same areas as the gastric-brooding frog, has not been seen in the wild since 1979. Three other species in southeastern Queensland have declined by more than 90 percent. Further north, in the tropical forests near Eungella, two species declined sharply in 1985; one of them has not been seen since. And in the tropical rain forests of northern Queensland, large-scale declines began in 1989—seven species dipped sharply; four can no longer be found in the wild. All these species were locally endemic: they did not exist anywhere else in the world.[4]

Several hypotheses about Australian amphibian declines have been proposed, but there are no widely accepted, definitive answers. And the mystery spreads beyond Australia. By the end of the 1980s, reports of declines and disappearances emerged from most other regions of the world where amphibians were reasonably well monitored—in North America, parts of South America, and Europe. In Costa Rica's Monteverde National Park, populations of 20 of the 50 native amphibian species have declined or disappeared since 1987. And in California's Sierra Nevada, five out of seven native amphibians have disappeared or declined sharply since the early 1900s. (See Table 4–1.)[5]

As news of the declines began to surface around the world, scientists wondered if they were simply anecdotal accounts that reflected natural population fluctuations. But there were some uncanny similarities—patterns that suggested something more ominous could be taking place. The declines were, in the first place, very rapid. They sometimes involved whole assemblages of species, rather than just one or two. And they were occurring not just in areas that were obviously disturbed, but in some of the world's most carefully protected parks, such as Costa Rica's Monteverde Cloud Forest Reserve and Yosemite National Park in the western United States. These were not the kinds of losses that could be readily predicted—or explained. Something peculiar was happening to *Amphibia*—something bad enough to dis-

Table 4–1. Selected Large-Scale Losses of Amphibians

Location	Species of Concern	Status	Possible Cause
Montane areas of eastern Australia	14 species of frogs, including the southern day frog and the gastric-brooding frog.	Sharp population declines since the late 1970s. Four species are thought to have become extinct.	Parasitic fungus, possibly introduced through international trade in aquarium fish and amphibians.
Monteverde region of Costa Rica	20 species of frogs and toads (40 percent of total frog and toad fauna), including the golden toad.	Disappeared after simultaneous population crashes in 1987. Missing throughout 1990–94 surveys.	Climate change, possibly combined with other factors such as parasites.
Las Tablas, Costa Rica, and Fortuna, Panama	More than 15 species.	Severe declines since the early 1990s.	Chytrid fungus infection, possibly combined with other factors.
Yosemite region of California	Five out of the region's seven frog and toad species—including the mountain yellow-legged frog and the foothill yellow-legged frog.	Severe declines—one species has disappeared entirely, another has declined to a few small populations.	Overall cause unknown. Introduced predatory fish combined with drought-induced loss of habitat contributed to the decline of some species.
Montane areas of Puerto Rico	12 of 18 endemic amphibian species.	Three may be extinct, the others are in decline or at risk.	Unknown. Possibly climate change.
Cordillera de Mérida, Venezuela	Five species.	Sharp declines since the 1970s.	Unknown. Possibly deforestation, floods, roadkill, pollution.
Reserva Atlântica, Brazil	8 out of the reserve's 13 native frogs.	Disappeared or declined in the 1980s.	Unknown. Possibly unusually dry winters, pollution, or a combination of both.

SOURCE: See endnote 5.

tinguish it from the broader tragedy known as the biodiversity crisis.

More than a decade since amphibian declines first emerged as a major scientific mystery, there have been significant advances in our understanding of the issue. While there are still many unanswered questions, there is now little doubt that the problem is real, and that a complex mixture of causes is involved. More important than the many details, however, is the realization that it is not just the loss of frogs and salamanders that should concern us, but what their loss tells us about the state of the environment. Stemming the loss of amphibians will involve changes far beyond the protection of wetlands or forests. It will require fundamental changes in the way we live.

Why *Amphibia*?

The decline of amphibians seems particularly surprising, given that these creatures have been around for about 350 million years and were able to survive three mass extinctions. These events are thought to have killed off approximately half of all animal species, including the dinosaurs that disappeared 65 million years ago. Yet overall, amphibians persisted when so many other groups of organisms did not.[6]

When amphibians first appeared, Earth's terrestrial area was essentially one giant landmass inhabited by plants and insects. Amphibians were the first vertebrates to make the transition from water to land. Somehow, a type of bony fish evolved into a creature that had four legs, could breathe atmospheric oxygen instead of dissolved oxygen, and had a body structure that allowed it to maneuver without the support of water. It was these kinds of traits that laid the groundwork for other classes of vertebrates—such as reptiles, birds, and mammals—to develop.[7]

Since the first amphibian emerged from the water, the class has evolved into three distinct groups (or, to use the scientific term, orders). *Anura*, which includes frogs and toads, is the largest group. The order *Caudata* includes salamanders and newts. And *Gymnophiona*, the least-known group, which are commonly referred to as caecilians, are legless, subterranean creatures that are found only in tropical and subtropical regions of the world. (See Table 4–2.)

Scientists have thus far identified nearly 5,000 species of amphibians—more than the known number of mammals. And compared with other groups of organisms, new species of amphibians are being described at a phenomenal rate—the number of described species is increasing by about 1–1.5 percent a year. And there are still hundreds, perhaps thousands, that have yet to be discovered. Amphibians' collective domain includes every continent except Antarctica, and probably most of the world's major islands. They achieve their greatest variety in tropical and warm temperate forests. The neotropical realm is by far the richest; Central and South America house close to half of the world's amphibian species. But amphibians also live in deserts, grasslands, northern bogs—even tundra, in the case of the North American wood frog, one of four species of North American frogs that can survive in a practically frozen solid state.[8]

Given their durability and ubiquity, the rapid decline of so many species is particularly unsettling. Why are amphibians disappearing now? Many scientists claim that amphibians are important bioindicators—a sort of barometer of Earth's health, since they are more sensitive to environmental stress than other organisms. One trait that gives them this distinction is their "amphibious" nature—the typical amphibian lifecycle is partly aquatic and partly terrestrial. That can make them doubly vulnerable: disturbance of either water or land can affect them. In water, for example, some species have fairly narrow temperature requirements. Some do best in still water; others need flowing water.[9]

And many amphibians are particular about where they will breed. In southwestern California, the endangered arroyo toad does not reproduce well unless it lays its

Table 4–2. Global Distribution of Amphibians

Area	Frogs and Toads	Salamanders	Caecilians	Total Amphibia	Share of World's Amphibian Species
		(number of known species)			(percent)
United States and Canada	90	151	—	241	5
Central and South America and the Caribbean	2,135	248	82	2,465	50
Eurasia and North Africa	103	89	—	192	4
Sub-Saharan Africa	761	—	29	790	16
India, Southeast Asia	745	29	44	818	17
Australia, New Zealand, Papua New Guinea	443	—	—	443	9
Total	4,277	517	155	4,949	

SOURCE: William E. Duellman, "Global Distribution of Amphibians: Patterns, Conservation, and Future Challenges," in William E. Duellman, ed., *Patterns and Distribution of Amphibians: A Global Perspective* (Baltimore, MD: The Johns Hopkins University Press, 1999), 3. For specific boundaries among areas, see page 6.

eggs on the sandy bottom of a slow-moving stream. Some frogs and salamanders will lay eggs only in the shallow "vernal pools" that appear with the spring rains and disappear with the summer heat. This is a kind of evolutionary gamble with the weather: the young are safe from predatory fish in a vernal pool, but they must reach their terrestrial phase before the pool dries up.[10]

Given such preferences, it is no surprise that habitat degradation is the leading cause of amphibian decline. Many amphibians, for example, are forest animals—and the world is currently losing approximately 14 million hectares of natural forest each year, an area larger than Greece. Even when the result is not outright deforestation, logging can devastate amphibian populations. If forests are allowed to regenerate, amphibians can return, although estimates of how long that will take vary; in the southeastern United States, for example, the estimates range from 20 to 70 years.[11]

Consider the logging boom in the U.S. Southeast, where forests shelter one of the world's richest assemblage of salamanders. More than 65 percent of all salamanders belong to a lineage called the *Plethodontidae*, which lack lungs. These creatures breathe through their skin, which must remain moist at all times to facilitate gas exchange or they will suffocate. Plethodontids are consequently extremely sensitive to changes in temperature and humidity. Even selective logging is likely to reduce a population, because it opens up the canopy and dries out the forest floor. In the U.S. Southeast, the logging of mature hardwood forest, the primary salamander habitat, is expected to overtake the hardwood growth rate by the close of the decade.[12]

The Oregon spotted frog has largely disappeared from its historical range in the heavily farmed Willamette River Valley.

Deforestation-induced losses are almost certainly far greater in the tropics, although generally far less is known about them. An extreme case appears to be Sri Lanka. As recently as 1993, the island's amphibian fauna was thought to comprise only 38 species, but a recent five-year survey of the remaining rainforest may have turned up as many as 200 additional amphibian species that are apparently endemic (though these have not yet been formally described, nor has the estimate been published in a peer-reviewed journal). Yet Sri Lanka has only 750 square kilometers of rainforest left; over the past 150 years or so, the island has lost approximately 96 percent of its original rainforest cover. When survey researchers checked the records of naturalists who were exploring Sri Lanka before 1900, they found that more than half of the amphibians mentioned were no longer present. Most of Sri Lanka's surviving natural forests are legally protected, but they continue to dwindle in the face of illegal logging, primarily for fuelwood.[13]

It is true that in many parts of the world, large tracts of new "forests" are being created, but tree plantations do not generally provide an ecological substitute for natural forest. In Florida, for example, approximately 80 percent of the natural longleaf pine forest has been lost over the last 50 years, and the majority of what was cleared has been converted to commercial slash pine plantations. This is considered to be a primary reason why populations of the threatened flatwoods salamander have declined or disappeared in recent decades. The longleaf pine forest was the primary habitat for the salamander, which does not do well in the drier, more uniform conditions of the planted slash pine forest. In South Africa, tree plantations have also come at a cost to local amphibians. The Matatiele wetland near Mtunzini was one of the last remaining natural coastal wetlands in the country and home to more than 25 species of frogs. But when an exotic tree plantation was installed next to it, the water table dropped and the wetland dried up.[14]

Amphibians also depend on many other types of habitat, and unfortunately the grasslands, wetlands, and streambeds of the world are not faring much better than the forests. In California, the conversion of native grasslands to farms and suburban developments is a leading reason for the endangered status of the California tiger salamander—at least 75 percent of the original grassland habitat has been irretrievably lost. And in the United Kingdom, the loss of breeding ponds is a primary reason for the decline of all six native species. Most of the world's rivers are now under the control of dams that cause flooding, scouring, and siltation—the kind of forces that are largely responsible for the disappearance of the endangered Arroyo toad from 75 percent of its former range in California.[15]

Although habitat loss is the leading cause of amphibian decline, it obviously cannot account for the many large-scale disappearances that have taken place in protected areas like Monteverde and Yosemite—places that would seem to be about as close to pristine as possible. And yet the amphibians in these areas are apparently reacting to dramatic changes. But these are changes that are hard to see, or that just do not seem "unnatural."

Toxics are one threat that is difficult to see, and there's no question that amphibians are highly vulnerable to all sorts of pollutants. Because they breathe and even drink through it, amphibians have thin, permeable skin that readily absorbs contaminants. Their eggs lack protective shells and are highly permeable as well. So it is hardly surprising that in heavily industrialized areas, pollution is frequently invoked as a cause for local declines. In some older centers of heavy industry, the pollution is so intense and pervasive that it's a wonder there are any amphibians left to study. Amphibian populations in much of Ukraine, for example, appear to be under attack from heavy metals, pesticides, aromatic hydrocarbons, acid rain, and radioactive waste.[16]

But pollution is taking a toll in healthy-looking landscapes as well. In Britain, the acidification of ponds is a contributing factor in the endangerment of the Natterjack toad. This toad is now nearly extinct in British lowland heath, a habitat that used to support about half the species' population in that country. (The toad is faring poorly in Scandinavia too, but it seems to be in better condition farther south.) California's Sierra Nevada range is losing many of its amphibians, and pesticide contamination could be a factor. The pesticides malathion, chlorpyrifos, and diazinon have been detected in precipitation at altitudes as high as 2,200 meters. The pesticides are not used in the mountains themselves; they are presumably drifting up from the state's heavily farmed lowlands.[17]

Fertilizers, which are used in far greater quantities than pesticides, may be creating problems we are even less prepared to counter. Some amphibians are very sensitive to the nitrogen compounds that typically leach out of artificially fertilized fields. For example, researchers have discovered that tadpoles of the Oregon spotted frog are poisoned by water with nitrate and nitrite levels low enough to pass the drinking water standards set by the U.S. Environmental Protection Agency. (Nitrate and nitrite are compounds that soil microorganisms make from synthetic nitrogen fertilizer.) Many water bodies in the United States contain levels of nitrate that violate government standards. If these are not taken seriously as a matter of public health, it seems unlikely—to say the least—that more stringent standards will be mandated for the welfare of frogs. The Oregon spotted frog has largely disappeared from its historical range in the heavily farmed Willamette River valley.[18]

Pollution, like habitat loss, is thus clearly a major factor in amphibian decline. But even considering both stresses, many declines remain unexplained. Not far from the Willamette valley, in the Cascades Range of Oregon, for example, the Cascades frog and the western toad are disappearing, even though their habitat has not been significantly disturbed or polluted. Researchers have found that these species are the victims of another stress: increased exposure to ultraviolet (UV) light, a consequence of the depleted stratospheric ozone layer, which filters much of the UV wavelength out of incoming sunlight. UV light can damage DNA and even kill cells. Amphibians, with their naked skins and eggs, are particularly at risk. The Cascades species may be losing their eggs to the extra UV exposure.[19]

It is possible that increased UV levels are injuring other amphibians as well, particularly at higher latitudes, where the ozone layer tends to be thinner. Unfortunately, seasonal fluctuations probably increase amphibian vulnerability: in either hemisphere, the ozone layer tends to be at its weakest during winter and spring, a period that overlaps with the egg-laying season for most species. Amphibians at higher elevations could be especially susceptible as well, since the higher you go, the less atmosphere there is to filter out the UV. But researchers have found that not all amphibians are especially sensitive to UV light, nor are all of them exposed to appreciable quantities of it.

Amphibians face another major pressure that, like UV light and pollution, is hard to see. Introduced, non-native species are often in plain view, but because they usually look perfectly "natural," it can be difficult to see them as a threat. Yet certain non-native species frequently prey on amphibians or out-compete them for food. In New Zealand, introduced rats are thought to have caused the extinction of several species of *Leiopelma* frogs. And in the Yosemite region of California's Sierra Nevada, large-scale trout introductions have played a major role in the disappearance or severe decline of five of the region's seven native amphibians. Further south, in California's Santa Monica Mountains, the recent spread of introduced mosquitofish (used widely to combat mosquitoes, but unfortunately they eat more than that) may be depressing local populations of the California newt and the Pacific treefrog.[20]

One invader that frequently injures amphibians is itself an amphibian: the bullfrog, native to the eastern United States. An aggressive, fast-growing species that can reach a length of 15 centimeters, the bullfrog has been introduced into ponds and marshes in many places around the world, as food and fishbait. It is not very particular about where it lives or what it eats. It will try to swallow almost anything it can fit in its mouth—including other frogs. It can also compete aggressively with other species for food. In California, where it was introduced in the early 1900s, the bullfrog may have contributed to the declines of several species. In South Korea, where it was imported for food in the early 1970s, the subsequent loss of native frogs and other small creatures inspired a government-sponsored anti-bullfrog campaign, with hunting contests and bounty prizes.[21]

Another organism that is apparently lurking in many forests and swamps may be consuming far more amphibian species than bullfrogs or trout do. In the mid-1990s, an unknown pathogenic fungus was found to be infecting frogs in Costa Rica, Panama, Australia, and the United States. Researchers in these places discovered the pathogen separately, but as they began to compare notes they realized that they were dealing with the same suspect.

Between 1993 and 1997, Karen Lips, a herpetologist studying amphibians in Costa Rica and Panama, discovered several dead and dying frogs at her research sites—a rare find, since dead frogs are usually snapped up quickly by scavengers. During this same period, she also observed population collapses in several species, and in five out of the seven Panamanian streams she monitored, frogs simply vanished altogether.[22]

Meanwhile, in Australia, researchers were beginning to wonder if an epidemic disease might be responsible for the series of declines and extinctions that occurred along the east coast of the country in the 1970s and 1980s. As with the Central American declines, the victims were stream dwellers and had succumbed very rapidly—traits suggestive of a highly virulent, waterborne pathogen. When researchers compared skin samples from Panamanian and Australian victims, they found them infected by the same type of organism: one or more fungi of the phylum *Chytridiomycota*. That came as a surprise, since chytrid fungi are common pathogens of plants and insects but had never before been reported to attack vertebrates.[23]

In the United States, the fungus was first noticed in 1991, when it started killing captive arroyo toads in California. In 1996, the toads' disease was identified as the chytrid fungus, which has since been found in the wild in various places around the country. In Illinois and Maryland, it has been encountered as an apparently benign infection of stable frog populations: the infected animals seemed perfectly healthy. But in 1999 and 2000, wildlife officials in the U.S. Southwest discovered the fungus in lowland leopard frogs outside Phoenix, Arizona, and in boreal toads near Denver, Colorado. In both cases, they found large numbers of dead and dying animals. Both species have been in sharp decline, and the fungus has become a prime suspect.[24]

Scientists are now wondering how many other declines the fungus might explain. Could it have been behind some of the older die-offs—events later marshaled as evidence for global amphibian decline? In many parts of the world, scientists are uncorking old specimen bottles, and in a few cases the chytrid has turned up. In Colorado, the chytrid has been found in a few preserved leopard frogs from the 1970s. It has also been detected in museum specimens of the Yosemite toad and the Northern leopard frog, collected from the California Sierra Nevada and the Colorado Rockies, respectively, during the 1970s when both species were in decline. In Australia, it has been traced back to 1978—to a treefrog collected in southern Queensland.[25]

In the mid-1990s, an unknown pathogenic fungus was found to be infecting frogs in Costa Rica, Panama, Australia, and the United States.

Even though our understanding of the fungus has grown tremendously in the last few years, there seem to be more unanswered questions now than ever before. For one thing, scientists do not yet know if there is more than one new species of fungus. Nor do they know how it kills its victims. It may suffocate them by causing their skin to thicken (amphibians breathe partly through their skin), or it may produce toxins. And there are plenty of other questions. Where, for instance, did it come from? It may have been widespread for a long time, and only recently detected, or it may have had a restricted range but recently spread to new regions of the world, where it is finding new host species.[26]

If the chytrid has been expanding its range to such distant parts of the world, what could be moving it around? Some herpetologists think it may have been brought into new terrain on the boots of tourists. Others see a likely conduit in the trade in aquarium fish, and in amphibians themselves. In Australia, some scientists suspect that the chytrid may have been transported across the continent on the skin of an infected frog that stowed itself away in a box of fruit. And once the fungus arrives in an area, there are many ways it could spread: it may be dispersed by local people, for example, or by cattle. It can even be carried by birds and insects.[27]

Another hypothesis is that the fungus could be a well-established pathogen in many parts of world, and that something is upsetting amphibian immune responses, making them more vulnerable to attack. After all, the chytrid is not the only pathogen that has been implicated in amphibian declines. A group of viruses called iridoviruses may have caused the tiger salamander die-offs in Arizona, Utah, Maine, and Saskatchewan. Iridoviruses have also triggered declines of the common frog and the common toad in the United Kingdom.[28]

In Costa Rica, 20 species have declined or disappeared since the late 1980s—including the famous golden toad.

Why so much disease in so many places in so short a time? It is possible that scientists are just finding more disease because they are getting better at looking for it. Or it might be evidence that some widespread stresses are throwing amphibian immune systems out of kilter.

By the late 1990s, evidence for one such stress turned up in Costa Rica, where 20 species have declined or disappeared since the late 1980s—including the famous golden toad. Working on the initial hunch that the declines may have had something to do with the weather, a group of climatologists and biologists led by Alan Pounds, head of Monteverde's Golden Toad Laboratory for Conservation, found strong evidence that the reserve's cloud forest is losing its clouds. According to the researchers, local sea surface temperatures have risen since the mid-1970s, and that has tended to push the cloud bank higher. The mountain tops are bathed in the clouds less frequently, so the forest is now somewhat drier. This might account for the amphibian losses. The theory is corroborated by bird observations: some lowland, "cloud-forest-intolerant" birds have moved upslope, into areas they had never occupied before. But the golden toad lived on top of the range and had nowhere to go.[29]

It is possible that the changing climate led to the toad's demise, but some scientists see another, more ominous possibility in the team's findings. Climate change could be "overlapping" with disease: the stresses of a changing climate could make amphibians more susceptible to infection. It is possible the fungus haunts Monteverde as well. Maybe the Monteverde declines are the result of a kind of synergism between the pathogen and the warming seas.

Climate change and spreading disease: both of these forces have a global reach and they could overlap in a number of ways—either simultaneously or in sequence. A change in the moisture regime, as at Monteverde, or a change in water temperature might weaken amphibian immune systems. Warmer water might also affect a pathogen's virulence, or its capacity to move from one animal to another. Warmer air temperatures might increase the range of insects that carry it.

Infections are likely to combine with other types of stress as well. Excessive UV exposure, like climate stress, could suppress amphibian immune systems. So could some forms of pollution. And some diseases appear to have been spread through the introduction of infected game fish like trout; in such cases, a new predator overlaps with a new disease.[30]

Other overlaps are also cause for concern. Consider, for example, non-native species. Organisms that are successful invaders are often capable of tolerating very disturbed conditions. If some kind of disturbance injures the native amphibians but does an invader no harm, then the latter may gain a level of dominance it might not otherwise have achieved. In the Ural Mountains of

Russia, this mechanism has apparently allowed the introduced lake frog to displace some of the natives. The lake frog seems to tolerate industrial pollution much better than the natives do, so the local frogs have been hit by a pollution-invasion overlap. In the Yosemite region of California, a combination of introduced trout and a five-year drought is considered to be responsible for the decline or disappearance of several species. The trout reduced the amphibians' ranges to isolated patches, and that increased their vulnerability to the drought that hit the area from 1987 to 1992.[31]

Overlapping factors may also be partly responsible for the recent rash of amphibian deformities—although there is no evidence that suggests a relationship between deformities and the declines. In some places where deformities have been observed, scientists believe the cause is a kind of parasitic flatworm called a trematode. As the trematode is often carried by snails, there may be a connection between nutrient-rich waters (for example, water receiving runoff from agricultural areas) that attract snails and trematode infections.[32]

Habitat loss, pollution, UV exposure, non-native species, disease, and climatic instability—those are the stresses that we know or suspect are killing off so many of the world's amphibians. Perhaps there are other, as yet unidentified stresses as well. But a simple inventory like this, dismal as it is, still does not adequately represent the threat because it does not account for the overlap factor. Unfortunately, we have virtually no idea what the overlaps will eventually do.

Beyond the Declines

The loss of amphibians demands our attention not just because we need to know why these animals are dying, but because we need to know what their death will mean. Amphibian decline itself is a form of environmental degradation, since amphibians play critical roles in many ecosystems. Given their sometimes secretive and inconspicuous nature, it is easy to underestimate their importance. Many species are only seen in large numbers when they are breeding or emerging from streams or ponds after they have developed from their aquatic to terrestrial stage. And in many parts of the world amphibians stay buried in forest soils or desert sand for several months at a time—waiting out unfavorable weather.

Despite their cryptic nature, amphibians sometimes outnumber—and may even collectively outweigh—other classes of vertebrates in many temperate and tropical woodlands. In Shenandoah National Park, in the southeastern United States, the Red-backed salamander has been found at densities as high as 10 per square meter. Since the park is a hotspot for salamander diversity, each square meter could well contain other salamander species too. In a New Hampshire forest, scientists found that the total mass of salamanders alone was more than double the mass of birds (even during the birds' peak breeding season) and about equal to the mass of small mammals.[33]

Because they are so numerous, amphibians play key roles in many aquatic and terrestrial ecosystems. As tadpoles, they devour vast quantities of algae and other plants. A Marsh frog tadpole, for example, can filter algae out of about a liter of water each day. In some small lakes and ponds, amphibians may be the most important animal regulators of algae and other plant growth. Algal growth is itself a primary regulator of aquatic oxygen levels; the more algae a body of water has, the less oxygen it is likely to have.[34]

Increased algal growth has already been

observed in parts of the world where amphibians have declined. In Ecuador, for example, algal growth rose in the streams where a species of poison dart frog was once abundant. Karen Lips noticed the same phenomenon following the disappearance of several frog species from her study sites in Central America. This kind of subtle ecological change is presumably occurring in many streams and ponds around the world.[35]

On land, amphibians play an important role as consumers of vast quantities of insects and other invertebrates. Some species also eat small reptiles, birds, and mammals. A study of the Blanchard's cricket frog in Iowa concluded that a small pond population of 1,000 frogs would consume about 4.8 million insects and other arthropods per year.[36]

The loss of amphibians could mean the loss of potential cures for many common ailments.

The pest control benefits of amphibians have long been recognized. A *Farmer's Bulletin* of the U.S. Department of Agriculture from 1915, for example, contained a paper entitled "The Usefulness of the American Toad" by entomologist A. H. Kirkland. After examining the stomachs of 149 toads, Kirkland determined that "injurious insects" such as gypsy moths, tent caterpillars, and may beetles made up 62 percent of the toads' total diet. In one stomach he found the remains of 36 tent caterpillars; in another, 65 gypsy moth caterpillars. He estimated that over 90 days a single toad could consume nearly 10,000 noxious insects. It was Kirkland's hope that toads could receive better protection and use as a natural pesticide in farms and gardens. At the time, the toad's most formidable threat, according to Kirkland, was the "small boy."[37]

Today amphibians get little more credit than they did in Kirkland's time, even though their usefulness is now much more evident. In the 1970s and 1980s, for instance, India and Bangladesh were exporting large quantities of frog legs to Europe, the United States, and Japan as a culinary delicacy. These harvests were intensive enough to virtually empty many large marshes of their frog populations. As the frogs disappeared, the insect populations exploded. Crop damage increased and there was a rise in malaria infections. Frog exports were banned in India in 1979. In Bangladesh, a ban was passed in 1990, after farmers realized that it was cheaper to protect frogs than to buy pesticides. Unfortunately, however, there is still a substantial frog smuggling operation in the region.[38]

In addition to their role as consumers of algae and insects, amphibians also support the many different types of animals that prey on them. They are important in the diets of many birds, mammals, reptiles, and freshwater fish. Some bats and snakes live exclusively on amphibians. In California's Sierra Nevada, for instance, there is a type of garter snake that depends on three of the local amphibian species, two of which are in serious decline. If the garter snake goes into decline as well, that could affect the predatory birds that depend on it. The fate of raptors may be bound up in the fate of frogs.[39]

In both aquatic and terrestrial food webs, amphibians often play an important role in energy transfer and nutrient cycling. Some scientists refer to amphibians as important "conveyor belts" of energy through the ecosystems in which they occur. Compared with birds and mammals, amphibians are much more efficient in converting food into new tissue, so a substantial amount of the energy that enters an ecosystem through photosynthesis passes through amphibians, who then are eaten by foxes, snakes, and owls. Some studies have found that amphibians can also be a much better source of protein than birds and mammals. A predator would be better off eating a pound of frogs rather than a pound of mice. And given their dual roles—aquatic juvenile stage and terrestrial adult stage—amphibians are often an important "nutrient bridge" from water to land. Through them, the algae that might otherwise suffocate a pond becomes a part of the terrestrial fauna.[40]

But the role of amphibians is often far more subtle and complex than food web dynamics might suggest. Consider, for example, the tiny frogs that live inside the tank bromeliads in the Brazilian Atlantic forest. Bromeliads are a type of plant common in tropical American forests; most of them look rather like the rosette of leaves on the top of a pineapple, and most of them are epiphytes—that is, they live in trees instead of rooting directly in the soil. The rosettes of tank bromeliads have an especially large central cavity that holds a pool of water in which many different organisms can live. Some of these "arboreal ponds" are quite substantial; one species of tank bromeliad can hold as much as 30 liters of water. But each pool is isolated from all the others, and many of the inhabitants are strictly aquatic. An effective dispersal mechanism is therefore essential to the survival of these communities, and this is the role that the frogs play. Researchers studying these communities in the Atlantic forest have found a host of little animals—from crustaceans to various types of worms—attached to the skin of bromeliad-dwelling frogs. Apparently, the frogs are moving these creatures from one bromeliad to another. The bromeliads grow and die, but the community as a whole might remain stable because of the frogs.[41]

The loss of amphibians could also mean the loss of potential cures for many common ailments. Amphibians are an incredibly diverse group of organisms that we know relatively little about. We do know that they are living chemical factories; amphibians produce all sorts of powerful compounds, or they concentrate compounds found in their prey. This characteristic is frequently a form of defense. Because amphibian skin is thin and permeable, it offers little physical protection from attack or infection. So protection is often chemical instead. Many species produce antibiotics and fungicides. Some produce powerful poisons, which they advertise to predators with their bright coloring.[42]

These chemicals are a medical treasure, as many traditional cultures have recognized. Pulverized toads, for example, have been used in traditional Chinese medicine for a variety of ailments. And while traditional remedies doubtless vary greatly in their efficacy, modern chemistry is substantiating the power of many of the raw materials. Epibatidine, a compound with powerful painkilling properties, was discovered in a skin secretion from a small, brightly colored Ecuadorian frog known as *Epipedobates tricolor*. Today, U.S. pharmaceutical giant Abbott Laboratories is developing a synthetic compound related in structure to epibatidine that is 200 times more powerful than morphine and does not have that drug's negative side effects, such as constipation, respiratory failure, addiction, and resistance to the desired painkilling effects. Clinical trials of the compound are under way. If it fulfills expectations, the epibatidine analog would be a major advance for medicine—improving the lives of millions of people suffering chronic pain.[43]

Another frog of interest is a large green treefrog that is used in an extraordinary hunting ritual practiced by Indians in the Amazon Basin. The Indians create "hunting magic" by applying a secretion scraped from live frogs to self-inflicted burns on the arms or chest. Following a coma-like sleep, the hunters awake and find they have increased stamina and keener senses. Research on compounds from this frog have yielded several novel compounds that are able to lower blood pressure, reduce pain, and provide antibiotic effects. One compound interacts with a variety of receptor systems and has provided new insights into how the brain's message-reading receptor systems work. Whether or not these compounds are responsible for the "hunting magic" properties experienced by the Amazonian Indians remains unknown.[44]

We have yet to learn even the most obvious lessons from some of these creatures. The wood frog is able to tolerate temperatures low enough to turn up to 65 percent of its body water into ice. It produces some sort of natural antifreeze to keep the remaining water liquid, but how does this system work? And what might we have learned from the gastric-brooding frog if it had not disappeared from the rain forests of Australia? This and other secrets are now beyond reach.

Reconceiving the Science

It has been more than 10 years since the prospect of global amphibian decline first attracted widespread scientific attention. We now understand some species and some localized areas very well. We have discovered that a whole complex of threats, such as nitrogen pollution, climate change, and disease, is involved. But there are still some major gaps in our understanding of the issue.

To begin with, there is an incomplete picture of what is at stake. As eminent herpetologist William Duellman recently noted, "the amphibian faunas of many parts of the world remain essentially unknown, and the faunas in many other regions are far from completely documented." In South America, nearly one third of the amphibian species were discovered in just the last two decades. And new species continue to be discovered at a rapid rate. The parts of the world that contain the largest numbers of undiscovered amphibian species include the mountainous regions of South America and Southeast Asia and the islands in the Indo-Australian Archipelago (Sulawesi, Sumatra, and the Irian Jaya part of New Guinea). In New Guinea, the current estimate of the number of amphibian species is 200, but herpetologist Michael Tyler believes the total is probably twice that. The tropics may harbor thousands of amphibian species not yet known to science.[45]

In part, the gaps in our knowledge stem from the enormous geographical mismatch between the research capacity and the creatures themselves. Canada, for example, has plenty of amphibian specialists, but not a single endemic amphibian. On the other hand, Mexico, like most tropical countries, has a vast amphibian fauna and only a few herpetologists. Although a good deal is known about the amphibians of the United States, Western Europe, Costa Rica, and Australia, there is limited understanding of those of South America, Asia, or most African countries—places that house some of the richest diversity of amphibians in the world. (See Table 4–3.) And among the tropical species already identified, many are in the literature only by virtue of their orig-

Table 4–3. Major Known Hotspots of Amphibian Diversity

Hotspot	Total species	Share endemic	Portion of original area still intact	Amphibian diversity per unit area
	(number)	(percent)		(species per thousand square kilometers of remaining intact natural vegetation)
Tropical Andes	830	73	25.0	2.6
Mesoamerica	460	67	20.0	2.0
Chocó-Darién-Western Ecuador	350	60	24.2	5.6
Atlantic Forest Region	280	90	7.5	3.0
Sundaland (western half of the Indo-Malayan archipelago)	226	79	7.8	1.8
Indo-Burma	202	56	4.9	2.0
Madagascar and Indian Ocean Islands	189	99	9.9	3.2
Caribbean	189	87	11.3	6.3
Brazilian Cerrado	150	30	20.0	0.4
Western Ghats and Sri Lanka	146	80	6.8	11.7
Guinean Forests of West Africa	116	77	10.0	0.9
Mountains of South-Central China	85	60	8.0	1.3
Philippines	84	77	8.0	3.5
Eastern Arc Mountains and Coastal Forests of Tanzania and Kenya	63	52	6.7	31.5
Mediterranean Basin	62	52	4.7	0.6

SOURCE: Russell A. Mittermeier et al., *Hotspots: Earth's Biologically Richest and Most Endangered Terrestrial Ecoregions* (Mexico City: CEMEX, S.A. 1999), 34, 42, and 50.

inal descriptions; in such cases, virtually nothing is known of the animal's ecology or even whether it still exists.[46]

In many parts of the world, amphibian research is frequently constrained by a lack of funds—but not generally by a lack of interest. Consider the dire situation in Bangladesh, where amphibians were widely abundant up until a few decades ago, although today most species are threatened throughout the country. Amphibians in Bangladesh are being decimated by habitat loss, exposure to pesticides and fertilizers, and commercial exploitation. Some species are even being ground up for poultry feed. Yet researchers still have a very tenuous understanding of the country's fauna. As biologist Sohrab Sarker watches the amphibians of his country disappear, he bemoans the lack of funding for field studies. "Amphibians in Bangladesh are not studied scientifically," he says, "there are many species that still need to be investigated and identification of some specimens has been held up due to lack of relevant literature."[47]

From Canada to New Zealand, a growing army of volunteers is canvassing the world's swamps and forests, searching for amphibians.

In Ecuador, there is a similar need for additional research. A little frog called the jambato used to be so common you could find it in the backyards of Quito. But the last time a jambato was seen alive was in 1988. Perhaps the frog fell victim to a fungus, possibly the chytrid, but no one knows because there is no money to do the research.[48]

An article in an April 2000 issue of *Nature* provides an excellent demonstration of how skewed the research is. It contains the most exhaustive overview of amphibian population trends to date. To produce it, a team of researchers combed through the scientific literature and coaxed additional data out of more than 200 scientists in 37 countries. Of the 936 populations the team managed to cover, 87 percent were in Western Europe or North America, 5 percent were in Central and South America, 2 percent were in Asia, and less than 0.5 percent were in Africa and the Middle East.[49]

Moving into the second decade of investigating amphibian declines, it will be essential to fill the gaps in our understanding and support research and conservation efforts in understudied regions. There is also a need to coordinate research programs, make information more accessible, and build a global amphibian database.

A key organization that has already played a major role in informing many of the world's herpetologists and supporting research efforts in developing countries is the Declining Amphibian Populations Task Force (DAPTF). This was established in 1991 under the auspices of the World Conservation Union–IUCN's Species Survival Commission. It encompasses a network of over 3,000 scientists and conservationists in more than 90 countries and serves as a global "information clearinghouse" on amphibian decline. The Task Force produces the newsletter *Froglog*, which is an important vehicle for communicating updates on research and conservation endeavors from all over the world. Currently, the volunteer scientists who manage the Task Force are compiling a comprehensive summary of the declines and a CD-ROM database of all data available on amphibian populations, declining or otherwise. Both of these will be available in 2002.[50]

In 1999, the DAPTF helped organize a series of amphibian decline workshops in Mexico, Panama, and Ecuador. The organizers hoped the meetings would be the first in a series of efforts to coordinate and build amphibian research in the region. Given the level of participation—the workshops drew in 88 people from 13 countries—that seems like a reasonable expectation.[51]

The Internet is also providing new and speedy ways for researchers to communicate and contribute to global databases. The recently launched Amphibiaweb is a site dedicated to providing information on all the world's known amphibian species. Each species will have its own "home page," including species description, life history traits, conservation status, and literature references.[52]

Acquiring more information will get us only so far. Once the information is collected, it must be used in the most effective way possible. For this, it will be necessary to change the way the science is done. The declines cannot readily be contained in a single field of inquiry. They involve microscopic pathogens and global climate change; they are part of forestry economics and wildlife toxicology. Understanding them will require a much more interdisciplinary, integrative approach than is typical of conventional research. Instead of individual specialists working on problems independently, research programs will have to find ways to focus the efforts of many disciplines into a collaborative whole. Some of the most successful recent work is already moving in this direction—the climate research at Monteverde, for example, or the international investigation of the chytrid fungus.

That idea is beginning to resonate within some major scientific institutions, such as the National Science Foundation, the U.S. government body that is the chief source of federal funds for scientific research. In 1999, the agency awarded $3 million to a team of 24 scientists from fields as diverse as veterinary epidemiology and evolutionary ecology to study host-pathogen relationships as an aspect of amphibian decline. Jim Collins, the Arizona State University biologist who heads the team, explained the challenge this way: "As we went through thinking about how to answer the questions, we really had to think about how we did the science. And how we did the science had to change—it couldn't be just an individual investigator laboring away in an isolated laboratory." Collins emphasizes the need for interaction not just between different biological disciplines, but also with the social sciences and possibly even the humanities. "To understand this problem we have to do a better job of integrating humans into ecological and evolutionary theory," he argues. "The nature of science itself is going to have to change."[53]

But the new research paradigm cannot be just a matter of linking scientists together. Because the problem is so pervasive, science is going to need to build a community of serious amateurs—members of the general public who can become "citizen scientists." A volunteer amphibian monitoring movement is already gaining steam. From Canada to New Zealand, a growing army of volunteers equipped with amphibian identification tapes, field guides, flashlights, pencils, and waterproof notepads is canvassing the world's swamps and forests, searching for amphibians. From schoolchildren to retirees, these people are heading out in evening hours, listening for frog calls, and keeping careful records of what they hear and see.

One of the largest volunteer networks is the North American Amphibian Monitoring Program (NAAMP). The program began in the mid-1990s, and by 1999 hundreds of volunteers were working in 25 states. They are trained to identify their local amphibian fauna and follow a standardized monitoring protocol. The data they collect become part of a global pool of information that is being used to gain a better understanding of amphibian declines. Indicative of the extent of this effort, in Minnesota the program has sold more than 1,000 copies of its amphibian audio tape, and its well-tuned volunteers have covered 150 monitoring routes since 1993. In Maine, NAAMP volunteers monitor 60 routes; Texan volunteers cover 20 routes.[54]

In many cases, volunteers see animals they have never seen before. A participant in a Massachusetts program commented that she had lived in her town for her entire life and "had no idea that each spring hundreds of salamanders crossed through her backyard." Massachusetts volunteers are also involved in identifying vernal pools—the shallow, seasonally flooded wetlands that provide critical breeding sites for many amphibians. Since the late 1980s, local citizens have helped identify a good portion of the 1,200 vernal pools that are now protected by the state.[55]

Another monitoring network is under development in Central America. Modeled on NAAMP, the Maya Forest Anuran Monitoring Project (MAYANMON) was initiated in 1997 to monitor amphibians in the Maya Forest region. Today, the program is a tri-national effort, with sites in Belize, Mexico, and Guatemala. It plans to add additional countries soon. Like NAAMP, this is a volunteer program based on regular, standardized monitoring of numerous sites. The information collected will be critical for the development of baseline data for an area that has not been well studied. Thus far, all participants are biologists, but MAYANMON recently received a grant to train local Mayan villagers who have an interest in monitoring the ecological health of their surroundings. The MAYANMON project is the first of its kind to be undertaken in Central America, and it represents a unique example of international cooperation.[56]

Many European countries have encouraged volunteer efforts to reduce road traffic mortality. Beginning in the 1960s, conservationists initiated programs to reduce amphibian roadkills by informing motorists, installing "toad-crossing" traffic signs, and setting up detours during peak migration times. More recently, drift fences have been used to capture migrating amphibians as they head toward roads. Once the migrants have been detained at the base of the fence, volunteers capture them and ferry them to safety. A "toad patrol" squad at one road crossing in England rescued more than 4,500 toads in just one season. At the same crossing, more than a thousand toads were killed by motorists.[57]

In addition to providing crucial information about population trends, migration patterns, and breeding sites, the new trend in "citizen science" is providing something that may be even more significant. By getting people out into the woods and swamps and attuned to their local fauna, the volunteer programs could serve as a way to capture the public's imagination and develop bigger constituencies for conservation science. Amphibian decline is kind of a case study of unsustainability, touching on all the major environmental issues of our day. It could serve as an important catalyst for change on these issues.

Protecting Amphibians

There are clearly many impressive and promising initiatives under way that bode well for the future of amphibian research. But will the evidence ever translate into sweeping changes in conservation policy? Considering amphibian decline as a political issue, it is almost inconceivable to imagine an international treaty to prevent frog extinctions. Amphibian decline lacks the political "weight" of issues like climate change. True, some countries do have explicit policies protecting amphibians—but these vary greatly around the world, and as is clear from the current trend in amphibian populations, they are not entirely effective.

In the United States, the Endangered

Species Act (ESA) is the only law that provides direct protection for amphibians. Since its inception in 1973, 18 frog, toad, and salamander species—out of 194 identified in the country—have been listed as threatened or endangered under ESA. Species listed under ESA receive special protection, primarily in the form of a "take" prohibition that makes it unlawful for a person to kill or injure a listed species—this can include any significant habitat alteration or degradation that kills or injures the species. Other recovery strategies, such as captive breeding and reintroductions, are also frequently used.[58]

Over 27 years ESA has been successful in rescuing several amphibian species from the brink of extinction. And it has provided many indirect benefits through the protection of habitats—wetlands or forests that are saved for amphibians clearly provide many important services beyond being homes for frogs. But there are also some pitfalls associated with ESA. For one thing, it can take years for the listing to be approved once a petition has been filed. The Santa Barbara County population of the California tiger salamander, for example, finally received permanent listing as endangered under ESA in September 2000—eight years after the listing petition was submitted and 15 years after the species was first recognized as a candidate. In those 15 years, more than half of the population's known breeding sites were destroyed.[59]

Denmark, France, Germany, Luxembourg, and the Netherlands also have legislation protecting all amphibians. But in other European countries, there is no protection. As in the United States, formal protection of amphibian species usually entails some form of habitat protection.[60]

There is one international treaty that provides some specific protection for amphibians. The Convention on International Trade in Endangered Species of Wild Fauna and Flora (CITES) controls trade in numerous endangered plant and animal species. (See Chapter 9.) As of 2000, 81 amphibian species were protected under CITES. But the focus of CITES is on restricting international trade, not on providing on-the-ground protection within a country's borders.[61]

As laudable as ESA and CITES are, such species-specific approaches are not enough to slow the current pace of global amphibian declines. Beyond the fact that they apply to just a tiny fraction of the world's amphibian species, these strategies are not effective for several reasons. For one thing, many species are going extinct that are not on any lists at all. They are disappearing in remote areas of the world where the amphibian faunas are not well understood. In addition, many of the declines and disappearances have taken place in just a year or two—far too quickly to do the paperwork for listing under laws like ESA or CITES. And finally, even if species such as these had been protected under an endangered species law, it probably would have made no difference. They already lived in highly protected areas. The Monteverde Cloud Forest is one of the world's premier parks, and home to expert biologists who monitor the local fauna very closely. Aside from a last-ditch effort at captive breeding (which was attempted with the gastric-brooding frog), there was no obvious way these losses could have been prevented by the time they were detected.

Given the magnitude of the threats at play, it is obvious that a species-by-species approach is just not going to work. The problem of amphibian declines is teaching us an important lesson about the way we approach environmental issues in general.

While it is certainly reasonable to argue for more money and research and conservation, and to demand more powerful conservation policies, the painful fact is this: no amount of money or legislation is going to make this problem go away—if it is directed only at this problem. What is happening to amphibians reflects what is happening to the planet in general. If we don't address the underlying reasons for amphibian decline, there is no way we can save these species.

While efforts to protect particular species are important, amphibians show us that what really matters are the initiatives to protect forests, wetlands, and other essential habitat; to reduce the spread of non-native species; to improve farming practices so that farmers don't depend on heavy inputs of pesticides and fertilizers; and to reduce the emissions of greenhouse gases that contribute to climate change. Clearly, the benefits of these changes will be felt far beyond frogs and salamanders. They will have far-reaching impacts on ecosystems and human health in general. Ultimately, the survival of amphibians and other creatures will depend on our willingness to confront the major, systemic environmental issues of our day.

Chapter 5

Decarbonizing the Energy Economy

Seth Dunn

The human harnessing of energy has long included the release of carbon atoms, dating at least as far back as a wood fire in the Escale cave near Marseilles, France, more than 750,000 years ago. Reliance on wood was a common feature of energy use in most settled parts of the world until the mid-nineteenth century. Growing population density and energy use in Great Britain meant that bulky and awkward-to-carry wood—in spite of its abundance—gradually lost out to coal, which was likewise abundant but more concentrated and easily transported. Though it was not noted at the time, the new fuel also happened to produce less carbon and more hydrogen per unit of energy—with one or two molecules of carbon per each one of hydrogen, versus a 10-to-1 ratio for wood. Thus began the first wave of "decarbonization."[1]

Despite its negative health and environmental effects, coal remained King of the energy world for the remainder of the nineteenth century and well into the twentieth. But the automotive revolution eventually favored another new fuel. Oil boasted an even higher energy density, could be transported more easily and through pipelines, and emitted less soot. By the 1960s, it surpassed coal. With only one molecule of carbon for every two of hydrogen, oil marked the second wave of decarbonization.[2]

At the end of the twentieth century, oil was still the world's leading energy source. Like its predecessors, however, it faced an up-and-coming challenger. Natural gas, which burned more efficiently, used a distributed network of pipes that made petroleum pipelines appear clumsy, and benefited from being known as the cleanest fossil fuel, began its ascent in the final decades of the second millennium. With one unit of carbon to four units of hydrogen, natural gas represents the third wave of decarbonization.[3]

This molecular perspective on the energy economy is not typical: generally speaking, most experts prefer to discuss energy trends in terms of politics and prices, resources and reserves. Yet the trend toward decarbonization—progressively reducing the amount of carbon produced for a given amount of energy—is as illuminating and important as it is overlooked. Jesse Ausubel of Rockefeller University goes so far as to argue that this pattern lies "at the heart of understanding the evolution of the energy system." Exploring this trend is therefore critical if we are to grasp where the energy economy might be headed in the future.[4]

Yet even as the third wave begins its rise, a fourth has already appeared on the horizon. Initially building on the natural gas network for its distribution, and derived at first from natural gas to run high-efficiency fuel cells, hydrogen could within several decades displace natural gas. Eventually using its own full-fledged network, and created by the splitting of water into hydrogen and oxygen from solar, wind, and other forms of virtually limitless flows of renewable resources, the hydrogen economy will over time "free energy from carbon," as Nebojsa Nakićenović of the International Institute for Applied Systems Analysis (IIASA) has put it.[5]

The first three waves of decarbonizing were driven by the search for more abundant and easily harnessed energy sources; local and regional environmental factors played a limited role in aiding the ascents of oil and natural gas. But the past century's discovery of carbon's role in changing Earth's climate, the role of humans in adding this carbon to the atmosphere, and the potential risks that accompany climate change have all made global environmental concerns a major new factor in the approaching fourth wave. Present and future generations thus face the challenge of deliberate decarbonization.

A key barrier to enhancing the decarbonization process is the lingering misperception among many governments and businesses that limiting carbon emissions necessitates reductions in energy use and economic growth. Yet the past quarter-century provides strong evidence that carbon and economic output can be "decoupled." This chapter explores one particular dimension of the decarbonization challenge: reducing the carbon intensity of the energy economy, as measured in emissions per unit of economic output. Between 1950 and 1999, the number of tons of carbon produced for each million dollars of gross world product was reduced by 39 percent, from 250 to 150 tons per million dollars of output, averaging an annual intensity improvement rate of 1 percent—but with a 2-percent annual average during the 1990s. (See Figure 5–1.) Carbon intensity is becoming an important indicator of our movement toward a sustainable energy economy, particularly as evolving scientific understanding of climate change strengthens public concern and pressure for action.[6]

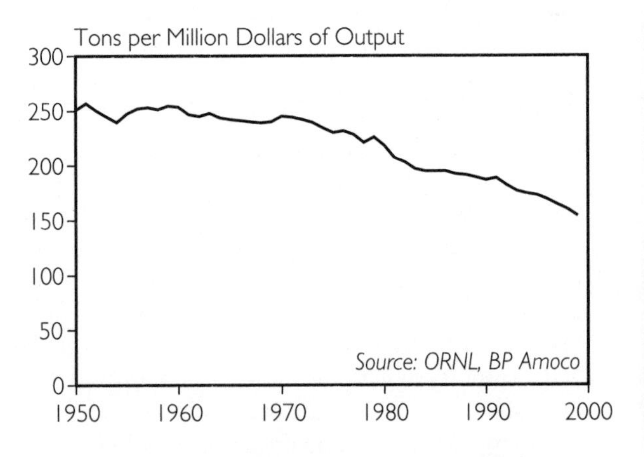

Figure 5–1. Carbon Intensity of the World Economy, 1950–99

The Climate Constraint

The global carbon cycle is among the most complex and least understood of the planet's large-scale natural processes. An estimated 42 trillion tons of carbon are either housed in or circulate between the atmosphere, oceans, and biosphere. The precise amount found in each of these reservoirs is subject to much uncertainty, however, and scientists believe that many of the important fluxes can vary significantly from year to year. But research into this dynamic process is advancing quickly, permitting a rough accounting of the global carbon budget. (See Figure 5–2.)[7]

One of the most conspicuous characteristics of the carbon cycle is the near-equilibrium of natural fluxes, or flows, between the atmospheric, oceanic, and terrestrial components. When these flows more or less offset each other, the size of the reservoirs changes little. Beginning in the late eighteenth century, however, a significant human component was added to the cycle. Since 1751, the dawn of the Industrial Revolution and the accompanying large-scale combustion of carbon-based fossil fuels, more than 271 billion tons of carbon have been added to the atmospheric reservoir through the burning of fossil fuels. (See Figure 5–3.) Today, annual carbon emissions total slightly more than 6.3 billion tons—second in mass only to water flows linked to human activities.[8]

A clear consequence of this addition to the global carbon cycle has been the elevation of atmospheric levels of carbon dioxide (CO_2), which is naturally present in the atmosphere but which also forms when fossil fuel burning releases carbon in the presence of oxygen. According to samplings of air bubbles trapped in the world's deepest ice core in Vostok, Antarctica, current CO_2 levels are "unprecedented" in relation to the last 420,000 years; analyses of fossilized plankton suggest that they may be at their

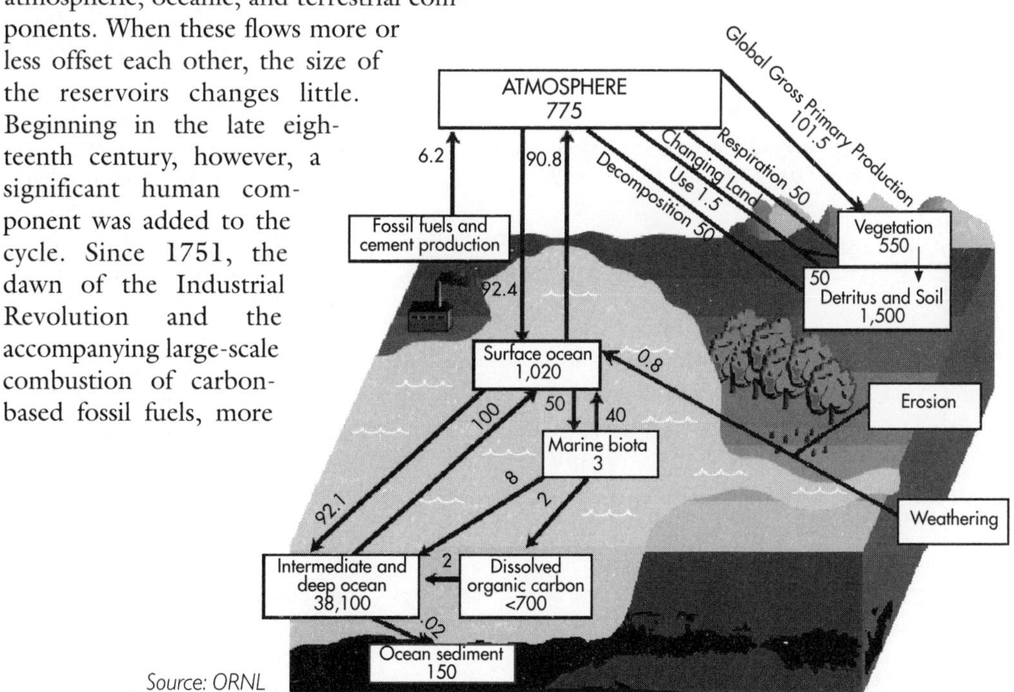

Figure 5–2. Global Carbon Cycle (in billion tons of carbon)

highest point in 20 million years. (See Figure 5–4.) Atmospheric CO_2 concentrations reached 368.4 parts per million volume (ppmv) in 1999, marking nearly a 32-percent increase from the pre-industrial level of 280 ppmv—and a 17-percent rise just since 1958.[9]

Carbon dioxide is one of the "greenhouse gases"—which also include methane, nitrous oxide, and halocarbons—that alter the planet's energy balance by trapping infrared radiation reflected from Earth's surface, preventing it from escaping to space and thus causing surface temperatures to rise. The Vostok ice core data suggest a strong correlation between greenhouse gas concentrations and temperature change, and that periods of CO_2 buildup have contributed to past global warming transitions between glacial and interglacial periods. Average surface temperature measurements suggest that another transition is under way. Land-based surveys from NASA's Goddard Institute of Space Studies reveal a temperature increase of 0.6 degrees Celsius since 1866. Researchers from the University of Massachusetts studying tree-ring samples have concluded that the 1990s were the warmest decade of the last millennium, with 1998 being the warmest year.[10]

Natural events, among them changes in solar variability and volcanic activity, have also contributed to past temperature changes such as warming during the medieval period and the Little Ice Age between the seventeenth and nineteenth centuries. But reconstructions of past climates suggest that human influences have had the dominant role in twentieth- century temperature trends. Thomas Crowley, a geologist at Texas A&M University, estimates that natural factors have accounted for only 25 percent of the warming since 1900, with the remainder attributable to increases in greenhouse gas emissions.[11]

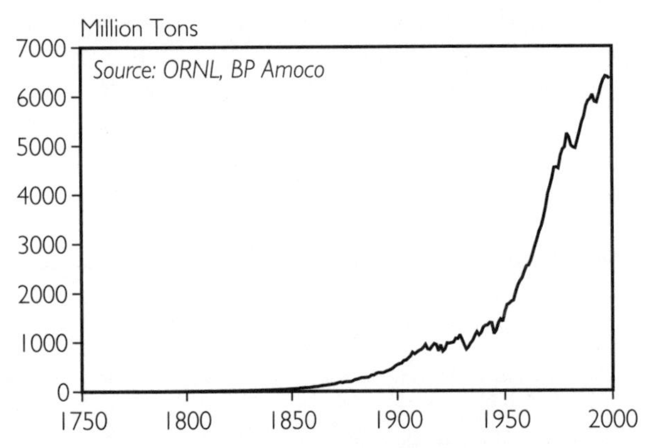

Figure 5–3. World Carbon Emissions from Fossil Fuel Burning, 1751–1999

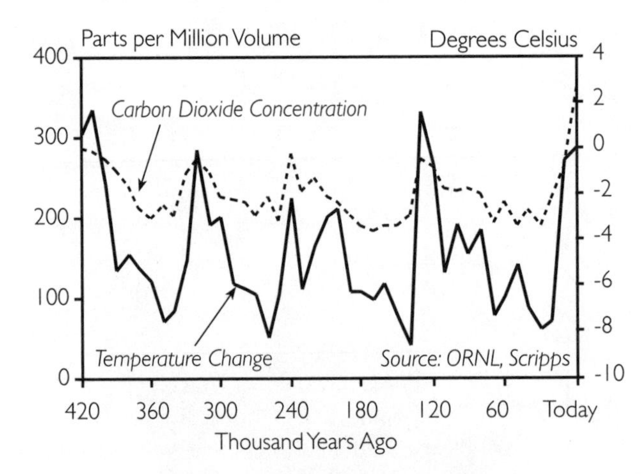

Figure 5–4. Atmospheric Carbon Dioxide Concentration and Temperature Change, 420,000 Years Ago–Present

The observed surface temperature rise is actually smaller than that projected by climate models—a "missing warming" is believed to be taking place in oceans, which have warmed dramatically during the last four decades. The oceanic heat storage—half of which, surprisingly, has taken place in deep water—implies that more atmospheric warming lies ahead, as increases in ocean temperature have generally preceded increases in atmospheric temperatures by approximately one decade. James Hansen of Goddard estimates that, because of this ocean-induced delay, surface temperatures will rise 0.5 degrees Celsius over the next century even if greenhouse gas concentrations are stabilized immediately.[12]

In a June 2000 article in the *Proceedings of the National Academy of Sciences*, Hansen and his colleagues present "an alternative scenario" that combines the reduction of non-CO_2 greenhouse gases and aerosols such as sulfates with success in slowing carbon emissions growth over the next 50 years. They note that these other gases have mainly driven the rapid warming of recent decades, and suggest that, because their growth rates have declined in the last decade, it may be more practical to focus on them rather than on carbon. The authors add, however, that the "climate forcing" by CO_2 remains by far the largest of any single natural or human factor, accounting for just under half of the human-made warming since 1850. They also point out that carbon dioxide will become even more dominant as a greenhouse gas as emissions of sulfur dioxide—aerosols that cause a temporary cooling—are reduced. "This interpretation," they assert, "does not alter the desirability of limiting CO_2 emissions."[13]

The Intergovernmental Panel on Climate Change (IPCC), a U.N.-designated body of more than 2,000 scientists, has projected a wide range of adverse impacts resulting from a doubling of pre-industrial atmospheric CO_2 concentrations, including sea level rise and coastal inundation, more frequent and intense weather extremes, stresses on water and agricultural systems, changing migration patterns and loss of biodiversity, and greater prevalence of infectious diseases. Perhaps the strongest agreement with IPCC climate models to date has been evidence of accelerated polar warming and diminishing sea ice and ice sheets in the northern hemisphere. Submarine probes suggest Arctic ice thickness has declined by 42 percent since the 1950s, and Norwegian researchers estimate that Arctic summers may be ice-free by 2050. Airborne surveys indicate that the Greenland Ice Sheet is losing 51 cubic kilometers of volume each year; complete loss of the sheet would raise sea level 7 meters. Such large influxes of fresh water could induce a slowdown of oceanic heat circulation, which has in the past led to abrupt cooling in the north Atlantic region.[14]

Norwegian researchers estimate that Arctic summers may be ice-free by 2050.

The IPCC has developed a new set of scenarios that explore how changes in demographics, social and economic development, and technology may affect future emissions of carbon dioxide and other greenhouse gases. In the six groups of scenarios, annual carbon emissions from fossil fuel combustion range from 9 billion to 12.1 billion tons in 2020; from 11.2 billion to 23.1 billion tons in 2050; and from 4.3 billion to 30.3 billion tons in 2100. These scenarios yield cumulative fossil fuel–related carbon dioxide emissions by 2100 spanning from 983 billion to 2.2 trillion tons.[15]

Tom Wigley of the U.S. National Center for Atmospheric Research estimates that these scenarios would lead to CO_2 levels of 558 to 825 ppmv by 2100—a doubling to tripling of pre-industrial levels. Mean surface temperature would increase by 1.9–2.9 degrees Celsius—three to five times the warming rate of the past century. Average sea level would rise by 46–58 centimeters—two to four times the rise of the last hundred years. These changes are influenced by all greenhouse gases but dominated by CO_2, which accounts for 66–74 percent of the warming.[16]

The Hadley Centre for Climate Prediction and Research, with the U.K. Meteorological Office, has produced scenarios suggesting that stabilizing CO_2 levels at a doubling of preindustrial levels (roughly 550 ppmv) would reduce the magnitude of—and in some cases avoid—serious impacts in many regions. By contrast, uncontrolled emissions would raise the temperature by 3 degrees Celsius and the sea level by 40 centimeters by 2080; cause substantial tropical forest and grassland dieback in Latin America and Africa; subject an additional 3 billion people to water stress, primarily in Africa, the Middle East, and India; and increase the annual number of people flooded from 13 million to 94 million, most of them in southern and southeast Asia. Significant rainforest loss and water resource stress would still result from stabilizing CO_2 at 750 ppmv, but would be avoided by a 550 ppmv stabilization.[17]

Although the objective of the U.N. Framework Convention on Climate Change is to stabilize atmospheric concentrations of CO_2 and other greenhouse gases at levels that will avoid "dangerous anthropogenic interference with the climate system," scientists have not reached a consensus on the stabilization level that would meet this objective. Because a doubling of CO_2 would entail serious dislocations, the IPCC has also considered a more aggressive stabilization target of 450 ppmv. According to the panel, achieving this goal would require cutting carbon emissions by roughly 60–70 percent—to about 2.5 billion tons annually—by 2100, and eventually down to less than 2 billion tons per year. These deep reductions imply the need for an aggressive strategy to accelerate the decoupling of carbon emissions and economic output. Like the carbon cycle itself, such a strategy would have four major components: switching to lower-carbon fossil fuels, improving energy intensity, increasing reliance on renewable energy, and developing an infrastructure for the use of hydrogen as an energy carrier.[18]

From Solid to Liquid to Gas

In 1850, wood accounted for nearly 90 percent of world energy use, but its share fell steadily and was surpassed by that of coal in the 1890s, when both accounted for roughly half of the global total. Coal's piece of the world energy pie expanded to about 60 percent in the 1910s and then shrank, but the fuel remained dominant until the 1960s, when it was displaced by petroleum as the leading source. And 1999 witnessed another milestone in the late fossil fuel era, with natural gas use moving past that of coal for the first time. (See Figure 5–5.) Today, these fuels supply roughly three quarters of world energy, with respective shares of 32, 22, and 21 percent for oil, natural gas, and coal.[19]

As noted earlier, this gradual move from solids to liquids to gases has resulted in the use of fuels with a lower carbon content. Since 1860, the amount of carbon per unit of energy has decreased by roughly 0.3 percent a year, a rate that is moving in the right direction but that scientists have recom-

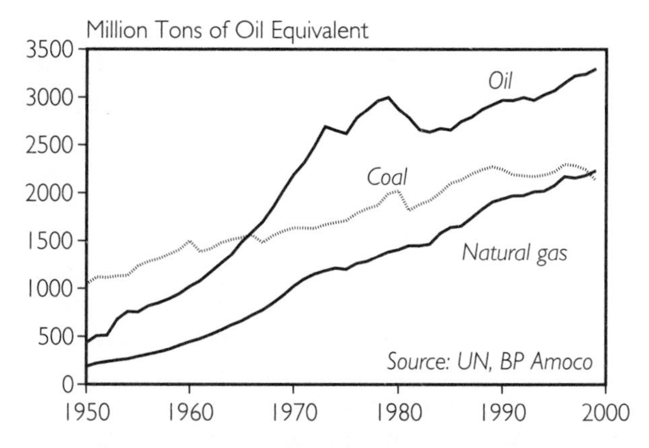

Figure 5–5. World Fossil Fuel Consumption by Source, 1950–99

mended be doubled or tripled to safely stabilize the climate. Partially driving the recent drop in carbon intensity has been a decline in worldwide use of coal, which is the most carbon-intensive fossil fuel and accounts for 36 percent of energy-related carbon emissions. Presently at its lowest point since 1985, coal consumption had an average annual growth rate of –0.6 percent in the 1990s. Usage dropped 7.3 percent just between 1997 and 1999.[20]

Half of global coal consumption takes place in the United States and China. The United States, with a 26-percent share, saw an 11-percent increase during the past decade. Much of this is due to increased burning of cheap coal in old power plants, some of it illegal, by utilities anticipating industry deregulation. China, in contrast, with 24 percent of world use, experienced a 4-percent drop over the decade and a 25-percent decline between 1997 and 1999. This has been attributed to reductions in coal production subsidies, a steep drop in demand for industry due to closings of inefficient state-owned industries, and shifts to natural gas for residential cooking and heating.[21]

Outside the two coal superpowers, consumption trends have varied widely. Usage dropped 24 percent in Europe in the 1990s and 44 percent in the former Eastern bloc, mostly due to subsidy cuts in the United Kingdom and Russia. In the Asia-Pacific region, India and Japan saw increases of 29 and 17 percent, respectively. South Africa and Australia, which together account for 6 percent of coal use, registered increases of 15 percent each.[22]

Use of oil, which contains 23 percent less carbon per unit of energy than coal and accounts for some 43 percent of energy-related carbon emissions, is still growing, although more slowly than in the postwar era. World oil consumption grew by about 1.2 percent on average each year during the 1990s. More than a quarter of global petroleum use occurs in the United States, which boosted use by 11 percent in the last decade. Japan, with an 8-percent share, increased oil use by 4 percent, while China, with a 6-percent share, raised consumption by 45 percent. Russia, accounting for 4 percent of world oil use, halved its consumption in the 1990s.[23]

The Asia-Pacific region has experienced the most rapid recent growth in petroleum use, with a 42-percent increase in the past decade. South Korea roughly doubled oil use during this period, while India had a 64-percent increase. Exploding use of motor vehicles with internal combustion engines in these countries is driving these jumps, just as steam turbine–powered railroads fueled coal use in an earlier stage of the fossil energy era.[24]

Natural gas—which accounts for 21 percent of energy-related carbon emissions, and on a per-energy unit basis releases 28 percent less carbon than oil and 44 percent less than coal—is now the fastest growing of the three fossil fuels. Consumption of natural gas averaged an annual growth rate of 1.9 percent during the 1990s. In industrial nations, natural gas has become the fuel of choice in turbines for power generation, replacing coal. The United States, the leading consumer with a 27-percent share of the global total, expanded consumption by 14 percent in the 1990s. Europe, with a 20-percent share, saw consumption rise by 34 percent. Leading this trend was a "dash for gas" in the United Kingdom—a 75-percent increase enabled by the removal of subsidies to the coal industry.[25]

Globally, fossil fuel subsidies total more than $120 billion a year, with natural gas receiving the least support.

Natural use has declined, on the other hand, in former Eastern bloc nations, which use close to a quarter of the world total but where economic troubles have dramatically slowed exploitation of the region's reserves. Collectively these countries saw gas consumption fall 19 percent in the 1990s, with the dominant user, Russia, experiencing a 13-percent drop. Ukraine, another major consumer, saw consumption plummet 43 percent.[26]

Developing regions have been the major contributors to recent growth in natural gas use, though they still account for only a quarter of the world total. Central and South America—sites of some of the most recently discovered reserves—have boosted use by 58 percent over the last 10 years, with the largest increases occurring in Argentina and Venezuela. The Middle East raised consumption by 85 percent, with Iran and the United Arab Emirates doubling use. In the Asia-Pacific region, consumption has swollen 70 percent; India and Malaysia have doubled use, Taiwan and Thailand have tripled use, and South Korea has quintupled consumption. As in industrial nations, growing reliance on gas turbines for power generation is the main factor in this trend, which may in turn be contributing to coal's slowing growth.[27]

Looking ahead, future policies to reduce carbon emissions may significantly affect fossil fuel markets. A near-term scenario by the Oxford Institute for Energy Studies sketches the effect on fossil fuels of the Kyoto Protocol, which commits industrial and former Eastern bloc nations to bring greenhouse gas emissions to 5 percent below 1990 levels by 2008–12. The study, *Fossil Fuels in a Changing Climate*, assumes that virtually all countries achieve their targets primarily through domestic action with carbon/energy taxes, voluntary measures from industry, and energy efficiency policies. By 2010, coal production is cut 4.4 percent from a "business-as-usual" trajectory, natural gas by 4 percent, and oil by 3 percent.[28]

Fossil fuel use is affected more in six long-term scenarios constructed by IIASA and the World Energy Council (WEC) in their book *Global Energy Perspectives.* In all the scenarios, which use vantage points of 2050 and 2100, "the peak of the fossil era has passed." In two ecologically driven scenarios, the global energy shares of natural gas, oil, and coal decline to roughly 11, 6, and 3 percent by 2100, and the overall fossil fuel portion to below 20 percent—the share some 140 years ago. Absolute consumption eventually declines for all three, lowering annual carbon emissions to 2 billion tons by 2100 and stabilizing atmospheric CO_2 levels at 450 ppmv. This scenario is predicated on "ambitious policy measures" that accelerate energy efficiency improvements and promote environmentally benign, decentralized energy technologies.[29]

Subsidy removal and carbon taxes are the most ambitious ways to hasten the transition to and beyond lower-carbon fossil fuels. Globally, fossil fuel subsidies—price supports, favorable tax rates, direct financial aid—now total more than $120 billion a year, with the lowest-carbon fuel (natural gas) receiving the least support. These subsidies can be redirected to social programs, such as worker retraining for affected industries. Similarly, carbon tax revenues can be used to reduce taxes on wages and employment to minimize negative economic impacts. Six countries—Denmark, Finland, Italy, the Netherlands, Norway, and Sweden—have implemented carbon taxes to date, while several others are considering them.[30]

Improving Energy Intensity

Another important element of decarbonization is reducing energy intensity, the amount of energy required per unit of economic output. Since the Industrial Revolution, energy intensity has averaged annual improvements of 1–2 percent. This has been responsible for roughly two thirds of carbon intensity reductions since 1950, with the remaining third due to fuel switching. In the United States, one unit of gross domestic product (GDP) requires less than one fifth as much energy as two centuries ago. Between 1973 and 1986, U.S. energy consumption remained level while the gross national product rose over 35 percent, although intensity gains have since slackened.[31]

Energy intensities have generally declined over time for a number of reasons. These include technical efficiency improvements in power plants and appliances, structural economic shifts to less energy-intensive activities, and changing patterns of energy use and lifestyles. But national circumstances and history are also important. Differences in energy resources, population density, and land area go far in explaining, for example, why Australia has a much higher energy and carbon intensity than Japan. Industrial-nation energy and carbon intensities have fallen over the last two decades, due to energy efficiency improvements, although these gains have recently slowed with plateaus in automobile efficiency and low fossil fuel prices. (See Figure 5–6.)[32]

As with fossil fuel use, patterns of energy intensity vary regionally. Much of this is due to differing economic structures: as nations move from industrial to "post-industrial" or service economies, energy use tends to rise at a slower rate or to fall. Scientists have also observed a correlation between rising incomes and falling energy intensity, as more-efficient technologies become available at higher income levels. The energy

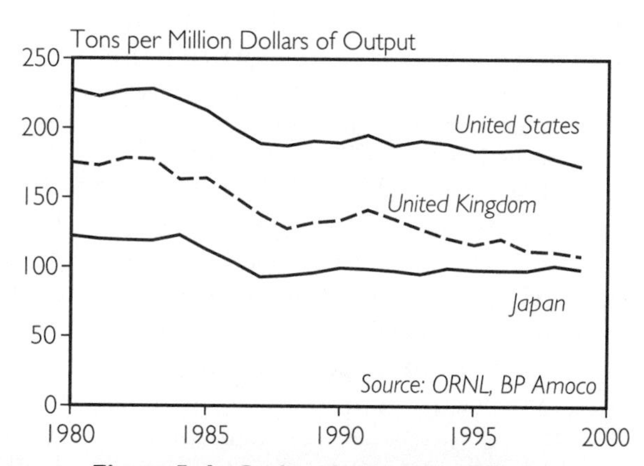

Figure 5–6. Carbon Intensities, Selected Industrial Economies, 1980–99

intensity paths of many of today's developing countries resemble those of industrial nations at low levels of per capita income. Most of these nations experience a surge in energy intensity as they begin to industrialize, as did China in the 1970s, paralleling India in the 1960s and the United States in the first decade of last century.[33]

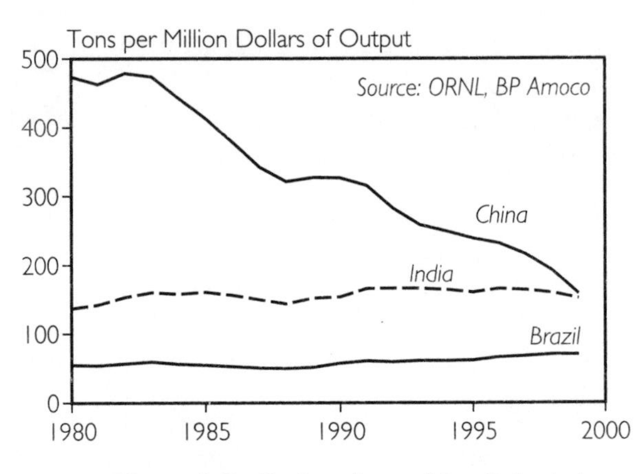

Figure 5–7. Carbon Intensities, Selected Developing Economies, 1980–99

China's experience since the late 1970s demonstrates the opportunity to make major energy intensity strides. As part of the move toward a more market-oriented economy, the government enacted an ambitious set of energy efficiency policies—creating corporations that focused investments on equipment upgrades and efficiency in new construction, building and appliance efficiency standards, and energy conservation centers. As a result, energy demand growth rates have been cut to half the rate of economic growth. Had Chinese energy intensities remained at 1980 levels, by now the nation would be emitting as much carbon as the United States does. Instead, China now has carbon intensities below the U.S. level, as do India and Brazil. (See Figure 5–7.)[34]

Although energy use has fallen sharply in former Eastern bloc nations over the past decade, steeper drops in economic output have pushed the region's carbon intensity upward, with the exception of Kazakhstan. (See Figure 5–8.) As its economy recovers, the region is projected to halve its energy intensity over the next 20 years—albeit to levels that will still be two to five times higher than in developing and industrial countries.[35]

Industrial nations are expected to average annual improvements of 1.1 percent, compared with 1.3 percent since 1970. A 1-percent annual improvement is also projected for developing nations, though national trends may vary widely. In general, the faster the economic growth and turnover of capital and equipment, the greater the energy intensity improvements. Gains are thus expected to be largest in countries with currently high intensities that are experiencing rapid economic growth and capital turnover.[36]

Opportunities for deploying energy-efficient technologies in buildings, transportation, and industry are significant. Buildings account for roughly 36 percent of energy use and have lifetimes ranging from five to eight decades, meaning that delays in improving efficiency will lock in waste for many years. It is estimated that improved building designs and wider use of appliances that are already available could cut industrial-nation energy consumption in half; even greater savings are possible in other regions. Policies that can help overcome barriers to efficiency investments in buildings and appliances include mandatory codes and standards, financial incentives, and design competitions. Current U.S. mandatory appliance standards are projected to save

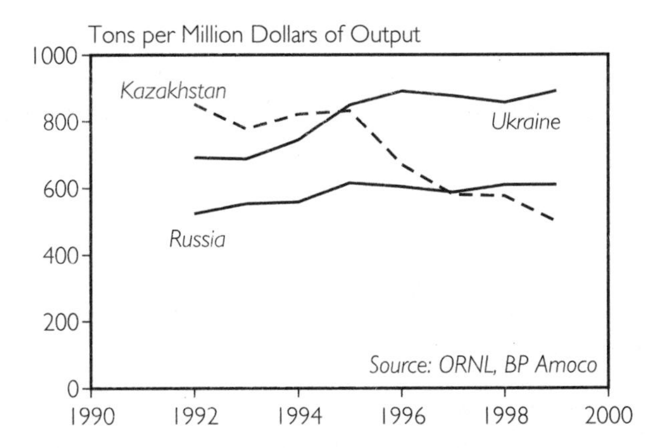

Figure 5–8. Carbon Intensities, Selected Former Eastern Bloc Economies, 1991–99

more than $160 billion and 60 million tons of carbon by 2010: the average electricity use of a refrigerator today is one third what it was in 1974. Brazil, China, Mexico, and South Korea are among those working to develop similar codes and standards.[37]

One quarter of world energy use goes to transportation, 70 percent of it in industrial nations. The failure to strengthen fuel economy standards has caused efficiency to level off in the United States, while most emerging automobile markets use highly inefficient engines. Considerable potential exists for improving the fuel economy of the existing internal combustion model and for introducing far more efficient hybrid-electric cars. The recent hybrids introduced by Toyota and Honda achieve 48–64 miles per gallon, compared with the U.S. fleet average of 27.5. Standards for manufacturers and incentives for buyers can help put these cars on the road, while fuel taxes can encourage greater use of public transit. (See Chapter 6.)[38]

Industry consumes close to 40 percent of world energy, with half of this amount coming from the production of energy-intensive materials such as steel, chemicals, cement, and paper. In transitional and developing economies, industry consumes more energy than do buildings and transport, although economic difficulties have recently slowed growth in some of these nations. Industrial energy use can be cut in half or more through more-efficient process technologies such as electronic controls, heat-recovery boilers, and motor drives (which consume half to two thirds of industrial power in Brazil, China, and the United States). Industries that use power on-site rely today on steam turbines with electrical efficiencies of 33 percent, but advanced gas turbines offer 55 percent. By generating power on-site in tandem with the steam and heat used in production—a process known as cogeneration—these turbines, along with other "micro-power" devices such as microturbines and Stirling engines, can achieve overall efficiencies of 80 percent and higher. Cogeneration is already a major source of electricity in Europe, providing 30, 34, and 40 percent of the national totals in the Netherlands, Finland, and Denmark, respectively.[39]

Structural economic changes also offer potential. As noted earlier, the gradual move toward less energy-intensive, service-based economies has considerably lowered energy use in industrial nations. But the impact of the information-based or "Internet economy" is hotly debated. While some commentators emphasize direct energy increases through greater use of electronic equipment, others argue that the Internet will indirectly temper energy use by reducing the need for energy-intensive manufacturing, retail space, and transportation.[40]

Energy analyst Joseph Romm notes that between 1996 and 1999, U.S. energy intensity fell by more than 3 percent a year, compared with average annual rates of less than 1 percent in the previous 10 years. One third of this change was structural—growth in services and information technology—with the remaining two thirds due to efficiency improvements. Romm projects the growing role of e-commerce will lead to U.S. energy intensity gains averaging 1.5–2 percent annually over the next 10 years. Further improvements, however, will require new policies to enhance energy efficiency.[41]

Beyond Fossil Fuels

"New" renewable energy sources are in the position of petroleum about a century ago: accounting for a fraction of of world energy, but gaining footholds in certain regions and markets. It was concern about the future of oil that initially sparked the first modern wave of interest in renewable energy technologies in the 1970s and 1980s. The dynamic growth phase that began in the 1990s, however, features significant technological improvements and is driven in part by policies designed to address carbon emission reduction commitments.[42]

The most spectacular recent growth in renewable energy use has occurred with wind power, which averaged a 24-percent annual increase in the 1990s and is now a $4-billion industry. Advances in wind turbine systems have dramatically lowered the generation cost of wind power over the last two decades, to the point where it is becoming cost-competitive with fossil-fuel-fired power generation in some regions. But its strong market entry also owes much to policy support, particularly electricity "in-feed laws" in Europe that provide generous fixed payments to project developers. Seven of the top 10 nations using wind power are European.[43]

Solar photovoltaics (PVs), which convert sunlight into electricity, have also witnessed significant cost declines and market growth—a 17-percent annual average during the past decade. The global solar industry is estimated at $2.2 billion; BP Solar—the leading manufacturer, with a 20-percent share of the world market and estimated annual revenues of $200 million—has its cells in use in more than 150 countries and manufacturing facilities in the United States, Spain, India, and Australia. The PV market is experiencing a shift from primarily off-grid uses to grid-connected applications, which are the fastest-growing sector due to subsidy programs in Japan, Germany, and the United States. Also poised for growth is the off-grid rural market in developing nations, which is forecast to expand more than fivefold over the next 10 years.[44]

Hydropower, geothermal power, and biomass energy have experienced slower but steady growth over the last decade, ranging from 1 to 4 percent annually. Increasingly, new hydropower systems in North America and East Asia are oriented toward small-scale applications. Geothermal power is on the rise in parts of Asia-Pacific and Latin America Use of biomass energy—primarily agricultural and forestry residues—which accounts for as much as 14 percent of world energy and has both residential and commercial uses, is benefiting from modern applications in turbines and factories. Less-established technologies, such as harnessing wave and tidal energy, may yet prove viable.[45]

Several recent publications have explored the possibility of renewable energy providing a significant share of world energy by mid-century. In the Shell Group Planning "Sustained Growth" scenario, renewables first capture niche markets and "by 2020 become fully competitive with conventional energy sources." Solar PV technologies experience cost reductions similar to those for oil in the 1890s, and between 2020 and 2030 developing countries turn aggressively to renewable energy.[46]

During the next two decades, in the Shell study forecast, these technologies become widely commercial as fossil fuels plateau, with wind, biomass, and solar PVs achieving market penetration rates analogous to those of coal, oil, and gas in the past. By 2050, over 50 percent of primary energy supply comes from renewables, with 10 sources each holding a market share of 5–15 percent. The company responded to its own scenario by establishing in 1997 a Shell Renewables core business, which has earmarked $500 million over five years and has projects under way in solar, biomass, and wind energy in Europe, South America, the Middle East, Africa, and the Asia Pacific.[47]

Renewable energy is prominent in other studies as well. The IIASA/WEC ecologically driven scenarios show renewables reaching a 40-percent share in 2050 and 80 percent in 2100. They note a "changing geography of renewables," as developing nations take a leadership role in harnessing the resources by the 2020s and account for two thirds of renewable energy use by 2050. In *Bending the Curve: Toward Global Sustainability*, the Stockholm Environment Institute (SEI) describes a 25-percent renewables scenario that "requires neither heroic technological assumptions nor economic disruption." Renewables promotion may, it argues, further the goals of economic development and job stimulation; the primary constraints to achieving the energy goal are institutional and political.[48]

The SEI study lays out specific policies that could make its scenario a reality. Carbon taxes are coupled with reductions in other levies, as in the tax shifts of several European countries. Fossil fuel subsidies are phased out. New financing initiatives and economic incentives spur investment in renewable technologies. Expanded research, development, and demonstration create new technologies. Better information, capacity building, and institutional frameworks overcome barriers to investing in renewables. And global initiatives to transfer technologies and know-how make these sources the foundation of developing nations' energy economies.[49]

Renewables promotion may further the goals of economic development and job stimulation; the primary constraints are institutional and political.

A common question in discussions of the long-term role of renewable energy is whether these resources could conceivably meet worldwide energy requirements. Bent Sørensen at the Roskilde University in Denmark has delved into this question, using an array of economic, population, and energy data and projections to create scenarios that consider whether solar, wind, biomass, geothermal, and hydropower collectively could meet global energy demand by 2050. His scenarios achieve near-zero carbon emissions, and are more expensive than the current system only when environmental costs are neglected.[50]

The Roskilde study concludes that a combination of dispersed and more centralized applications—placing solar PVs and fuel cells in buildings and vehicles, and wind turbines adjacent to buildings and on farmland, plus a number of larger solar arrays, offshore wind parks, hydro installations—would create a "robust" system capable of meeting the world's entire energy demand. But the study also stresses that significant additional technological and policy development would be required to realize the scenario.

The policy preconditions for a wholesale shift to renewable energy include a mix of free market competition and regulation, with environmental taxes correcting marketplace distortions; temporary subsidies to support the market entry of renewables; and the removal of hidden subsidies to conventional sources. Taxes would need to be synchronized internationally to avoid differential treatment of energy sources among countries, and to be adjusted if the market does not respond enough to the initial price change. And the energy transition would have to be kept on course by continuous "goal-setting" and monitoring.[51]

In fact, the European Union (EU) has already established a target for renewables of 12 percent of total energy by 2010. National goals in Germany and Denmark have helped stimulate rapid wind power development that has in turn led to more ambitious goals. The Danish company BTM Consult, which projects that wind power could supply 10 percent of global electricity by 2020, argues that nations should set wind-specific goals, backed up by legally enforced mechanisms such as those now popular in Europe. The United States aims to increase wind power's share of electricity to 10 percent by 2010—compared with about 0.1 percent today—but it has yet to provide such policy support.[52]

One of the variables shaping how fast an energy economy based on renewable resources emerges is the extent to which storage systems are developed that can harness the intermittent flows of these sources and store them for later use. Viable energy storage is essential for turning renewables into mainstream sources, and engineers have experimented with a long list of candidates, including batteries, flywheels, superconductors, ultracapacitors, pumped hydropower, and compressed gas. But the most versatile energy storage system, and the best "energy carrier," is hydrogen.[53]

Entering the Hydrogen Age

The ultimate step in the decarbonization process is the production and use of pure hydrogen. As noted earlier, the gradual displacement of carbon by hydrogen in energy sources is well under way. Between 1860 and 1990, the ratio of hydrogen to carbon in the world energy mix increased more than sixfold.[54]

Hydrogen—the universe's lightest and most abundant element—is known most commonly for its use as a rocket fuel. It is produced today primarily from the steam reformation of natural gas for a variety of industrial applications, such as the production of fertilizers, resins, plastics, and solvents. Hydrogen is transported by rail, truck, and pipeline and stored in liquid or gaseous form. Though it costs considerably more to produce than petroleum today, the prospect of hydrogen becoming a major carrier of energy has been revived due to advances in another space-age technology: the fuel cell.[55]

An electrochemical device that combines hydrogen and oxygen to produce electricity and water, the fuel cell was first used widely in the U.S. space program and later in a number of defense applications such as submarines and jeeps. While these cells were traditionally bulky and expensive, technical advances and size and cost reductions have sparked interest in using them in place of internal combustion engines (ICEs), central power plants, and even portable electronics.

Their initial costs are several times higher than these conventional systems, but are anticipated to drop sharply with mass production.[56]

Fuel cells are nearing the market for both stationary and transportation uses. Ballard Power Systems and FuelCell Energy plan to deliver their first commercial 250-kilowatt units in 2001. DaimlerChrysler, which is devoting $1.5 billion to fuel cell efforts over the next several years, aims to sell 20–30 of its fuel cell buses to transit systems in Europe by 2002, and to mass-produce 100,000 fuel cell cars and begin selling them by 2004. Toyota and Honda have set 2003 commercialization dates for their fuel cell vehicles.[57]

An important stimulus of the fuel cell market has been the state of California's requirement that 2 percent of new cars sold in 2003 be zero-emissions vehicles. The mandate has spurred new fuel cell investments and collaborations. The California Fuel Cell Partnership, composed of major car manufacturers, energy companies, and government agencies, intends to test 70 fuel cell vehicles by 2003, with energy companies delivering hydrogen and other fuels to refueling stations. In November 2000, the partnership unveiled its headquarters, which includes a refueling station and public education center, and the first fleet of vehicles in Sacramento.[58]

Another region at the vanguard of the hydrogen transition is Iceland, where in February 1999 a $1-million joint venture to create the world's first hydrogen economy was launched by the government and other Icelandic institutions, DaimlerChrysler, Shell Hydrogen, and Norsk Hydro. The joint venture, Icelandic New Energy, emerged from a parliament-appointed study commission that recommended the initiative; it is now official government policy to promote the increased use of renewable resources—geothermal and hydroelectric resources provide 70 percent of the nation's energy—to produce hydrogen. The strategy is to begin with buses, followed by passenger cars and fishing vessels, with the goal of completing the transition between 2030 and 2040.[59]

In Iceland, it is official government policy to promote the increased use of renewable resources to produce hydrogen.

Hydrogen-powered buses are a logical first step because they can handle larger and heavier fuel cells and do not need to be refueled as often. Ballard has demonstrated fuel cell transit buses in Vancouver and Chicago, running on compressed hydrogen gas stored in tanks onboard the vehicles. Hydrogen refueling stations for buses and vans are also appearing in Germany, in the Munich airport and Hamburg, although these will initially supply vehicles with ICEs that use the fuel directly. The Hamburg station intends to eventually import hydrogen from Iceland.[60]

The introduction of fuel cell cars faces three tough technical challenges: integrating small, inexpensive, and efficient fuel cells into the vehicles; designing tanks that can store hydrogen onboard; and developing a hydrogen refueling infrastructure. The design issue is being overcome by improvements in power density and reduced platinum requirements. The storage issue is being addressed through vehicle efficiency gains, tank redesign, and progress in storage technologies such as carbon nanotubes and metal hydrides. Although the direct hydrogen fuel cell vehicle is the simplest and most elegant approach, industry is devoting substantial research to having cars use onboard reformers that strip hydrogen from gasoline, natural gas, or methanol. From their perspective this approach may appear preferable to spending the money needed to develop a new refueling infrastructure, but the economics and environmental effects are less straightforward.[61]

Studies indicate that by the time a critical mass of infrastructure and vehicles are in place, direct hydrogen will be more cost-effective than onboard reformers. Reformer-based fuel cell cars, furthermore, are unlikely to achieve the environmental performance of those using direct hydrogen. The Canada-based Pembina Institute, comparing the "well-to-wheel" greenhouse gas emissions of various hydrogen vehicle production systems over 1,000 kilometers of travel, found that reforming hydrogen from hydrocarbon fuels does provide an improvement over a gasoline-powered internal combustion engine, but the improvement varies widely, depending on the fuel. Gasoline and methanol, which are the preferred fuels for many transport and energy companies, offered the least improvement, with reductions of 22–35 percent. The hydrocarbon demonstrating the greatest climate benefits—natural gas, whose life-cycle emissions were 68–72 percent below that of the gasoline ICE—has been relatively ignored by industry.[62]

One near-term solution to the "chicken-and-egg" infrastructure problem in many countries would be to use small-scale natural gas reformers at fueling stations, relying on existing natural gas pipelines to distribute the fuel. Marc Jensen and Marc Ross of the University of Michigan estimate that building 10,000 such stations—covering 10–15 percent of U.S. filling stations—would be enough to motivate vehicle manufacturers to pursue mass production of direct hydrogen fuel cell vehicles. This would require $3–15 billion in capital investment, which "can be weighed against the social and environmental benefits that will be gained as a fleet of hydrogen fueled vehicles grows." Ultimately, hundreds of billions of dollars will need to be invested over decades in a network of underground pipelines engineered specifically for hydrogen.[63]

While natural gas is currently the most common source of hydrogen, and the reformation of coal and oil is also being explored, renewable energy sources are likely to eventually produce hydrogen most economically. Electrolysis of water can convert solar and wind energy into hydrogen; scientists have recently boosted the efficiency of solar-powered hydrogen extraction by 50 percent. Biomass can be gasified to produce the fuel. Other potential renewable hydrogen sources include photolysis—the splitting of water with direct sunlight—and common algae, which produces hydrogen when deprived of sunlight. Coupling renewable energy systems with hydrogen will address their inherent intermittency and easily meet energy demand, provided there are continued technical advances and cost reductions in fuel cells and electrolyzers.[64]

Wise decisions made in today's early hydrogen economy could yield enormous economic and environmental benefits. Wrong turns toward an interim infrastructure, on the other hand, could strand millions of dollars in financial assets, lock in fleets of obsolete fuel cell cars, and add millions of extra tons of carbon emissions. There is an appropriate role for governments to play in collaborating with transport and energy companies to develop a direct hydrogen infrastructure through greater research into storage technologies and the identification of barriers and strategies to surmount them. A January 2000 report from the U.S. National Renewable Energy Laboratory concluded that "there are no technical showstoppers to implementing a near-term hydrogen fuel infrastructure for direct hydrogen fuel cell vehicles." The study did, though, point out engineering challenges and institutional issues, such as the need for codes and standards for hydrogen use.[65]

If governments and industries can come up with forward-looking roadmaps, optimistic scenarios for hydrogen may materialize. In the Oxford Institute's Kyoto scenario, hydrogen becomes more competitive due to emissions policies that cause oil prices to rise, undercutting oil supply and reaching production of 3.2 million barrels per day of oil equivalent by 2010, and 9.5 million barrels by 2020. In their 1999 book *The Long Boom*, former Shell executive Peter Schwartz and colleagues describe a scenario in which fuel cells have displaced the internal combustion engine within two decades, and "by 2050 the world is running on hydrogen, or close enough to call it the Hydrogen Age"—with climate change concerns a major driver of the transition.[66]

Deliberate Decarbonization

A wider understanding of the progressive squeezing of carbon from the energy economy could help resolve differences in the international climate debate. There is little dispute among economists, for example, that international emissions trading—the buying and selling of greenhouse gas credits among nations and companies to meet abatement goals—could lower the overall cost of reducing emissions. There is concern among a number of analysts and politicians, however, that some nations will reach their targets primarily by purchasing credits from other nations with large emissions surpluses—due to relatively lax Kyoto targets—rather than by seizing domestic carbon-cutting opportunities. One solution might be to encourage nations taking part in an international emissions trading regime to implement a set of "structural decarbonization" policies—such as fossil fuel subsidy removal, carbon taxation, energy efficiency standards, voluntary covenants with industry, and incentives for renewable energy.[67]

Greater dialogue on decarbonization policies can improve their design. Participants in "best practices" workshops hosted by the Climate Change Secretariat have concluded that comprehensive strategies are needed that use a mix of policies and measures across all sectors. While European nations appear to have the most progressive policies, even these have had limited effect. In a review of programs in five EU member states, former U.K. environment secretary John Gummer illustrated that only his home nation is on track to meet its Kyoto target. Gummer contends that the EU will only meet its goal if member states rely more on eco-taxes, voluntary agreements with industry, and efforts to increase use of renewable energy and cogeneration.[68]

One obstacle to the adoption of new climate policies is that modeling estimates of their economic impact range widely, from substantial savings to dire losses. John Weyant of Stanford University notes that several key variables account for the majority of these differences: the extent to which energy price rises stimulate the development of low-emission technologies and changes in consumption patterns; the expected emissions path under current policies; the type of policy regime adopted; and whether the benefits of avoided climate impacts are accounted for. These varying assumptions can lead to a 10-fold difference in cost projections for an individual model.[69]

Some governments are striving to surpass Kyoto goals and take a longer-term perspective. The United Kingdom has a goal of cutting its carbon emissions to 20 percent below 1990 levels by 2010. In March 2000, the government unveiled a new package of policies including a climate change levy, negotiated agreements with energy-intensive sectors and car manufacturers, and a requirement that electricity suppliers generate at least 10 percent of their power with renewable energy sources. An expert committee appointed by the Swedish government has looked even further ahead and mapped out a long-term national strategy for halving greenhouse gas emissions by 2050. Key parts of the plan include tighter efficiency regulations for housing, industry, and offices; the introduction of a car tax; and the expansion of subsidies for wind power.[70]

Decarbonization may be an especially useful concept for developing countries. As experience with industrial nations has shown, emission "baselines"—the starting point against which trends are measured—using net emissions can be highly problematic because they are subject to changes in economic growth rates. This is even more so for developing nations, where emissions trends have been volatile in recent years. A 1999 U.N. Development Programme (UNDP) study asserts that "baselines constructed using carbon emissions per unit GDP...tend to be less variable and less subject to factors beyond the control of policymakers, and are thus a better indicator of whether a country is continuing to make progress in de-coupling emissions from economic development."[71]

The UNDP study, which contains case studies of Argentina, Brazil, China, India, Mexico, and several African nations, finds that these countries are slowing emissions growth through a range of policies adopted more for their social, economic, and environmental benefits—energy savings, lower air pollution, reduced oil imports—than for their carbon cuts. This suggests a seemingly counterintuitive lesson for all nations: that it may be easier to implement decarbonization policies when their non-climate benefits are emphasized.[72]

Decarbonization implies major change in international institutions such as the World Bank and export credit agencies. During the 1990s, the latter funded up to $100 billion in energy and carbon-intensive activities in developing countries. But reforms are under way: the Bank is setting up a Prototype Carbon Fund that will support the Clean Development Mechanism established under the Kyoto Protocol to steer public and private capital to carbon-saving projects in developing nations.[73]

The decarbonization challenge is rising on the radar of business executives. In March 2000, participants at the World Economic Forum in Davos voted climate change as the most significant global problem they would face during the twenty-first century, identifying it as an issue where business as well as government must play more of a leadership role. The hunt for carbon savings may become a major new source of competitive advantage, as companies that adopt energy-saving practices and develop low or no-carbon technologies gain an edge in the marketplace.[74]

Kimberly O'Neill Packard and Forest Reinhardt argue in the *Harvard Business Review* that "business leaders need to inform themselves about climate change and think systematically about its effects on their companies' strategies, asset values, and investments." They contrast the forward-looking attitude of BP—whose CEO, John Browne, believes that confronting the problem is unleashing creativity within his company—with the status-quo strategy of ExxonMobil—which continues to challenge the scientific evidence and attempt to slow international negotiations The authors conclude that "companies that calculate the risks and opportunities effectively—as they would for any other part of the business—will be able to survive the coming storms."[75]

A first step for companies in identifying decarbonization opportunities is to perform inventories and report their carbon and other greenhouse gas emissions, as national governments have done. A study prepared for the Pew Center on Global Climate Change documents a growing trend toward corporate reporting of carbon and other greenhouse gas emissions. It notes that beyond considerations of environmental concerns and corporate citizenship, inventories enable companies to identify ways to enhance productivity and energy efficiency, improve relations with stakeholders, take advantage of emissions trading, and position themselves to count their voluntary reductions when future regulations kick in. Dupont has committed to reducing its greenhouse gas emissions by 65 percent between 1990 and 2010. Shell and BP aim to cut emissions 10 percent below 1990 levels by 2002 and 2010, respectively, and have set up internal emission trading systems to reach these goals most cost-effectively.[76]

Nongovernmental groups can be valuable partners in corporate climate efforts, lending advice, legitimacy, and public oversight to the process. The U.S. group Environmental Defense is partnering with BP, Shell, Dupont, Alcan, and three other corporations to measure and publicly report their emissions and meet their internal targets. The World Resources Institute and World Business Council for Sustainable Development are working to develop internationally accepted standards for corporate greenhouse gas emissions reporting.[77]

The hunt for carbon savings may become a major new source of competitive advantage.

Joseph Romm's Center for Energy and Climate Solutions and the World Wildlife Fund have launched a Climate Savers program to help firms like IBM and Johnson and Johnson implement corporate climate strategies. Romm's 1999 book, *Cool Companies*, provides an overview of existing efforts to cut carbon; many of these investments—building upgrades, industrial process improvements, on-site cogeneration—quickly pay for themselves through productivity gains and energy savings. The Pembina Institute has developed an online database of success stories and information on reducing greenhouse gas emissions not only for industry and commerce, but also for agriculture, public institutions, municipalities, and families and individuals.[78]

Decarbonization also merits greater attention as an indicator of sustainability, since political and business leaders are wary of steps that might have negative effects on economic growth. Lee Schipper of the International Energy Agency observes that "the goal is to widen the historical carbon/GDP gap." One avenue for mainstreaming this indicator is the Pilot Environmental Sustainability Index developed by the World Economic Forum's Global Leaders for Tomorrow Task Force, which compares 58 nations on a range of variables. These include total and per capita emissions of carbon and energy efficiency; adding a carbon/GDP indicator might be a logical next step in the further development of this index. At the corporate level, the amount of carbon emissions per unit of product is another criterion by which companies could be assessed and compared in the recently created Dow Jones Sustainability Index.[79]

Finally, scientists could propose specific goals for improving the carbon intensity of the energy economy, and circulate their proposals widely within the policymaking community. A 1999 report from the U.S. National Research Council, entitled *Our Common Journey: A Transition Toward Sustainability*, recommends a doubling of the historical rate of improvement as a reasonable goal. It has taken roughly 50 years to reduce global carbon intensity from 250 to 150 tons per million dollars of output; doubling this rate of improvement over the next half-century—or sustaining the rate of the past decade—would pull the ratio down to 125 by 2007, to 100 by 2018, to 75 by 2032, and to 50 by 2052. A tripling of the past half-century's decarbonization rate may eventually be necessary to achieve the most ambitious CO_2 stabilization goals, and thus to potentially stave off some of the more adverse impacts of climate change. But a doubling would be an important first step—and a measure of our progress on the long journey toward a sustainable energy economy.[80]

Chapter 6

Making Better Transportation Choices

Molly O'Meara Sheehan

Transportation planners and policymakers in Central and Eastern Europe are at a critical juncture. Soviet-era railways are in decline, car sales are booming, North American–style shopping malls are opening on the outskirts of towns, and drivers are clogging deteriorating roads to reach new "hypermarkets." Should these nations build more highways, going farther down the road toward car reliance and urban sprawl? Or should they instead funnel new investments toward rail, bus, and bicycle infrastructure, so that people have a variety of attractive, non-car choices?

Although the situation in Central and Eastern Europe is unique, all countries have important transportation decisions to make. Russia, which covers more land than any other country, must improve its deteriorating transport links. Although the former Soviet republics have less money than their neighbors in Central and Eastern Europe, they too have growing car fleets and crumbling rail tracks. Most of the world's people live in developing countries that have inadequate road and rail networks, so they also have the opportunity, in theory, to build a more balanced transportation system. And many industrial countries have both the money and the chance to add choices such as public transit and bicycling to existing communities and to channel future construction around such transportation networks. The investments that societies make today will determine which of two very different transportation futures will unfold.

By leaving bus and rail networks to disintegrate and focusing instead on a new generation of roads and highways, governments could build themselves into a corner. Cities in North America, Western Europe, and elsewhere have experienced the dangers of heavy reliance on fuel-guzzling motor vehicles. Local problems include accidents, air pollution, noise, and loss of wildlife habitat and open space; globally, the transportation sector is the fastest-growing source of the carbon emissions that are changing the climate. Increased reliance on airplanes cannot solve these problems, as air travel also contributes to air pollution, noise, and climate change.

But governments could choose to go in a different direction, to create transportation systems that give people more choice with less damage to the environment. This would require a concerted effort to build and revitalize railway and transit systems and to steer new development to locations easily reached by a variety of transportation means—not only car, but also bicycle, bus, and rail.

The path toward cleaner and more equitable transportation systems, however, is the path of greater resistance. In Central and Eastern Europe, the collapse of centrally planned economies has left state-owned rail agencies weak and inefficient. At the same time, western construction and car companies are eager to expand into new markets. Car advertisements promising individual freedom can be especially appealing to people newly freed from communist rule. Moreover, money is flowing from the European Union to upgrade transportation in Central and East European countries that hope to join the Union—and these funds are aimed at building long-distance links between East and West, such as highways, rather than at reviving urban transit services.

Many of the challenges to reforming transportation systems are found in similar forms around the world. One common obstacle is the division of responsibility in government and international lending agencies between different departments for transportation policy, land use planning, and environmental protection. Another barrier is the political sway and advertising power of today's motor vehicle and construction industries.

Decisions about transportation have far-reaching effects. Transportation links shape the villages, towns, and cities where people live. Throughout history—from compact cities well suited to walking, to radial towns stretched into spokes by trolley lines, to sprawling metropolises dominated by cars—transportation modes have dictated urban form. They also allow economies to function by bringing raw materials together to create products and services, and by connecting goods and services to markets.

Existing technologies can help us improve our transportation systems so they better serve the world's cities, economies, people, and the environment. This chapter explores how people can get better access to goods and services while reducing the environmental and social costs associated with transportation today.

From Sledge to Jet

Changes in our current transportation systems are certainly possible. Economist Joseph Schumpeter has described the history of transportation as "a history of revolutions." The quest for faster movement, the development of new fuels, and the response to environmental problems are some of the forces that have contributed to the rise and fall of various transport modes.[1]

Beginning with the earliest wooden sledges, people have sought ways to move themselves, messages, and goods from place to place. The wheel, a key advance in transportation, may have been invented around 4000 B.C. in Mesopotamia. Societies in China and Persia built some of the earliest highways for wheeled vehicles, and by the first century A.D., Romans had built the world's most sophisticated paved roads.[2]

Transportation technology has developed rapidly in the last two-and-a-half centuries. Although advances in the eighteenth century included steamboats and hot-air balloons, by 1800 people were still traveling at speeds comparable to those of ancient Rome. But by 1900, speeds of up to 100 miles per hour were possible on land; the spread of steamships, the introduction of railways—including the first subway—and the development of bicycles were among the transforming innovations of the nineteenth century. The scope of transportation in industrial countries was further widened in the twentieth century with the mass production of the automobile and the development of air travel.[3]

When new transportation technologies are introduced, they tend to grow rapidly and spread to other countries before declining. England began to build canals, the first infrastructure to overcome some of the shortcomings of pre-industrial transportation, around 1760, with feverish construction occurring around 1795 and canal mania extending to continental Europe and the United States. The pace of canal building slowed down after 1830, with network length peaking in most countries around 1850. Railways, introduced in the mid-nineteenth century, grew so rapidly that by 1900 they dominated the transportation system, accounting for close to 90 percent of all passenger traffic and about 70 percent of freight traffic in Europe and the United States. Railway networks reached their maximum size in most of these countries by the 1930s.[4]

Whereas rails rose to dominance by displacing canals for long-distance transportation, the automobile first gained ground as an urban vehicle that took the place of horse-drawn vehicles for relatively short trips. Some 4 million draft horses were used for transportation in the United States in 1900, for instance, compared with 400,000 motor vehicles; by the end of the 1930s, the number of horses had fallen to below 400,000, and the number of cars had risen to over 20 million. Automobiles initially complemented long-distance rail transport. It was not until after 1940 that cars and trucks began to overtake the railroads in western industrial nations.[5]

Changes in fuel sources and energy-related technologies have helped drive the evolution of transport systems. Arnulf Grübler of the International Institute for Applied Systems Analysis has identified a series of technology clusters in the history of industrial nations. The era of canals, which extended until about 1870, was associated with steam and iron. The age of railways, which lasted until 1940, coincided with the rise of coal, steel, and industrial electrification. The automobile rode to its present dominance along with oil, plastics, and consumer electrification.[6]

Environmental concerns have also shaped transportation choices. When the automobile was introduced, its fumes were regarded as less offensive than the solid and liquid wastes from horses. (At their peak, horses in nineteenth-century urban England produced an estimated 6 million tons of manure a year.) But the automobile held the environmental advantage only until car fleets began to exceed the horse populations they displaced.[7]

The next era of industrial and economic development may similarly see changes in transportation that take advantage of cleaner energy sources, from natural gas to renewable fuels (see Chapter 5), along with the growth of information technologies. Cities might boast a diversity of transportation options, transit systems strengthened by computer technology, and seamless connections between different modes.

Current Transport Trends

In many parts of the world, recent trends do not favor diverse, next-generation transportation systems. Rather, cars, trucks, and planes are moving more people and goods, while rails, bicycles, and other less environmentally damaging transport modes are declining. Increasing incomes, sprawling cities, and globalizing companies, among other factors, are sustaining these shifts. Fuel use for transportation is rising as a result.

Since the end of World War II, motor vehicle production has risen almost linearly. As a result, the global car fleet now numbers more than 500 million. In the United States, the number of household vehicles increased at six times the rate of the population between 1969 and 1995. In Western Europe, passenger car traffic more than doubled between 1970 and 1995, and road freight traffic tripled, growing faster than rail or water. The World Bank estimates that the demand for freight and passenger transport in most developing and transition countries is growing 1.5–2.0 times faster than gross domestic product (GDP), fueled by a demand for road transport. In many African countries, however, political instability, government corruption, and low economic growth have made it difficult to maintain roads.[8]

While road traffic dominates the transportation system, air transportation is the fastest-growing segment. Transportation volume is often measured in "passenger-kilometers" or "ton-kilometers," units that encompass the number of passengers or the weight of cargo goods multiplied by the distance they are moved. Global air passenger transportation has risen from 28 billion passenger-kilometers in 1950 to more than 2.6 trillion passenger-kilometers in 1998. Since 1960, global air passenger traffic has grown at an average yearly rate of 9 percent, air freight shipments have grown an average 11 percent a year, and mail by air, 7 percent. As with road traffic, however, the African continent is largely missing from the air transport picture. Between 1982 and 1992, air traffic grew 11.4 percent on routes between Europe and Asia, but increased hardly at all between Europe and Africa.[9]

As road and air travel have grown, rail has become relatively less important. Since the 1980s, road networks have expanded more than rail networks in a number of countries. (See Table 6–1.) Although railways serve a much larger number of passengers than airlines do, the volume of rail traffic has fallen to less than 70 percent that of air. (See Figure 6–1.)[10]

The top five rail networks—in the United States, Canada, the former Soviet Union, India, and China—account for about half of the total tracks but carry more

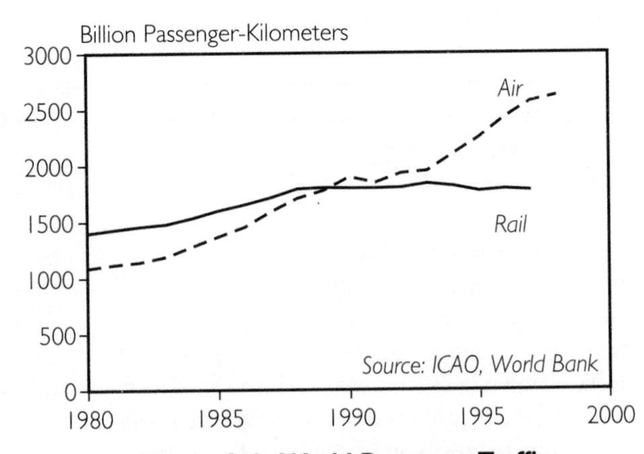

Figure 6–1. World Passenger Traffic, Rail Versus Air, 1980–97

Table 6–1. Road and Rail Networks, Selected Countries, Mid-1980s and Mid-1990s

	Road			Rail		
	Mid-1980s	Mid-1990s	Change	Mid-1980s	Mid-1990s	Change
	(kilometers)		(percent)	(kilometers)		(percent)
United States	6,213,852	6,308,086	+ 2	235,103	170,200	–28
India	2,962,470[1]	3,319,644	+12	61,836	62,725	+ 1
Brazil	1,583,172	1,980,000	+25	—	—	—
China	—	1,526,389	—	52,119	54,616	+ 5
Japan	1,127,501	1,152,207	+ 2	22,801	20,134	–12
France	804,650	892,500	+11	34,676	31,939	– 8
Russia	—	570,719		82,600[2]	87,543	+ 6
Indonesia	205,030	342,700	+67	6,458	—	—
South Africa	183,728	331,265	+80	23,821	25,555	+ 7

[1]1989. [2]1980.

SOURCE: International Road Federation, *World Road Statistics* (Geneva: various editions); World Bank, *World Bank's Railways Database* (Washington, DC: Transportation, Water, and Urban Development Department, August 1998).

than 90 percent of the world's ton-kilometers of freight and account for 56 percent of passenger-kilometers. The World Bank railways database shows that rail passenger traffic has been decreasing in many countries and that rail freight traffic has changed little except in the United States, where it has increased. In Europe, however, railways have become less important for moving freight, while road transport of freight has greatly expanded.[11]

Just as cars and trucks appear to be taking long-distance customers away from railways, motorcycles and cars are supplanting bicycles for shorter trips. Bicycle production fell to 79 million units in 1998, which was 25 percent below the peak of 107 million bicycles in 1995. Bicycles are on the decline even in Asia, the world's leader in production and use.[12]

Many factors are contributing to the growth in road and air travel. As people get richer, they often buy motor vehicles. For instance, between 1980 and 1995 a more than tripling of GDP in South Korea corresponded to a 16-fold increase in the nation's vehicles. In Japan, GDP more than doubled between 1970 and 1990, while the number of vehicles more than tripled. Rising incomes have also supported air travel.[13]

However, researchers at Australia's Murdoch University have demonstrated that income alone does not dictate differences in travel. Studying 47 major metropolitan areas in Asia, Australia, Europe, and North America, they found that car use in cities of North America and Australia is higher than could be explained by wealth alone. For instance, the U.S. cities surveyed had, on average, 141 percent more car use than the European cities, but had 15 percent less per capita income. Lower car use is linked not simply to lower incomes but to greater support for public transit and bicycling, land use rules that promote development along public transport routes, and higher fuel taxes.[14]

With just 19 percent of the world's population, industrial nations use 59 percent of all energy that goes to transportation.

Cars have helped to change the structure of cities, which in turn has contributed to greater vehicle use. Cities have spread out over larger expanses of land as builders have constructed wide expressways and ample parking to accommodate motor vehicles. Asked in a survey to identify the top influences shaping the American metropolis, a sampling of urban historians, social scientists, and architects chose the highway system and dominance of the automobile as the number one influence. Even U.S. metropolitan areas with stable or declining populations over the past 30 years—places like Detroit or Cleveland—have actually grown in land area. As cities sprawl, cars become essential while transit, bicycling, and walking become less practical. In the Czech Republic, for instance, car use has surged and public transit use has fallen as the number of suburban hypermarkets has ballooned from 1 to 53 between 1997 and 2000.[15]

Bikes are much cheaper than motor vehicles and are well suited to short trips, but pollution, unsafe roads, and a lack of safe bike parking often keep cyclists off the road. For instance, the U.S.-based Institute for Transportation and Development Policy found that 60 percent of very short, "bikable" trips in Surabaya, Indonesia, are made by motor vehicle; in Germany, where streets are more inviting for pedestrians and cyclists, only 15 percent of such trips are by motor vehicle.[16]

In industry, adoption of "just-in-time" delivery has increased the importance of trucks for hauling freight. Popularized by the Japanese automobile industry in the 1980s, this practice allows companies to cut down on inventory by requiring raw materials or intermediate products to arrive at a manufacturing facility at a precise location and time to fit seamlessly into a production schedule. Rather than pay for warehouse storage, companies keep goods in motion, using trucks as rolling warehouses.[17]

With the most energy-intensive transportation methods—cars, trucks, and planes—being used more often, fuel use is increasing. One measure of energy use is emissions of the carbon dioxide released when fossil fuels are burned. Looking at a group of major industrial countries, Lee Schipper and his colleagues at the Lawrence Berkeley Laboratory in California found that between 1980 and 1994, carbon emissions per unit of GDP fell in all sectors of the economy except transportation. Between 1973 and 1992, Japan's primary mode of passenger travel shifted from rail; increased reliance on autos and airplanes raised total carbon emissions from transport by 20 percent. Denmark was the only industrial country during these 20 years in which a shift toward buses and rail led to a clear decline in travel energy use and emissions.[18]

Schipper identifies a trend in energy use in industrial countries from "production to pleasure." While efficiency gains have lowered energy use in manufacturing, trends in personal mobility toward higher levels of motor vehicle use with fewer passengers per vehicle have actually raised energy use in the transportation sector. Personal travel now outweighs freight transport in industrial countries. Passenger trips generally account for 60–70 percent of energy use and emissions from transportation. And the key component in these countries is automobile travel. (See Figure 6–2.) According to travel surveys in the United States and a number of West European countries, work trips account for 20–30 percent of personal travel, shopping and other "service" trips account for about 25 percent (except in the United States, where the share was higher), and leisure for the rest.[19]

Behind the growth in automobile energy use is a boom in larger, more energy-consuming "light trucks" such as minivans and sport utility vehicles (SUVs). In the United States, these vehicles now dominate new sales, and in the 1990s this reversed gains made in the 1980s in the fuel economy of the total passenger fleet. Car makers are now introducing smaller SUVs into Western Europe, where sales of the vehicles increased by 26 percent between 1998 and 1999.[20]

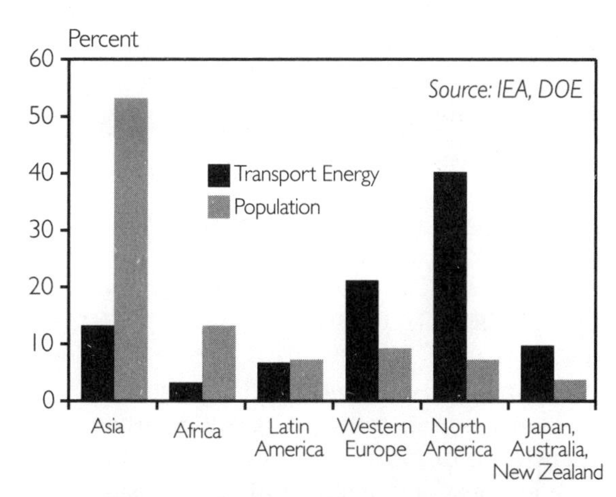

Figure 6–3. Global Transport Energy and Population, by Region

Industrial nations have the bulk of the world's automobiles, trucks, and airplanes, and accordingly use a disproportionate share of transport fuels. (See Figure 6–3.) With just 19 percent of the world's population, industrial nations use 59 percent of all energy that goes to transportation. The United States alone uses more than one third of the world's transport energy. Americans travel almost twice as far per person as Europeans do. In 1997, annual oil use for all transportation averaged 18 barrels per person in the United States, 13 barrels in Canada, and 6 barrels in Western Europe, Japan, Australia, and New Zealand.[21]

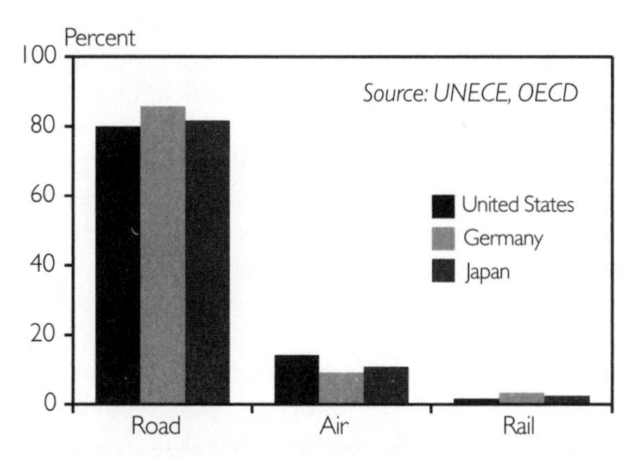

Figure 6–2. Transportation Energy Use by Mode in the United States, Germany, and Japan, 1995

In the developing world, the fastest growth in transportation energy use is in the urban areas of Asia and Latin America. While average transportation fuel consumption in the industrial world increased 1 percent between 1996 and 1997, the equivalent figure for developing Asia was 6 percent and for Latin America, 5 percent. In Africa, transportation energy use was unchanged.[22]

The Costs of Mobility

Advances in transportation technology have brought benefits, but growing vehicle fleets and escalating fuel use have also created problems. These include social inequities, accidents, smog, traffic congestion, urban sprawl, loss of forests and farms, noise, and carbon emissions that contribute to rising global temperatures.

First on this list is that a lack of transportation choices worsens existing social inequities. Although car travel is the only viable means of transportation in some U.S. cities, for instance, roughly one third of the nation is too young, too old, or too poor to drive. In rural areas of developing countries, lack of transportation alternatives, such as bicycles, takes a disproportionate toll on the health of women, who are often responsible for transporting fuel, food, and water. In a study of village-level travel and transport in sub-Saharan Africa, World Bank–sponsored researchers found that adults spend on average an hour and a quarter each day walking for essential trips, carrying some 20 kilograms (45 pounds) over 2 kilometers; most of the burden falls on women.[23]

Beyond this equity impact, motor vehicles do all sorts of other damage to people and the environment. Traffic accidents kill and injure many people. Researchers estimate that nearly a million people are killed on the world's roads each year, and most of them are pedestrians. According to the World Bank, the annual cost to the global economy in lost productivity from traffic-related deaths and injuries is roughly $500 billion.[24]

Motor vehicles also contribute to local and regional air pollution. Toxic ingredients in motorcycle and car fumes include carbon monoxide, sulfur dioxide, nitrogen oxides, fine particles, and sometimes lead. Nitrogen and sulfur that travel beyond urban areas acidify lakes, forests, and farms. Emissions from many vehicles are harder to reduce than emissions from relatively few stationary sources. In the United States, for instance, while sulfur and nitrogen have been reduced from smokestacks as a result of Clean Air Act amendments, nitrogen emissions from cars have proved more difficult to abate. As a result, acid rain continues to eat away at forests.[25]

In some parts of the world, vehicular air pollution actually kills more people than traffic accidents do. In Austria, France, and Switzerland, in 1996 the premature mortality brought about by particulate emissions from vehicles stood at about twice the number from traffic accidents, according to a recent World Health Organization study published in the *Lancet*.[26]

Vehicle emissions are a growing source of air pollution in many urban centers of the developing world as well. Highly polluting two- and three-wheeled vehicles are prevalent in many of these cities. Most of these are powered by simple but dirty "two-stroke" engines, in which much of the fuel goes unburned and is released with the exhaust. A World Bank report suggests that vehicles with these engines emit more than 10 times the amount of fine particulate matter per vehicle-kilometer as a modern car and only slightly less than a diesel truck.[27]

Motor vehicles impede other forms of traffic and cause delays. Cities such as Bangkok and Jakarta are less densely populated than Paris, Moscow, or Shanghai, yet they suffer worse traffic delays because they have neither effective public transit systems nor adequate facilities for bicyclists and pedestrians. Congested roads in São Paolo have prompted the wealthiest residents to take to the skies, boosting the city's helicopter fleet to the third largest in the world, after New York and Tokyo.[28]

Roads also cause profound changes in ecosystems. A great deal of land in car-dependent cities is lost to roads and parking lots. Water quality and quantity both suffer in proportion to the amount of paved roads and parking that cover a watershed. Plants and animals are killed during road construction, as well as by vehicles. And roads promote the dispersal of species that are not native to a given area, alter the physical and chemical environment, block wildlife corridors, and divide populations of various species into smaller, less stable subpopulations. In the United States, 6.2 million kilometers of public roads cover about 1 percent of the nation's surface area, but a rough estimate suggests that this network has effects on wildlife on up to 20 percent of the surface area.[29]

Noise is perceived by many urban residents as one of the greatest problems associated with road traffic. It contributes to stress disturbances, cardiovascular disease, and hearing loss. The problem is particularly acute in Japan, where 30 percent of people experience noise levels greater than 65 acoustically weighted decibels—a level considered unacceptable in many countries—and in Europe, where the figure is 17 percent.[30]

Transportation is a growing contributor to climate change. According to Lee Schipper of Lawrence Berkeley Lab, "travel is emerging as the primary leader of growth in carbon emissions in the wealthy, industrialized countries." Worldwide, the share of carbon dioxide gas from transportation climbed from 17 percent in 1971 to 23 percent in 1997. Road traffic is spurring this jump: motor vehicles accounted for 58 percent of worldwide transportation carbon emissions in 1990, but their contribution grew to 73 percent by 1997.[31]

Nearly a million people are killed on the world's roads each year, and most of them are pedestrians.

The environmental toll of air travel is increasingly coming under scrutiny as well. Airplanes can be especially fuel-inefficient over short distances. They are the primary source of heat-trapping greenhouse gas emissions from humans that are deposited directly in the upper atmosphere, and scientists have noted that these emissions have a greater warming effect than they would have if they were released at the surface. Some researchers have found an increase in cirrus clouds associated with the trails of condensation created by aircraft, which may affect climate patterns. Jets release not only carbon dioxide but also water vapor, nitrogen oxides, soot, and sulfate, and account for some 3.5 percent of global emissions of greenhouse gases. Scientists on the Intergovernmental Panel on Climate Change estimate that this share could increase fourfold within 50 years, even with substantial improvements in fuel efficiency.[32]

Over the past decade, analysts have tried to quantify all these problems associated with transportation—and with motor vehicles in particular. Estimates of the social costs of transportation differ from country to country and even within countries because there is no single framework for determining which costs to include and how to derive them. Nevertheless, clearly there are many costs to society that are not covered by fuel taxes, vehicle taxes, or fees for road use. For example, a study of road transportation in the United Kingdom found that the costs associated with air emissions, noise, congestion, road damage, and accidents outweighed the taxes paid by drivers by three to one. And a recent U.S. study similarly found that the cost of driving-related air and water pollution, climate change, noise, and accidents—at $125 billion—was nearly three times greater than the $42 billion paid by car drivers in highway user fees. Using different methods, various researchers have estimated the costs of road transport not covered by drivers to be around 5 percent of GDP in industrial countries, and even higher in some developing-country cities. (See Table 6–2).[33]

Clearing the Air

The single largest contributor to the costs of transportation borne by society in many countries is illnesses and deaths from air pollution. One challenge, therefore, is to tackle immediate health threats from the most polluting vehicles. By adopting policies that promote cleaner technologies, governments can take one important step toward clearing the air.

Cleaner fuels and engines can reduce local air pollution. Historically, the United States has set emissions standards to force vehicle makers to find cleaner technologies. Better engine design and catalytic converters have reduced emissions from individual cars to well below 5 percent of the levels in the 1960s. Other countries have adopted U.S. standards, but there is still much room for improvement. A recent study by the Organisation for Economic Co-operation and Development (OECD) suggested that combinations of advanced emissions control technologies now exist that could allow cars and light trucks to meet tailpipe emissions standards 50–80 percent more stringent than the strictest rules in any industrial country. Moreover, effective inspections and maintenance programs could significantly cut pollution by the existing fleet. Much-needed targets for cleaner diesel-fueled trucks and buses are on the horizon in Japan and the United States.[34]

A number of policies can reduce pollution from motorized two- and three-wheel vehicles, which are among the worst polluters on the roads of the developing world. Analysts at the World Bank have identified

Table 6–2. Estimates of Societal Costs of Road Transport as Share of GDP

Country, Region, or City	Year	Share of GDP (percent)	Costs Included in Study: Accidents	Smog	Climate	Noise	Roads	Land & Parking	Congestion
United States	1989	5.5	x	x	x	x	x	x	
European Union	early 1990s	4.63	x	x	x	x	x		x
United Kingdom	1993	4.68–5.79	x	x	x	x	x		x
Mexico City	1993	5.6	x	x				x	x
Santiago, Chile	1994	6.71	x	x		x	x	x	x

SOURCE: Christopher Willoughby, *Managing Motorization*, Discussion Paper, Transport Division (Washington, DC: World Bank, April 2000), 5.

several key short-term steps: public information campaigns on vehicle maintenance, regular inspections, minimum standards for fuel and motor oil, and tax incentives for low-polluting vehicles. Over the long term, it is often cheaper for vehicle owners to replace two-stroke engines. Two- and three-wheelers are popular in cities of the developing world because they are relatively inexpensive to buy, but their inefficient two-stroke engines require expensive fuel and motor oil to operate.[35]

Whereas some technical solutions allow the combustion engine to release less pollutants when it burns petroleum or diesel fuel, another set of solutions replaces the fuel or the engine with a cleaner alternative. For instance, the two-stroke engine on a small, three-wheeled cart could be replaced with an electric battery. In the early 1990s, environmental groups in Katmandu, Nepal, were pointing to the pollution caused by diesel-fueled vehicles that was threatening the city's tourism industry. The city government invited a nonprofit group from Eugene, Oregon, to design an electric three-wheeler. Upon learning of the design, the Agency for International Development pledged $500,000 to promote the vehicles. By the time that funding ended in 1996, two local businesses in Katmandu had started assembling and operating electric vehicles.[36]

Elsewhere, governments are spurring the deployment of different sorts of cleaner-fuel vehicles. London has begun converting taxis from diesel fuel to cleaner-burning liquefied petroleum gas. Government agencies in Mexico City have contracted with a private company to convert thousands of public transportation vehicles from gasoline to liquid propane gas. Most public buses in the United States run on diesel fuel, but the number of less polluting natural gas buses is on the rise, with some cities announcing that they will no longer purchase diesel buses. The state of California announced in 1990 that it would eventually require a certain percentage of cars sold there to be "zero-emissions" vehicles, which might run entirely on electric batteries; the original mandate has been postponed and modified to include very low emissions vehicles such as the hybrids now on the market by Toyota and Honda, which are powered by a small internal combustion engine combined with an electric motor.[37]

One possible successor to the combustion engine is the fuel cell, which can be used to generate electricity from hydrogen with fewer carbon emissions. (See Chapter 5.) DaimlerChrysler announced in 2000 that it would bring fuel-cell public buses to the market within two years, and passenger cars in four years. Toyota, Honda, and General Motors have pledged to have competing vehicles in roughly the same time period. Iceland is interested in converting its bus, car, and fishing fleets to fuel cell vehicles within 40 years.[38]

While such innovations are promising, history has shown that technical solutions to one problem can exacerbate other problems or even create new ones. Catalytic converters reduce carbon monoxide and volatile organic compounds but increase nitrogen oxides. Three-way catalysts also remove nitrogen oxides but decrease engine efficiency, resulting in more carbon dioxide. And these catalysts are destroyed by lead in gasoline. When catalytic converters were first introduced in the United States, leaded gasoline had not yet been phased out; in fact, it was cheaper than unleaded gasoline. As a result, many pollution-control devices were ruined when people opted to fill their tanks with the cheapest fuel. The story of methyl tertiary butyl ether (MTBE) provides a similar lesson. For the past decade, this has been added to about one third of the gasoline in the United States, mainly in smoggy areas, to reduce emissions of carbon monoxide and toxic pollutants. The additive has led to cleaner air but dirtier water, from leaking underground pipes and storage tanks. (See Chapter 2.)[39]

About 20 car-free communities are in various stages of development in Germany.

Even the best technologies alone are insufficient. Emissions reductions can be swallowed by greater vehicle use, so governments must bolster improvements in fuels and vehicles with efforts to reduce the volume of vehicle use and to shift transportation activity to less fuel-intensive modes. Moreover, fuel and vehicle fixes tend only to address air pollution, leaving other transportation problems—from accidents to congestion to climate change—unchecked.

Diversifying Our Options

The broad challenge is to ensure that transportation systems do not endanger public health or ecosystems and that they meet people's needs for access to places. An OECD team of transportation experts recently advanced a set of principles that would characterize such an environmentally and socially "sustainable" transportation system. (See Table 6–3.) To meet these targets, societies will have to not only shift to cleaner motor vehicles but also work to diversify transportation choices.

With a diversity of transportation alternatives, people could choose the best option for a given purpose. Forty years ago, U.S. urban critic Lewis Mumford noted, "what an effective network requires is the largest number of alternative modes of transportation, at varying speeds and volumes, for different functions and purposes....By pushing all forms of traffic onto high speed motor ways, we burden them with a load guaranteed to slow down peak traffic to a crawl."[40]

Today transportation planners increasingly recognize that building more roads does not necessarily solve traffic problems. A study in the Washington, DC, area found that despite new road capacity, severe congestion quickly overtook regional highways between 1996 and 1999. Michael Replogle, a transportation specialist at the U.S.-based advocacy group Environmental Defense, came up with this analogy: "Adding highway capacity to solve traffic congestion is like buying larger pants to deal with your weight problem."[41]

As new roads attract more cars, regions that have invested heavily in roads have fared no better at easing traffic than those that have invested less. Measures to diversify transportation options include regulations to curb car traffic, price incentives to reduce motor vehicle use and boost alternatives, and changes in urban design that enhance the viability of cycling, walking, and public transit.[42]

Some European cities explicitly ban private cars from central areas. Munich, Vienna, and Copenhagen, for instance, boast popular commercial centers that restrict motor vehicle traffic to ambulances, delivery trucks, and cars owned by local residents. A new residential neighborhood in Freiburg, Germany, bans cars entirely; about half of the residents own cars, but they keep them just outside the development. In fact, about 20 car-free communities are in various stages of development in Germany. And since 1994, more than 70

Table 6–3. Suggested Principles for Sustainable Transportation

Principle	Description
Access	People have reasonable access to places, goods, and services.
Equity	Transportation meets the needs of different groups within society and across generations.
Health and Safety	Transport systems are designed and operated in a way that protects the health and safety of all people.
Education and Participation	People and communities are fully engaged in transportation decisionmaking.
Integrated Planning	Transportation planning involves people from various fields: environment, health, energy, urban design.
Land and Resource Use	Transport systems make efficient use of land and other natural resources while preserving habitats and biodiversity.
Environmental Integrity	Transportation does not threaten public health, global climate, or essential ecological processes.
Economic Well-Being	Taxation and economic policies promote equitable and clean transportation.

SOURCE: Organisation for Economic Co-operation and Development, *Environment and Transport: Synthesis of OECD Work on Environment and Transport and Survey of Related OECD, IEA and ECMT Activities* (Paris: November 1999), 28–29.

European cities have joined a Car Free Cities Network to exchange policy ideas.[43]

Based on these examples in Europe, the city of Bogotá in Colombia tried a "car-free" experiment in February 2000, banning private vehicles from 6:30 a.m. to 7:30 p.m. on a weekday. According to a news report, the air was cleaner, with lower levels of carbon monoxide and nitrogen dioxide recorded. No traffic fatalities were reported, and polls showed that some 65 percent of residents wanted to repeat the exercise.[44]

For a car ban to work, however, there must be other effective means of transportation, such as public transit or cycling networks. A well-known example of a policy failure is Mexico City's "hoy no circulan" policy that banned use of certain cars on one day each week. Without viable transportation choices, many residents just bought an extra car—often an old, cheap, and highly polluting one—to circumvent the ban. Thus a policy aimed at reducing pollution ended up worsening the problem. To avoid a similar result, the mayor of Bogotá is proposing a long-term program that by 2015 would develop the city's bus service and build a subway and bicycle paths so that people could realistically restrict their private car use for three hours in the morning and three hours in the afternoon.[45]

In response to the Mexico City fiasco, some analysts have argued that if a local air cleanup strategy complements new regulations with a taxation and pricing strategy, it has a chance of both reducing the use of vehicles and cleaning them up. An increase in the "variable cost" of operating a vehicle—the cost of each trip—could deter excessive use. In contrast, increasing the "fixed cost" through vehicle taxes may actually promote use because once people pay for a car, they want to use it as much as possible to get their money's worth. In the United States, unfortunately, the fixed costs of automobile ownership have risen while the variable costs, such as fuel, have fallen. One way to increase the variable cost of driving is to raise the cost of fuel with a tax, as many countries in Western Europe have done. Governments could make a fuel tax hike more palatable by accompanying it with a reduction in income tax, or by using it to fund transportation alternatives.[46]

Reflecting increasing concern about the emissions associated with air travel, a number of citizens groups and governments in Europe have recommended introducing some form of a jet fuel tax. Norway was the first to impose a carbon tax on jet fuel in 1998, but then revoked it under pressure from governments that argued that international agreements protected jet fuel from national taxation. In 1999, a Dutch environmental group convinced a number of companies to voluntarily introduce a flight "tax" of 17.5 percent of the ticket price, which is routed to a fund within each company for investments in environmentally friendly energy resources. In 2000, the European Commission recommended introducing a jet fuel tax, but European Union finance ministers would only sign it if an agreement at the global level could be forged.[47]

A fuel tax alone, however, is a blunt instrument. In the case of road traffic, targeted incentives—charging heavy trucks for the extra wear they take on roads, introducing fees for driving on congested roads at peak travel times, and substituting transit tickets for parking privileges—can more precisely address different concerns.

Some governments have accordingly begun to adjust road and parking policies to reflect the high cost of vehicle use and to limit unnecessary trips. Iceland, New Zealand, Norway, and Sweden levy a charge on diesel trucks that rises with the weight of the truck and the distance traveled. Singapore leads the world in using tolls to curb traffic. For more than 20 years, downtown-bound drivers have paid a fee that rises during rush hours; since 1998, the fee has been automatically deducted from an electronic card. In the United States, parking provided by employers is tax-exempt, which can be an incentive to drive. But the state of Maryland passed a law in 2000 that extends tax credits to nonprofit groups such as schools and hospitals if they pay for employee transit benefits or benefits equal to the cost of a parking space that the employee will not use.[48]

Removing some of the cost incentive for excessive driving can be done without taxes through private initiatives that decouple car ownership from car use. Privately run car-sharing networks, popular in Europe and recently introduced to a few North American cities, give people easy access to a car without the high costs of owning or the hassles of renting. Members are more likely to use a car only for the trips where it has a significant advantage over other modes. In addition, they have access to a wide range of vehicles, so they can choose the best one for a given trip: a small car for city use, a larger car for a family vacation, or a truck for moving furniture.[49]

Coordinated transportation and land use policies can lessen the need for travel and boost transportation options. Since 1947,

Copenhagen has guided its development along five rail corridors extending toward smaller population centers. Although government agencies and planning fashions have come and gone, a vision has endured of the region shaped as a hand, with the central city as the palm, urban development spreading out along the rail "fingers," and green space preserved between the fingers. This structure helps to make urban transit and cycling viable. A recent government directive states that new development should be within one kilometer of a transit station. Even though the road network is bigger than it was in 1970, the kilometers driven by motor vehicles is 10 percent below the figure then.[50]

Perhaps the best example of integrated planning on a budget comes from Curitiba, Brazil. In the early 1970s, the city designated several main roadways radiating from the city center as structural axes for busways, which are much less expensive than underground railways. Through zoning laws, the city encouraged construction of high-density buildings along the bus corridors. Since then, innovations such as extra-large buses for popular routes and inviting, tube-shaped bus stops where passengers pay their fares in advance have added to the system's speed and convenience. The bus stations also link to a 150-kilometer network of bike paths. Although Curitiba has one car for every three people, two thirds of all trips in the city are made by bus.[51]

Developing countries without extensive road systems may have the opportunity to develop mobility with lower dependence on the car, by setting aside rights-of-way for bicycles, buses, and other forms of public transportation. In cities such as Copenhagen and Curitiba, convenient connections between cycling and public transit make these options more attractive. Bicycles are often not convenient for long trips, and buses and trains are limited to fixed routes. But bicycles and transit can complement each other when people are able to carry their bikes aboard buses or trains, or to park them at stations.

Privately run car-sharing networks give people easy access to a car without the high costs of owning or the hassles of renting.

Cycling and pedestrian pathways provide other benefits beyond easing congestion and clearing the air. For instance, the exercise they stimulate decreases the risk of heart disease and diabetes. Studies in the United Kingdom have shown that the health benefits of cycling far outweigh the risks of accidents. The mayor of Bogotá is promoting bikes not only for their environmental benefits but also because of their potential to reduce crime: "The bike path ...is the safest place in the city because it creates a kind of solidarity—people help one another."[52]

The possibility for boosting cycling and public transit options exists even in countries with the most car-reliant cities. In the United States, public transit ridership is lower than it was 50 years ago, even though the population has doubled. But with increased government spending on infrastructure, public transportation is on the rise for the first time in decades. In 1999, ridership reached its highest level in nearly four decades, growing at a faster rate (4.5 percent) than car use (2 percent). The trend continued into 2000, with major increases found across the board—in small towns, suburbs, and large cities.[53]

Information technology may help enhance existing and new public transportation systems. For instance, computer-aided paratransit services using vans, taxis, shuttle buses, and jitneys with flexible pick-up and delivery stops may be better suited to spread-out metropolitan areas than conventional fixed-route services. A central switchboard and computer could match drivers and riders. Cities may need to reform laws governing bus and taxi routes in order to allow these new forms of public transportation to take root.[54]

While information technologies may support transportation initiatives, they will not do away with the need for integrated transportation and land use. Computers and electronic information transfer—e-mail, e-commerce, video conferencing—are transforming many aspects of business and personal life, and may reduce the need for some people to travel during peak traffic hours. But transportation and land use planning will be critical to prevent "tele-sprawl"—people living even farther away from others because they rely on machines for communication.[55]

Denmark, the Netherlands, and the United Kingdom each have one integrated national agency for transportation, regional planning, and the environment.

Moreover, by connecting more far-flung people, communications technologies may actually induce more transportation. The Internet now allows a person in Europe, for instance, to easily order a book to be air-shipped from the United States before it is available locally. History suggests that transportation and communication trends are complementary. Analysis of data on France, Germany, and the United Kingdom shows that since 1900 there has been a relatively constant ratio of transportation to communication, with around six kilometers traveled per message sent. Today's growth in communications may similarly be accompanied by an increase in transportation.[56]

Identifying Bottlenecks

Over the past several decades, a growing vehicle fleet has spurred escalating fuel use, air pollutants in traffic fumes have sickened and killed millions of people, and highway-led urban development has devoured farms and forests. Scholars, citizens groups, and governments have increasingly recognized the need for cleaner fuels and vehicles and for diverse transportation choices. The Victoria Transport Policy Institute in Canada has documented part of this knowledge base by listing more than 2,600 reports on transportation problems and remedies. What, then, is preventing these policy solutions from being put into action?[57]

Many bottlenecks are thwarting progress toward cleaner and more equitable transportation systems. From Copenhagen to Bogotá, every place has a unique history, economy, and political arena in which government, industry, and citizens interact to make transportation decisions. In government, key actors include agencies responsible for transportation planning, politicians and lawmakers who have the ability to change policies, and officials in charge of land use regulations and zoning. Private-sector players include the automotive, aviation, construction, and real estate industries. Individual citizens who make daily transportation choices and citizens groups that pressure government and industry also wield power.

Government structure profoundly affects the recognition of transportation needs and the establishment of priorities in this sector. A 1997 study of seven Central and East European nations found that transport policies often contradicted environmental policies. And as more established transportation ministries held more power than their environmental counterparts, transport policies were able to take precedence.[58]

Transportation ministries themselves are often segmented. For instance, within the U.S. transportation bureaucracy, the highway agency gets more money than the agencies for railways and urban transit. The availability of federal funds for U.S. highway projects has historically influenced the transportation priorities of states and cities, which themselves tend to have completely separate transportation and land use planning agencies. In contrast, countries such as Denmark and the Netherlands, and more recently the United Kingdom, have one integrated national agency for transportation, regional planning, and the environment.[59]

Given the parochial interest of certain government agencies in maintaining the status quo, historical trends in public spending for transportation are hard to overcome. In the United States, the landmark Intermodal Surface Transportation Equity Act (ISTEA) and its successor, the Transportation Equity Act for the 21st Century (TEA-21), allowed money to be diverted from highways to other transportation projects. But a recent study found that out of $33.8 billion in "flexible" funds that could be diverted to transit projects between 1992 and 1999, only $4.2 billion was actually reallocated.[60]

The U.S. citizens group Sustainable Transportation Policy Project (STPP) has documented backsliding on the initial progress of ISTEA toward diversifying transportation options. Between 1992 and 1997, an average of 22 percent of the federal transportation funds covered by the new legislation went to providing alternatives to driving; by 1999, the share had fallen to 17 percent. One reason, says STPP's Don Chen, is that state transportation departments have not really restructured themselves to seize the new opportunities.[61]

In many countries, international financial and development institutions sway transportation decisions. For example, a new source of transport funds for Central and East European countries is the European Union's Instrument for Structural Policies for Pre-Accession, which will cofinance with multilateral banks and national governments the extension of highways and high-speed transit from Western Europe. "By targeting improvements in long-distance links at the expense of urban transit, this money will distort transportation priorities" in Central and Eastern Europe, according to Magda Stoczkiewicz of CEE Bankwatch Network.[62]

Cozy relationships between industry and politicians often distort transportation and land use decisions. Citizens groups working to stop sprawl in Central and Eastern Europe observe that city councilors may be bribed to approve a new shopping mall, or may resign after approving a project to become head of its development company. Yaakov Garb, head of the Anti-Sprawl Campaign of the Institute for Transportation and Development Policy, notes that in Prague, "the regional planner's goal of finding harmony between economy, culture, and nature has receded, replaced by the growing power of investors."[63]

In many cases, "bribery" is legal. Large contributors to political campaigns expect to wield influence. In the 1998 U.S. congressional races, industries with a stake in transportation and land use decisions contributed some $128 million to political parties and candidates, which was 18 percent of the total spent by major industry groups. And during the 2000 U.S. presidential election campaign, a construction industry lobbyist in favor of expanded highway building told the *Wall Street Journal* that he looked forward to having the President's ear if the candidate he supported with a large contribution won: "I'm assuming, since we've supported him throughout his gubernatorial career, that we'd have an opportunity to go visit if we wanted to."[64]

In the United States, traffic growth and urban sprawl have become major concerns.

Politicians are often convinced by industry groups to oppose legislation that would be unfavorable to the industry. An example would be the efforts of the auto industry, which successfully lobbied the U.S. Congress to halt increases in fuel efficiency standards in the 1990s. Industries also persuade legislators to pass laws that benefit companies. The U.S. oil industry, for instance, has been successful at gaining a variety of tax advantages.[65]

Part of the increase in U.S. highway spending in recent years can be attributed to politicians eager to bring large road-building contracts to construction companies in their districts. In 1998, the largest expansion in U.S. highway spending since the 1950s was pushed through the U.S. Congress by transportation committee chairman Bud Shuster, who, in the words of one reporter, "has so many public works projects named after him you might think he was dead." To entice colleagues to support his transportation spending bill, he offered them their own public works projects.[66]

Once a transportation technology is widespread, it is often politically difficult to impose new rules that restrict its use or raise its costs. In Bangkok, for instance, replacing two-stroke motorcycles with ones that have four-stroke engines could dramatically reduce air pollution. But local manufacturers have been reluctant to change—as they would incur costs in retooling their plants and getting new licenses. In Germany, the Green Party has long encouraged less car use, but in May 2000 the party released a strategy paper acknowledging that its efforts have been unsuccessful—even though it has been part of the coalition government.[67]

Individuals make choices every day about different transportation modes based, in part, on perceptions about safety and comfort. Poorly maintained cycling and public transport networks are therefore less appealing than the private car. High-income residents of Mexico City, surveyed in a recent study of travel behavior, said they feared robberies on buses and that they would use public transportation only if it were safe, well organized, and comfortable. In Delhi, riders of privately operated "chartered" buses were asked what made these more attractive than other modes; comfort was the most common reason given (58 percent), followed by regularity (18 percent).[68]

People are also influenced by the car's image of freedom, power, and modernity. To many young people, getting a driver's license is a rite of passage. In one survey in England, young adults were asked: "Imagine you are only able to have one of the following two rights—the right to vote in an election, or the right to obtain a driving license—which would you choose?" Some 72 percent chose the license. The automotive industry spends more money than any other industry to perpetuate this image.

The United States is the largest market for car companies, but the automotive sector also dominates advertising spending elsewhere. (See Table 6–4.) By studying the emotional and instinctual reasons that people buy cars, companies are able to target their advertisements more effectively.[69]

But views and behaviors may change, as congested roads thwart the car's promise of individual freedom and power. In the United States, traffic growth and urban sprawl have become major concerns. In February 2000, the Pew Center for Civic Journalism found that the category "urban sprawl and traffic congestion" was the major concern of both urban and suburban Americans, and that nationwide it tied with "crime and violence" as the top worry, according to five public-opinion surveys around the country.[70]

Citizens have an important role in leaning on industry to change transportation priorities. Automotive and oil industries have shown signs of being sensitive to criticism. In 2000, the Ford Motor Company announced that it would voluntarily raise the average fuel efficiency of its sport utility vehicles by 25 percent by 2005, stating that it was responding to its customers' environmental concerns. And BP revamped its image to appeal to environmentalists with a campaign to install solar-powered Internet access at its gasoline stations.[71]

To foster innovation by industry—and not just a public relations "greenwash"—citizens will have to pressure governments too. Already, nongovernmental organizations have played a key role in pushing policy reform. The U.S. group STPP, for instance, was instrumental in drafting the ISTEA legislation that opened new funding options for alternatives to highway building and has acted as a watchdog group on ISTEA's shortcomings. And in response to advice from citizens groups such as the Institute for Transportation and Development Policy, the World Bank has rethought its approach to transportation lending.[72]

The automobile came to dominate the world's roads in the last century, the age of oil. Motor vehicles still consume the bulk of the world's transportation fuel, polluting

Table 6–4. Automotive Sector Ranking in Advertising Spending, United States and World, 1998

Rank	Category	United States Ad Spending	Rank	Category	World Ad Spending (excluding United States)
		(million dollars)			(million dollars)
1	Automotive	14,074	1	Automotive	9,904
2	Retail	11,572	2	Personal care	9,558
3	Movies and media	4,122	3	Food	5,225
4	Financial	3,850	4	Movies and media	2,449
5	Medicines	3,564	5	Medicines	1,573

SOURCE: "Total Measured U.S. Ad Spending by Category & Media in 1998" and "Measured Ad Spending by Category," Advertising Age, <adage.com/dataplace>, viewed 19 July 2000.

local communities and changing the global climate. Today, with the environmental and social costs of road traffic well documented, and with natural gas and renewable sources of energy beginning to replace oil, we can envision a new generation of transportation systems. Cars, trucks, motorized carts, and motorcycles could be cleaner, and cities and towns could be made more attractive and functional with integrated networks for bicycles, bus, rail, and new types of transit. People will need to work together to build this future, and to confront those in government and industry with vested interests in transportation systems that belong to the last century.

Chapter 7

Averting Unnatural Disasters

Janet N. Abramovitz

In October 1998, Hurricane Mitch slammed into Central America, pummeling Honduras, Nicaragua, El Salvador, and Guatemala for more than a week. As the powerful storm hung over the region, it dumped as much as two meters (80 inches) of rain. By the time it turned back out to sea, some 10,000 people had died, making Mitch the deadliest Atlantic storm in 200 years. Conservative estimates place the cost of damage to the region at around $8.5 billion—higher than the combined gross domestic product (GDP) of Honduras and Nicaragua, the two nations hardest hit. The storm set back development in the region by decades.[1]

But Central America is not the only region to experience such devastation in recent years. In fact, the 1990s set a new record for disasters worldwide. During the decade just over $608 billion in economic losses was chalked up to natural catastrophes, more than during the previous four decades combined.[2]

In 1998–99 alone, over 120,000 people were killed and millions were displaced from their homes. In India, 10,000 people lost their lives in a 1998 cyclone in Gujarat; the following year as many as 50,000 died when a "supercyclone" hit Orissa. Vast forest fires raged out of control in Brazil, Indonesia, and Siberia. Devastating landslides in Venezuela caused over $3 billion in losses and took more than 30,000 lives, capping off the deadly decade.[3]

Ironically, the United Nations had designated the 1990s as the International Decade for Natural Disaster Reduction, hoping to stem the rising toll taken by natural disasters. Instead, the 1990s may go down in history as the International Decade *of* Disasters, as the world experienced the most costly spate of floods, storms, earthquakes, and fires ever.

Around the world, a growing share of the devastation triggered by "natural" disasters stems from ecologically destructive practices and from putting ourselves in harm's way. Many ecosystems have been frayed to the point where they are no longer resilient and able to withstand natural disturbances, setting the stage for "unnatural disasters"—those made more frequent or more severe due to human actions. By destroying forests, damming rivers, filling in wetlands, and destabilizing the climate, we are unraveling the strands of a complex ecological safety net. We are beginning to understand just how valuable that safety net is.

The enormous expansion of the human population and our built environment in the twentieth century means that more people and more economic activities are vulnerable. The migration of people to cities and coasts also increases our vulnerability to the full array of natural hazards. And these human-exacerbated disasters often take their heaviest toll on those who can least afford it—the poor.

Ecologically, socially, and economically, many regions are now vulnerable and ill prepared for the onslaught of storms, floods, and other hazards. Hurricane Mitch washed away hillsides, sweeping up homes, farms, roads, bridges, and people in massive mudslides and floods. Given that Central America has some of the highest rates of deforestation in the world—each year it loses 2–4 percent of its remaining forest cover, and Honduras alone has already cleared half its forested land—the tragedy should not really be all that surprising. The pressures of poverty, population growth, and inequitable land rights had forced more and more people into vulnerable areas such as steep hillsides and unprotected riverbanks. Further, when crippling debt burdens consume most of a nation's budget and stall development, few resources remain to address these problems.[4]

To date, much of the response to disasters has focused on improving weather predictions before the events and providing cleanup and humanitarian relief afterwards, both of which have without doubt helped save many lives. Yet much more can be done. On average, $1 invested in mitigation saves $7 in disaster recovery costs. Nature provides many valuable services for free; healthy and resilient ecosystems are shock absorbers that protect against coastal storms and sponges that soak up floodwaters, for instance. We should take advantage of these free services rather than undermine them. In order to stem the ever rising social and economic costs of disasters, we need to focus on how to mitigate disasters by understanding our own culpability, taking steps to reduce our vulnerability, and managing nature more wisely.[5]

Counting Disasters

During the twentieth century, more than 10 million people died from natural catastrophes, according to Munich Re, a reinsurer that undertakes global data collection and analysis of these trends. Its natural catastrophe data include floods, storms, earthquakes, fires, and the like. Excluded are industrial or technological disasters (such as oil spills and nuclear accidents), insect infestations, epidemics, and most droughts.[6]

While some 500–700 natural disaster events are recorded every year, only a few are classified by Munich Re as "great"—natural catastrophes that result in deaths or losses so high as to require outside assistance. Over the past 50 years there has been a dramatic increase in this type of disaster. In the 1950s there were 20 "great" catastrophes, in the 1970s there were 47, and by the 1990s there were 86. (See Figure 7–1.)[7]

During the last 15 years, nearly 561,000 people died in natural disasters. Only 4 percent of the fatalities were in industrial countries. Half of all deaths were due to floods.

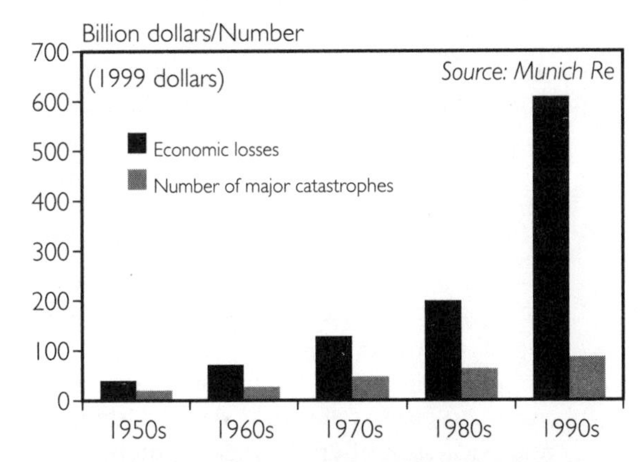

Figure 7–1. Rising Tide of Major Disasters, by Decade

(See Figure 7–2). Earthquakes were the second biggest killer, claiming 169,000 lives. Between 1985 and 1999, 37 percent of the events were windstorms, 28 percent floods, and 15 percent earthquakes. Events such as fires and landslides accounted for the remaining 20 percent.[8]

Asia has been especially hard hit. The region is large and heavily populated, particularly in dangerous coastal areas. There is frequent seismic and tropical storm activity. Its natural and social vulnerability is borne out by the statistics. Between 1985 and 1999, Asia suffered 77 percent of all deaths, 90 percent of all homelessness, and 45 percent of all recorded economic losses due to disasters.[9]

As tragic as the death toll of recent years is, in earlier decades and centuries it was not uncommon to lose hundreds of thousands of lives in a single great catastrophe. In the last 20 years, however, there has been only one such event—the cyclone and storm surge that hit Bangladesh in 1991 and took 139,000 lives. Still, in the last decade over 2 billion people worldwide have been affected by disasters.[10]

Early warnings and disaster preparedness have been a significant factor in keeping the death toll of recent decades from reaching even higher. So, too, have advances in basic services, such as clean water and sanitation. Following disasters, the life-saving benefits are apparent. According to the Chinese government, 90 percent of the 30,000 deaths from floods in 1954 were a result of communicable diseases like dysentery, typhoid, and cholera that struck in the following weeks and months. After the 1998 Yangtze flood, in contrast, no such epidemics were reported (although diarrheal diseases remained a problem).[11]

Worldwide, floods cause nearly one third of all economic losses, half of all deaths, and 70 percent of all homelessness. Damaging floods have become more frequent and more severe. They are the type of disaster that people have the greatest hand in exacerbating. In China's Hunan province, for instance, historical records show that

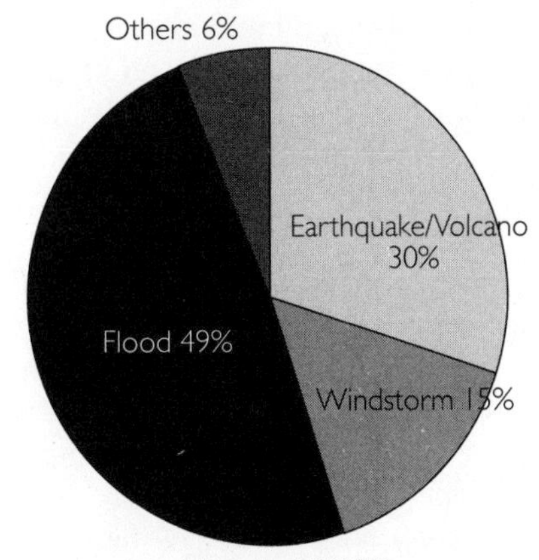

Figure 7–2. Global Deaths by Disaster Type, 1985–99

whereas in early centuries flooding occurred once every 20 years or so, it now occurs 9 out of every 10 years. In Europe, flooding on the Rhine River has worsened as a result of changes in the way the river is managed. At the German border town of Karlsruhe, prior to 1977 the Rhine rose 7.62 meters above flood level only four times since 1900. Between 1977 and 1995 it reached that level 10 times.[12]

Although there has been some success in reducing the death toll, the financial toll of disasters has reached catastrophic proportions. Measured in 1999 dollars, the $608 billion in economic losses during the 1990s was more than three times the figure in the 1980s, almost nine times that in the 1960s, and more than 15 times the total in the 1950s. The biggest single year for losses in history was 1995, when damages reached $157 billion. An earthquake in Kobe, Japan, accounted for more than two thirds of that total. For weather-related disasters, 1998 was the biggest year on record, at nearly $93 billion in recorded losses, with China's Yangtze river flood absorbing more than a third of this total.[13]

The economic losses measured usually include insured property losses, the costs of repairing public infrastructure like roads and power, and some crop losses. Such direct losses are the easiest to measure. But the tally rarely includes indirect or secondary impacts, such as the costs of business failures or interruptions, suicide due to despair, domestic violence, human health effects, or lost human and development potential. Losses in developing countries are particularly undercounted. Damage figures also exclude the destruction of natural resources.

During the last 15 years, Asia sustained 45 percent of the world's economic losses to disasters, North America 33 percent, and Europe 12 percent. (See Figure 7–3.) Rural areas and developing nations are in general underrepresented in global disaster data, as reporting systems tend to be weaker. Africa is particularly underrepresented because it is rarely hit by major storms or earthquakes. Most of the disasters in Africa are smaller, or are slow-onset disasters, like droughts, that are not counted in the global tallies. The region also has less infrastructure and capital exposure.[14]

Economic losses can be especially devastating to poor countries. As in Honduras and Nicaragua after Hurricane Mitch, disaster losses often represent a large share of the national economy. While the wealthiest countries sustained 57.3 percent of the measured economic losses to disasters between 1985 and 1999, this represented only 2.5 percent of their GDP. (See Figure 7–4.) In contrast, the poorest countries endured 24.4 percent of the economic toll of disasters, which added up to a whopping

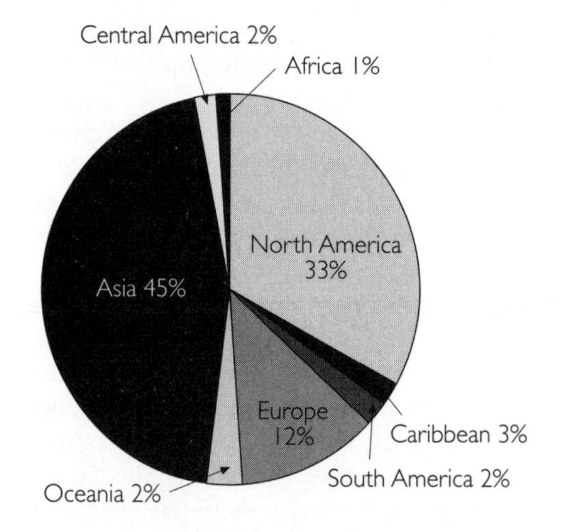

Figure 7–3. Global Economic Losses from Disasters, by Region, 1985–99

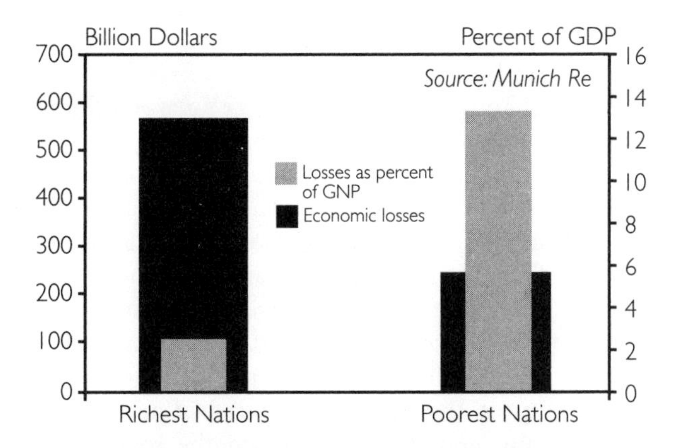

Figure 7–4. Disaster Losses, Total and as Share of GNP, in the Richest and Poorest Nations, 1985–99

13.4 percent of their GDP, further increasing their vulnerability to future disasters. And in the poorest countries, little if any of the losses are insured. Worldwide, only one fifth of all disaster losses were insured. The vast majority of insured losses, some 92 percent, were in industrial nations. Finding a way to provide a financial safety net for developing countries is of critical importance.[15]

The quickly rising economic toll and the troubling increase in the number of major catastrophes that overwhelm nations provide clear evidence that a new way of managing nature and ourselves is in order.

Ecological Vulnerability

There is an important distinction between natural and unnatural disasters. Many ecosystems and species are adapted to natural disturbance, and indeed disturbances are necessary to maintain their health and vitality, and even their continued existence. Many forests and grasslands, for instance, are adapted to periodic natural fires, and need them to burn off dead vegetation, restore soil fertility, and release seeds.

Likewise, river systems need periodic flooding, and plants and animals across the landscape are adapted to this regime. Fish use the floodplain as a spawning ground and nursery for their young. Some fish consume and disburse seeds, which can sustain them for an entire year. Many plants need the flood period to germinate and absorb newly available dissolved nutrients. Migratory birds also rely on the bounty that floods bring. Soils, too, benefit from the regular addition of nutrients and organic matter, and underground aquifers are refilled as floodwaters are slowly absorbed into the ground. By disrupting the natural flooding regime, we cut off the interactions between a river and its surrounding landscape—interactions that make them more diverse and productive. Indeed, natural flooding is so beneficial that some of the biggest fish and crop harvests come the year after a flood. Little wonder that floodplains and deltas have attracted human settlement for millennia and been the cradles of civilizations.[16]

Not every natural disturbance is a disaster, and not every disaster is completely natural. We have altered so many natural systems so dramatically that their ability to bounce back from disturbance has been greatly diminished. Deforestation impairs watersheds, raises the risk of fires, and contributes to climate change. Destruction of coastal wetlands, dunes, and mangroves eliminates nature's shock absorbers for coastal storms. Such human-made changes end up making naturally vulnerable areas—such as hillsides, rivers, coastal zones, and low-lying islands—even more vulnerable to extreme weather events.

Droughts, and the famines that often follow, may be the most widely understood example of an unnatural disaster. They are triggered partly by global climate variability and partly by resource mismanagement such as deforestation, overgrazing, and the overtapping of rivers and wells for irrigation. Considered slow-onset events, droughts are not as well reported as rapid-onset events like storms and floods, nor are they usually included in disasters data. Yet they affect major portions of Africa and Asia and are likely to continue worsening in the coming years.

Human settlements, too, have become less resilient as we put more structures, more economic activity, and more people in vulnerable places. Our usual approach to natural disturbances is to try to prevent them through short-sighted strategies using methods that all too often exacerbate them. Dams and levees, for example, change the flow of rivers and can increase the frequency and severity of floods and droughts.

China's Yangtze River dramatically shows the consequences of the loss of healthy ecosystems. The flooding in 1998 caused more than 4,000 deaths, affected 223 million people, inundated 25 million hectares of cropland, and cost well over $36 billion. Heavy summer rains are common in southern and central China, and flooding often ensues. But in 1998, as the floodwater continued to rise, it became clear that other factors besides heavy rains were at play. One influence was the extensive deforestation that had left many steep hillsides bare. Indeed, in the past few decades 85 percent of the forest cover in the Yangtze Basin has been cleared by logging and agriculture. The loss of forests, which normally intercept rainfall and allow it to be absorbed by the soil, permitted water to rush across the land, carrying valuable topsoil with it. As the runoff raced across the denuded landscape, it caused floods.[17]

In addition, the Yangtze's natural flood controls had been undermined by numerous dams and levees, and a large proportion of the basin's wetlands and lakes, which usually act as natural "sponges," had been filled in or drained. The areas previously left open to give floodwaters a place to go have filled instead with waves of human settlements. All these changes reduced the capacity of the Yangtze's watershed to absorb rain, and greatly increased the speed and severity of the resulting runoff.[18]

Chinese government officials initially denied that the Yangtze floods were anything but natural, claiming that the flooding was caused by El Niño. But as the disaster toll added up, the State Council finally recognized the human element. It banned logging in the upper Yangtze watershed, prohibited additional land reclamation projects in the river's floodplain, and stepped up efforts to reforest the watershed.[19]

Flooding and landslides following deforestation are not limited to developing countries. In the U.S. Pacific Northwest, where hundreds of landslides now occur annually, a study found that 94 percent of them originated from clearcuts and logging roads. The torrents of water and debris from degraded watersheds caused billions of dollars in damage in 1996 alone.[20]

Paradoxically, clearing forests also exacerbates drought in dry years by allowing the soil to dry out more quickly. Such droughts helped fuel the record-breaking fires in Indonesia and Brazil in 1997–98. These massive fires occurred in tropical forests that are normally too moist to burn. But when fragmented by logging and agricultural clearing, the forests dried to the point where fires set deliberately to clear land were quickly able to spread out of control. In Indonesia, industrial timber and palm oil plantation owners took advantage of a severe El Niño drought to expand their areas and in 1997–98 burned at least 9.8 million hectares, an area the size of South Korea.[21]

The smoke and haze from Indonesia's fires choked neighboring countries, affecting about 70 million people. The economic damage to the region has been conservatively estimated at about $9.3 billion. Schools, airports, and businesses were shut down. Many crops were lost to the drought and fires, and the haze impaired the pollination of other crops and wild plants, the ecological repercussions of which will unfold for many years. If harm to fisheries, biodiversity, orangutans, and long-term health were included, the damage figure would be far higher.[22]

Sumatra and Kalimantan, the provinces where most of the 1997–98 fires occurred, have lost up to 30 percent of their forest cover to exploitation and fire in just the last 15 years. One of the first smoke signals that indicated that the forests were in trouble due to forest exploitation policies was during another El Niño year, 1982–83, when 3.2 million hectares burned in Kalimantan. In 1991, another half-million hectares burned, and in 1994 almost 4.9 million hectares went up in smoke. As Charles Barber and James Schweithelm put it in *Trial by Fire*, a new study of Indonesia, "the fires of 1997 and 1998 were just the latest symptom of a destructive forest resource management system carried out by the Suharto regime over 30 years."[23]

In contrast to the human-made unnatural disasters that should be prevented but are not, considerable effort is spent trying to stop natural disturbances that are actually beneficial. The result is disasters of unnatural proportions. In the United States, for example, fire suppression has long been the policy, even in ecosystems that are fire-adapted. The result has been the buildup of debris that fuels very hot fires capable of destroying these ecosystems—and the homes that are increasingly built there. The well-publicized 2000 fire season is a telling example of the consequences of such wrongheaded policies.[24]

Considerable effort is spent trying to stop natural disturbances that are actually beneficial.

Likewise, a common response to floods is to try to prevent them by controlling rivers. But contrary to popular belief, containing a river in embankments, dams, channels, reservoirs, and other structures does not reduce flooding. Instead, it dramatically increases the rate of flow, and causes even worse flooding downstream. The Rhine River, for example, is cut off from 90 percent of its original floodplain in its upper reaches, and flows twice as fast as before. Flooding in the basin has grown significantly more frequent and severe due to increased urbanization, river engineering, and poor floodplain management.[25]

The Great Midwest Flood of the upper Mississippi and Missouri rivers in 1993 provided another dramatic and costly lesson on the effects of treating the natural flow of rivers as a pathological condition. The flood was the largest and most destructive in modern U.S. history. It set records for amounts of precipitation, upland runoff, river levels, flood duration, area of flooding, and economic loss. Financial costs were estimated at $19 billion. The floodwaters breached levees spanning nearly 10,000 kilometers. In hindsight, many now realize that the river was simply attempting to reclaim its floodplain. Not surprisingly, 1993 was a record spawning year for fish as the river was restored, temporarily, to more natural functioning.[26]

Bangladesh suffered its most extensive flood of the century in 1998; two thirds of the country was inundated for months.

Today's problems reflect the cumulative impacts of more than a century of actions by public and private interests to expand agriculture, facilitate navigation, and control flooding on the Mississippi and its tributaries. Nearly half of the 3,782-kilometer-long Mississippi flows through artificial channels. Records show that the 1973, 1982, and 1993 floods were substantially higher than they might have been before structural flood control began in 1927 after a major flood.[27]

Throughout the huge Mississippi River basin, the construction of thousands of levees, the creation of deep navigation channels, extensive farming in the floodplain, and the draining of more than 6.9 million hectares of wetlands (more than an 85-percent reduction in some states) have cut into the ability of the Mississippi's floodplains to absorb and slowly release rain, floodwater, nutrients, and sediments. Separating fish from their floodplain spawning grounds and upstream reaches has virtually eliminated some species and caused many others to decline. The commercial fish catch in the river has fallen 83 percent over the past 50 years.[28]

Flood control and navigation structures have also adversely affected the Mississippi Delta and the Gulf of Mexico. Because these structures trap sediments rather than allow them to be carried downstream to replenish the delta, as they have done for millennia, the coastal areas are actually subsiding as water inundates wetlands and threatens coastal communities and productive fisheries.[29]

The management and policy changes begun after the 1927 flood have had other perverse effects. One was to shift the cost and responsibility for flood control and relief from the local to the federal level. Another was to encourage people, farms, and businesses to settle in vulnerable areas with the knowledge that they would be bailed out of trouble at taxpayer expense.[30]

The government also fostered settlement in vulnerable areas by providing crop insurance and crop price guarantees, and by paying for most of the cost of levees. The net result is that farming the land in the former river channel is profitable only with regular federal payments for flood damage.[31]

In 1968, Congress created the National Flood Insurance Program (NFIP) to cover flood-prone areas that private insurers deemed too risky. Unfortunately, this led to rebuilding in many of these areas. Nearly half of the payments for flood claims went to the repeat flood victims who account for less than 1 percent of the policyholders. And for those without flood insurance, emergency relief aid was repeatedly provided, further contributing to the cycle of losses.[32]

The 1993 Mississippi flood's human and economic costs, combined with its benefits to the ecosystem's functions, inspired a rethinking of the way large rivers are managed. After the flood, a federal task force recommended ending the nation's over-reliance on engineering and structural means for flood control in favor of floodplain restoration and management. It emphasized managing the river as a whole ecosystem rather than as short segments.

Other reforms to the NFIP have been promoted by a wide range of groups (from floodplain managers to insurance companies and environmental groups) to reduce repeated flood losses, save taxpayer dollars, and restore the health of the Mississippi basin.[33]

On the other side of the globe, Bangladesh suffered its most extensive flood of the century in the summer of 1998, when two thirds of the country was inundated for months. Annual floods are a natural and beneficial cycle in this low-lying coastal nation, which encircles the meandering deltas of the Ganges, Brahmaputra, and Meghna Rivers. The people of Bangladesh have long adapted their housing, land use patterns, and economic activities to these "barsha" or beneficial floods. However, 1998 brought a "bonna" or devastating flood. Floodwaters reached near-record levels and did not recede for months. All told, 1,300 people died, 31 million people were left temporarily homeless, and 16,000 kilometers of roads were heavily damaged. Overall damage estimates exceed $3.4 billion—or 10 percent of the nation's GDP.[34]

A number of factors precipitated Bangladesh's bonna flood. Heavy rainfall upriver in the Himalayas of north India and Nepal, some of which fell on heavily logged areas, exacerbated the disaster, as did the runoff from extensive development upstream that helped clog the region's rivers and floodplains with silt and mud. In the future, rising sea levels due to climate change will make Bangladesh even more vulnerable to flooding. This problem will be made worse because large expanses of stabilizing mangroves have been removed from shores in recent years to make way for shrimp ponds, exposing the coast to more inundation.[35]

Further, a major reason that so much of Bangladesh was submerged for so long was that extensive embankments built in the last 10 years as part of the nation's Flood Action Plan actually prevented the drainage of water. (The structures also dried out the backwaters that once fertilized fields and provided fish after the floods receded.) While the Bangladeshi peasants look at most floods as beneficial, engineers and donors tend to see all flooding as a problem to be solved by technical measures. As researcher Thomas Hofer has noted, "when it comes to perception of floods and their danger, few heed the wisdom of villagers, even though it is they who have to (mostly) live with the flood."[36]

Social Vulnerability

A number of factors make some places and some people more vulnerable to natural hazards. Growing concentrations of people and infrastructure in vulnerable areas like coasts, floodplains, and unstable slopes mean that more people and economic activities are in harm's way. While poor countries are more vulnerable, in every nation some people and communities—notably the very poor, women, and ethnic minorities—are especially hard hit during and after disasters. For poorer countries and poorer people, disasters can take a disproportionately large share of income and resources. Misplaced development priorities and heavy debt burdens can exacerbate disasters and cripple recovery efforts, further hampering development.

Two major global social trends of recent decades have increased our vulnerability to natural hazards: the migration of people to coasts and cities, and the enormous expansion of the built environment. Approximately 37 percent of the world's population—more than 2 billion people—lives within 100 kilometers of a coastline. Coastal zones are especially vulnerable to storms, high winds, flooding, erosion, tidal waves, and the effects of inland flooding. In the U.S. Gulf and Atlantic coasts, the areas most vulnerable to hurricanes, 47 percent of the population lives in coastal counties. Between 1950 and 1991, a period of relatively few hurricanes, the population of South Florida exploded from under 3 million people to more than 13 million. And 80 percent of this growth occurred in coastal regions.[37]

A disproportionate number of the world's poor live on the front line of exposure to disasters.

Similarly, there has been explosive growth of cities. Since 1950, the world's urban population has increased nearly fourfold. Today, the urban population—almost half the people in the world—is growing three times faster than the rural population. Many cities are also in coastal areas, further compounding the risks. Of the world's 19 megacities—those with over 10 million inhabitants—13 are in coastal zones.[38]

As the built environment increases in amount and density, potential losses increase. As the *World Disasters Report* puts it, "growing cities concentrate risk." Urban areas are dense concentrations of not only people but also buildings, roads, rail lines, pipelines, communications systems, and water and sanitary services. The concentration of these "lifelines" means that a disruption in service can affect a very large share of a region's population and economic activity. The earthquake that rocked Kobe, Japan, in 1995 killed 6,350 people and cost over $100 billion, making it the most expensive natural disaster in history. It disrupted economic activity for months, including vital shipping and railway lines.[39]

Urbanization also increases the risk of flooding. When land is covered by impervious surfaces such as roads and roofs, the frequency and severity of flash floods increases. Urbanizing 50 percent of a watershed can increase the frequency of floods from once every 100 years to once every 5 years.[40]

In much of the developing world, urbanization has additional dangers. Up to half the people in the largest cities of the developing world live in unplanned squatter colonies, which are often sited in vulnerable areas such as floodplains and hillsides or even garbage dumps. These poorer communities are far less likely to have public services such as water, sanitation, storm drains, and health and emergency services. As a result, when disasters strike, the residents are even worse off. After disasters they have few, if any, resources to fall back on to survive and rebuild.[41]

Whether in urban or rural areas, the poorest and most marginalized suffer the most. A disproportionate number of the world's poor live on the front line of exposure to disasters. In Nicaragua, 80 percent of those who lost their homes during Hurricane Mitch were living at or below the poverty line even before the storm.[42]

In Central America, the nations most ravaged by Mitch—Guatemala, Honduras, and Nicaragua—have a history of highly inequitable distribution of land and wealth. Such extreme poverty invites disaster. In the Honduran capital of Tegucigalpa, one neighborhood that slid into the Choluteca River was home to vendors from the local market who had cobbled together shanties for lack of affordable housing. In the countryside, where prime agricultural land was being used mostly to produce export commodities such as bananas and coffee, subsistence farmers had been forced onto steep hillsides, where they were much more vulnerable to landslides.[43]

After the storm, half the people in Honduras had lost their homes or been evacuated and 70 percent were without clean water. More than 70 percent of the crops were destroyed—in a nation where two thirds of the workers are in agriculture, which accounts for half of export revenue. Nutrient-rich topsoil was also lost, and it will be years before many fields can be rehabilitated and crops can bear fruit. Thousands of land mines, planted during a decade of civil conflict, were washed to unknown locations.[44]

The United Nations estimated that Mitch set the region's development back by 20 years. The cost of rebuilding infrastructure in Honduras and Nicaragua alone was estimated at nearly $9 billion. But far from starting with a clean economic slate, Central American nations face the impossible task of rebuilding while paying the development debt of previous decades. Already over $10 billion in debt before the disaster, Honduras and Nicaragua were together paying $2.2 million a day simply to service their existing debts.[45]

After Mitch, the World Bank quickly arranged a large financial support package, including $1 billion in new interest-free credits for Nicaragua and Honduras, while some lender countries agreed to forgive all or part of their share of outstanding debt or to delay repayment. Yet with the destruction of much of the infrastructure and export capacity, these nations seem destined to slip further into debt unless there is more debt relief. (See Chapter 8.)[46]

In rich and poor nations alike, people living on the edges of society and the economy may be pushed over the edge when disaster strikes. Simply put, disasters make poverty worse. Community and family networks, which provide vital social security, may unravel. For subsistence farmers—both men and women—what little "insurance" they have is in the form of seeds, tools, and livestock, which are often lost along with their crops. Laborers lose their incomes. Squatters or illegal immigrants are usually in high-risk locations to begin with. After disasters, they often do not ask for help because they may fear being evicted from their settlement or deported. Illiterates cannot read disaster notices and instructions. Those who were homeless before the disaster have no resources or social networks to rely on, and are often invisible to government agencies. Indigenous people often have poor access to information and services before disasters, and are less likely to receive aid afterwards.[47]

Disasters can weaken the already vulnerable position of women and children. As one flood survivor put it, "life shatters along existing fault lines." Although needs may differ, relief efforts rarely make distinctions between men and women. Women may need special medical assistance when pregnant or lactating, or protection from the increased male violence and aggression that commonly occurs after disasters. Women usually bear the weight of responsibility for caring for children and the elderly, yet few emergency efforts provide assistance for these tasks. The disproportionate malnourishment of women and children worsens after disasters.[48]

As with development in general, men tend to be seen as the family providers and relief efforts focus on them to the exclusion of women. "Food for work" jobs and agricultural rebuilding often target men, despite evidence that the food does not always reach the home and is sometimes sold instead, whereas the food and money that a woman works for are almost universally dedicated to her family's needs. Most relief and rebuilding efforts focus on major infrastructure rather than on the priorities of local people, such as affordable housing or income-generating activities.[49]

Finally, planners rarely recognize that in pre- and post-disaster situations, women have different priorities and coping strategies. They generally have less tolerance for risk than men, so they are more likely to prepare for hazards and to heed disaster warnings and evacuation notices. After disasters they are more likely to mobilize social networks to find ways to meet the needs of their family and the community. Men, on the other hand, often cope by leaving the disaster zone to find employment, in some cases abandoning their families.[50]

Increasingly sophisticated engineering allows people to wrongly assume that nature can be controlled and they can be protected.

The tendency to view all disaster victims and their needs alike has a special danger for the disabled and the elderly. In the hurricane-vulnerable coastal communities of North Carolina, for example, 12 percent of residents have a physical or medical condition that impedes their ability to evacuate their homes—a reality that evacuation plans need to prepare for.[51]

While the "tyranny of the urgent" in disasters makes it easy to overlook gender and social issues, doing so makes efforts far less effective than they need be. Understanding social realities and vulnerabilities is as crucial for ensuring success of all phases of disaster management—from preparedness and response to recovery and mitigation—as it is for achieving truly sustainable development.[52]

The Politics and Psychology of Disasters

Responding to disasters is a genuine human reaction to the suffering of others. When tragedy strikes, there is an almost reflexive outpouring of help to try to feed, clothe, and house those in distress. Yet long-term rebuilding and disaster prevention efforts rarely elicit the same level of empathy and support. Among donors, governments, and even humanitarian organizations, there is a well-developed culture of response, but not an underlying culture of mitigation. Within the U.S. Office of Foreign Disaster Assistance, for instance, only 11 percent of its meager $155.4 million 1997 budget went to mitigation and preparedness activities.[53]

When people contemplate the future they "are typically unaware of all the risks and choices they face. They plan only for the immediate future, overestimate their ability to cope when disaster strikes, and rely heavily on emergency relief," according to Dennis Mileti, Director of the Natural Hazards Center and author of *Disasters by Design*. Even when they are aware of risks, people are generally less likely to expend effort and resources on something that might happen, perhaps sometime in the future, than they are to meet more immediate needs. For the very poor, these day-to-day needs are pressing indeed.[54]

While the improved accuracy and dissemination of warnings has saved countless lives, it can, ironically, foster a false sense of safety, and, along with insurance, can encourage people to build and live in risky places. Increasingly sophisticated engineering allows people to wrongly assume that nature can be controlled and thus they can be completely protected from hazards. In many wealthy countries, such as the United States, most people—rich and poor alike—who choose not to invest in mitigation measures (or even insurance) can do so with a near certain knowledge that they will be physically and financially rescued in the event of an emergency. All this can lead to unnecessary risk taking.

Just as individuals take calculated risks or risks out of ignorance, so too do governments. In many areas of government, including hazard management, short-term thinking prevails. Preparing for and mitigating hazards often takes a back seat to other priorities. Rescue and relief get much more financial support—and have more political appeal—than preparing for an event that may not happen during a politician's term in office.

And yet the adage "an ounce of prevention is worth a pound of cure" clearly applies to disasters. The World Bank and U.S. Geological Survey calculated that global economic losses from natural disasters could be reduced by $280 billion if just one seventh that amount were invested in preparedness and mitigation efforts. The costs of disaster preparedness and mitigation can be far less than the costs of disaster relief and recovery.[55]

Disasters can focus attention on the many failures in preparation and response. The aftermath of Hurricane Mitch, for instance, brought to light Central America's inadequate disaster preparedness. Despite the fact that the region has been repeatedly hit by hurricanes, earthquakes, and tidal waves, it seems that none of the lessons of those events were learned and applied before Mitch—or since. Nicaragua's government, especially the president, was criticized for failing to declare a state of emergency in the early days of the storm. National emergency planning did not start until days after the storm began, during which time the president repeatedly denied there was a crisis. Early warnings and evacuations could have saved people in the villages around the Las Casitas volcano. After seven days of pounding rains the side of the volcano slid away, killing more than 1,400 people—the worst single incident of Mitch.[56]

In India, the cyclone and tidal wave that hit the desert region of Gujarat and killed 10,000 people in 1998 was predicted by the federal government, but the warnings were not disseminated by local authorities. Some have even said that there was little political will to expend effort warning politically powerless people in the region.[57]

When a supercyclone hit Orissa, India, in late 1999, the official response was decidedly mixed. Though some sectors, such as public health, responded admirably, in general the government's reaction was disjointed and often ineffective. The confusion meant that the people hit hardest by the storm suffered for many days without relief. All told, as many as 50,000 died, 20 million were left homeless, and more than 1 million families lost their means of support. The lack of coastal management plans or an effective emergency communication network also help explain why this cyclone was so destructive compared with similar storms that strike elsewhere. Even a neighboring Indian state was more prepared than Orissa—just a month earlier, Andhra Pradesh managed to evacuate 1 million coastal dwellers to 1,000 cyclone shelters during another storm, while for the supercyclone Orissa evacuated only 150,000 people, and had only 21 shelters for evacuees. Andhra Pradesh had applied the lessons learned in three almost equally large cyclones: in 1974, 10,000 people died in a similar storm; in 1991, 1,000 people died; in 1996, just 60 people were killed.[58]

The failure of governments to develop or enforce adequate land use plans and building codes, even after multiple disasters, can also have devastating consequences. In earthquake-prone Turkey, as elsewhere, rapid urbanization in recent decades led to a housing crisis. To alleviate the crunch, 15 "building amnesties" were granted since 1950 that legalized illegal construction. Before the 1999 earthquakes, these amnesties were seen as a great populist gesture. Afterward, corrupt building contractors and local officials were denounced as "murderers" in newspaper headlines. While many poorly constructed apartment blocks, some as far away as 100 kilometers, turned into tombs, other properly constructed buildings at the quake's epicenter survived.[59]

Turkey is not alone in facing this type of problem. In many cities in developing countries, more than half of all homes are technically illegal. They are poorly constructed, sited, and served. In Honduras, the government has failed to enforce zoning laws introduced after Hurricane Mitch. Not all poorly located buildings are inhabited by the poor, either. In Venezuela, the 1999 landslides that claimed 30,000 lives hit luxury apartment high-rises built at the foot of landslide-prone slopes as well as more modest dwellings.[60]

Even in the industrial world, building in risky locations—from the cliffs of California to the barrier islands of the Carolinas and the mountains of Italy—is a widespread practice and problem. Sometimes it is even subsidized. Hazard mitigation codes can make buildings safer, but they must be enforced. If the State of Florida's codes had been upheld, for instance, more than 25 percent of the damage from Hurricane Andrew in 1992 could have been avoided. (Most of the damage from the hurricane was not from houses blowing away but from massive water damage due to broken roofs or windows.) For communities that lack the technical expertise to develop their own, model codes and standards can provide guidance.[61]

Ineffective development and enforcement of building codes are not the only governance problems faced by hazard-prone communities. According to the *World Disasters Report*: "Corruption and vested interests in and around government play a large role in many of the long-term precursors to disaster. Mafia organizations have been implicated in the widespread construction of illegal housing in disaster-prone areas of Italy. Timber smuggling cartels with political connections on the porous borders of Pakistan and Afghanistan are denuding and destabilizing mountain slopes in earthquake zones."[62]

In Indonesia, not only did former president Suharto's government turn a blind eye to timber and palm oil plantation owners (many of whom were his cronies) who were illegally using fire to clear forest to expand their operations, but some of the 1997–98 fires were set as part of the government's own misguided program to turn 1 million hectares of peat swamp into an agricultural settlement. Early on, the government tried to blame the rural poor for setting the fires that swept the country, despite satellite images tracing almost all the blazes to corporate plantations and timber concessions. When the government finally admitted who the real culprits were, little or nothing was done to stop them. Nor was anything done to help the millions who lost their homes and livelihoods or were sickened by the haze, while the nongovernmental organizations (NGOs) that stepped in to provide services were criticized.[63]

Governments should beware, as the failure to prepare for and respond to disasters can have political repercussions. In Indonesia, Suharto was finally ousted when outrage over the Asian financial crisis and the massive fires fanned the flames of widespread opposition to the regime's corrupt and authoritarian rule. In the elections following the Orissa disaster in India, the incumbent party was ousted by voters angry over the governments' apathy, bungling, and corruption.[64]

Fostering Resilience in Nature and Communities

The ever-rising human and economic toll of disasters provides clear evidence that a shift is needed in our coping strategies. This shift is all the more urgent if the current trends that make us vulnerable continue: the concentration of people and infrastructure in cities and along coasts, and growing pressure on ecosystems. The looming prospect of climate change and sea level rise can only exacerbate these troubling trends.

Scientists project that in the future the weather is likely to become more erratic and extreme as a result of climate change. Warmer ocean waters, for example, can fuel stronger storms. Many of today's disasters are also fueling climate change. The fires that ravaged Indonesia's forests and peat swamps in 1997–98 produced a third of the carbon pumped into the atmosphere by human activities during that time.[65]

It is already clear that sea levels are rising. During the last century, they rose about 20 centimeters and they are projected to rise another 50 centimeters by 2100. The British Meteorological Office and others have calculated that with uncontrolled climate change the number of people at risk of flooding "will increase ten-fold by 2080." Coastal cities, river deltas, and small islands will be especially vulnerable. Major river deltas like Bangladesh, the Amazon, and the Mississippi would be at risk. Some small island nations may see their national territory disappear. Rising sea levels could even flood the New York City subway system and turn parts of the metropolitan area into wetlands.[66]

Many like to blame "the weather" or "the climate," and use them as convenient excuses for inaction. But it is important to recognize that irrespective of any potential climate change dimension, we continue to put more people and more "stuff"—buildings, bridges, cities, and power plants—in harm's way and have weakened nature's ability to mitigate hazards. Equally important is understanding that just as our development choices have made the threats worse, we have the power to make better choices.

There is a growing awareness that disaster response and recovery—the traditional mainstays of past efforts—are not enough, and that mitigation actions are needed to reduce the impacts of natural disasters. The need for a new direction in policies toward disasters is evident in the rising costs of these events to government treasuries. In the United States, for example, between 1970 and 1981 domestic disaster assistance cost the federal government $3.8 billion. But for 1989–94, a period half as long, the bill topped $34 billion.[67]

While we cannot do away with natural hazards, we can eliminate those that we cause, minimize those we exacerbate, and reduce our vulnerability to most. Doing this requires healthy and resilient communities and ecosystems. Viewed in this light, disaster mitigation is clearly part of a broader strategy of sustainable development—making communities and nations socially, economically, and ecologically sustainable.

How can communities and nations begin to mitigate disasters and reduce the human and economic toll? They can make sure that they understand their risks and vulnerabilities. They can use this knowledge to ensure that their development efforts do not inadvertently increase the likelihood and severity of disasters. To the extent possible, people and structures should be located out of harm's way. When hazards are unavoidable, development can be made to withstand them—for example, buildings in earthquake zones should be designed to weather earthquakes. Disaster preparedness, too, is an integral part of saving lives and lowering the economic toll. And every segment of the community needs to be actively engaged in planning and implementing disaster mitigation efforts.

Instead of relying on structural engineering, the time has come to use the services provided by healthy and resilient ecosystems.

Identifying and delineating natural resources (like watersheds and floodplains), hazards (such as flood zones), vulnerable infrastructure (such as buildings, power lines, and bridges), as well as vulnerable communities and resources—and doing so at scales that are meaningful to communities and decisionmakers—is an essential step. Yet hazard mapping is incomplete, outdated, or non-existent in many communities and nations. Even most U.S. flood maps are more than 20 years old, and most other hazards are not mapped at all. Maps do not show the areas that would be flooded in the event of a dam or levee failure, or that are at risk from coastal erosion—despite the fact that in the next 60 years, 25 percent of homes within 500 feet of U.S. shorelines are projected to be lost to coastal erosion.[68]

A critical part of good land use planning is maintaining or restoring healthy ecosystems so they can provide valuable services. China, for example, now recognizes that the forests are more valuable for flood control and water supply than they are for timber. Ecosystem restoration and rehabilitation can be effective tools in hazard mitigation. An extensive study by the U.S. National Research Council recommended these tools to solve water quality, wildlife, and flooding problems at minimal cost and disruption. Restoring half of the wetlands lost in the upper Mississippi Basin would affect less than 3 percent of the agricultural, forest, or urban land, yet it could prevent a repeat of the flood that drowned the heartland in 1993. Allowing more of the natural floodplain to function can reduce the impact of future floods on human settlements and economic activities.[69]

In the past, making communities safe was seen as the job of engineers, who, for instance, would apply structural solutions to flood control and coastal storms—a costly and often unsuccessful approach. As noted earlier, many of these structures have ironically contributed to a false sense of security and to magnifying the hazard. Many of them are now reaching the end of their life span and should be decommissioned.

Instead of relying on structural engineering, the time has come to tap nature's engineering techniques—using the services provided by healthy and resilient ecosystems. Dunes, barrier islands, mangrove forests, and coastal wetlands are natural shock absorbers that protect against coastal storms. Wetlands, floodplains, and forests are sponges that absorb floodwaters. Nature provides these valuable services for free, and we should take advantage of them rather than undermining them.

There is still a role for traditional engineering. Buildings and bridges can be made to better withstand natural hazards. By ensuring that structures can withstand earthquakes of a certain magnitude, or winds of a certain speed, many lives and dollars could be saved.[70]

Making communities safer does not have to be high tech or high cost. In Maharastra, India, "barefoot" engineers and builders helped introduce new and safer building techniques during post-earthquake reconstruction. In many flood-adapted cultures—like in the Amazon or Mekong—houses sit on stilts above the high water mark or float up and down with the water levels. In Bangladesh, communities build and maintain raised mounds where they can go for safety during floods. The mound usually has a safe drinking-water well and a school or other community structure, providing a safe haven and an incentive for the community to maintain it. Active community participation in planning and implementation of all levels of disaster mitigation and recovery is essential.[71]

Basic community services have added benefits during disasters. As noted earlier, China credits improved sanitation with virtually eliminating the post-disaster epidemics of waterborne diseases that frequently used to kill more people than the disaster itself.

Communities can also act to reduce the "hidden hazards" that can create a "disaster after the disaster." After Hurricane Floyd hit North Carolina in 1999, for example, the contents of open waste ponds of industrial hog farms spread out over the landscape in the floodwaters. Chemical plants and other industrial sites also present special hazards during natural disasters. Ensuring safe containment of these facilities can save many lives and much money in post-disaster cleanup efforts. Among the most frightening and deadly hidden hazards are the land mines that are washed by floodwaters to new and unmapped locations, as has happened in Mozambique and Central America.[72]

In recent decades, great strides have been made in predicting extreme weather events and disseminating warnings. In 1992, warnings and timely evacuations were a major factor in limiting to 15 the number of deaths caused by Andrew, the costliest hurricane in U.S. history (at $30 billion). A comprehensive preparedness system has helped reduce the loss of life in Bangladesh, 90 percent of which is vulnerable to cyclones. Tens of thousands of community volunteers, working in teams of 10 men and 2 women, provide warnings, evacuation, search and rescue, and other emergency assistance—often at risk to their own lives. They are credited with saving 30,000 people in the powerful 1991 cyclone and countless others in recent events.[73]

Getting the right information to the right people at the right time remains an enormous challenge. Sometimes information is too technical to be useful or in the wrong language. Radio, television, satellites, computers, and the Internet can be very effective in expanding dissemination, yet much of the world is still without access to many of these technologies. Expanding effective early warning systems should continue to be a high priority.[74]

Sustainable mitigation must be an integral part of local and international development plans. Governments have a role to play in investing in hazard and risk assessments and in developing databases on losses, mitigation efforts, and social data. They can establish land use policies, limit subsidization of risk and destructive activities, use incentives to encourage sound land use and sustainable hazard mitigation, and encourage collaboration between agencies and civil society.[75]

Governments and civil society must also ensure the rule of law—without it, the social and ecological unraveling that precipitates and exacerbates disasters is far more likely. The fires in Indonesia provide a textbook case on the consequences of corruption and lawlessness. Russia may be unwittingly setting the stage for future disasters by allowing massive and poorly regulated logging in its Far East. Since China enacted its much needed logging ban in 1998 to restore the health of the flood-ravaged Yangtze basin, the impacts of logging have shifted to neighboring countries like Russia.[76]

Private and public insurers can help reduce hazard losses by providing information and education as well as incentives that encourage mitigation and disincentives to discourage building in hazard-prone places. Insurers have been active participants in the climate change debate, as they recognize the huge potential impacts of climate change on their industry. For most of the developing world, insurance is not available. Providing some sort of financial safety net is a large and unmet need.[77]

The publicly funded U.S. National Flood Insurance Program provides insurance in communities that adopt a set of minimum standards for floodplain management. Reduced insurance premiums are provided for communities that undertake activities (such as flood mapping, preparedness, public information, and so forth) that exceed minimum standards. While there have been some changes in the program, much more could be done. Currently, because erosion hazards are not mapped, homeowners in erosion-prone areas pay the same flood insurance rates as those in no-risk areas. The NFIP also reimburses communities for "beach nourishment": the costly, futile, and potentially destructive practice of regularly plowing sand from the ocean up to the beach. In the future, NFIP rates could be raised and coupled with land use controls such as mandatory set-backs from hazardous zones.[78]

Donors can provide leverage and resources to promote development policies that include disaster mitigation. As noted, every dollar spent on disaster preparedness saves $7 in disaster-related economic losses—a great return on investment. Considering the social and ecological losses that are also prevented, the return is far higher.[79]

Unfortunately, overall foreign aid budgets are small, and disaster prevention allocations are minuscule. At the 1992 Earth Summit, the Group of Seven industrial countries made a commitment to provide 0.7 percent of their GDP in aid, yet five years later they had managed to come up with only 0.2 percent. (If they had met their target, it would have added $155 billion to aid funds.) Of the aid that they do provide, what is spent for emergency assistance is painfully small. In 1997 it was less than 7 percent of bilateral aid. The amount spent for mitigation was far lower.[80]

Better coordination of emergency and development efforts within and among agencies is needed. In the United Nations, for instance, weather forecasting, humanitarian relief, food relief, and disaster preparedness and mitigation are each in separate agencies. Some donors are beginning to integrate these functions, a step that can help mainstream mitigation. The World Bank recently launched the ProVention consortium, in partnership with governments, intergovernmental organizations, private insurance companies, universities, and NGOs. Yet within the Bank, disaster and development are still largely segregated, and neither seems to influence the onerous debt demands of the World Bank, the International Monetary Fund (IMF), and other lenders on disaster-stricken countries.[81]

Donors and lenders also have the opportunity and the obligation to resolve the debt burden that cripples many nations. (See also Chapter 8.) The huge amount of money needed for both immediate disaster relief and long-term reconstruction in Central America after Hurricane Mitch and in Mozambique after Cyclone Eline focused attention on the growing problem of debt. Many question how these nations can realistically be expected to provide for their citizens and rebuild while repaying mounting foreign debt, especially since much of their capacity to generate revenue was wiped out by the storm. Before the disaster, Honduras owed $4.7 billion in external debt and Nicaragua owed $5.7 billion. In Nicaragua, per capita GNP was less than $400, while even before Mitch, each person's share of foreign debt was nearly three times that.[82]

A few months after floods and cyclones ravaged Mozambique, affecting nearly 5 million people, donor nations pledged $453 billion to fully fund its reconstruction. While Mozambique has received some measure of debt relief, debt elimination is what is needed.[83]

Much of the heralded post-Mitch "debt relief" involves simply postponing payments and supplying more loans (and therefore debt). The skepticism that met most creditor initiatives was summed up by the Roman Catholic Archbishop of Tegucigalpa, Oscar Andres Rodirigues, who likened the lender's moratorium on debt repayment to a "stay of execution."[84]

Indeed, the debt and structural adjustment programs of recent decades have forced extreme cutbacks in social services, such as health care and education, and in environmental and resource management programs—precisely the kinds of services that are needed to help prevent disasters and respond effectively when they occur. The new loans and structural adjustment programs are accelerating these cutbacks. One year after Hurricane Mitch, Nicaragua had spent almost as much on debt service ($170 million) as on reconstruction ($190 million). The IMF explicitly stated that Nicaragua must limit reconstruction spending to $190 million per year in 1999 and 2000.[85]

Every dollar spent on disaster preparedness saves $7 in disaster-related economic losses—a great return on investment.

What Central America needs for reconstruction, said Archbishop Rodirigues, "is debt cancellation, combined with adequate foreign assistance and with careful oversight by our civil society," an approach championed by the faith-based Jubilee 2000 coalition that applies equally as well in many disaster-stricken nations. Oxfam has proposed that no more than 10 percent of government revenues could be spent on debt payments. Such limits are not without precedent. After World War II, Germany's debt payments were limited to 3.5 percent of export revenues in order to spur peace and development. Yet today the IMF, World Bank, and the Paris Club of government creditors say that 20–25 percent is sustainable, a level far higher than industrial nations deemed sustainable for themselves in the past.[86]

The international community has additional avenues for action. The International Decade for Natural Disaster Reduction that ended in 1999 represented an important opportunity to raise the profile of hazards and disasters, advance science and policy, and inspire national action. Yet it may have been "a decade of missed opportunity," in the words of eminent geographer Gilbert White, as it focused on scientific and technical programs but failed to strengthen local capacity or to address slow-onset events such as those that plague Africa, among other important aspects of disaster reduction. To continue and expand the efforts of the decade, the United Nations has established a follow-up process, the International Strategy for Disaster Reduction. Unfortunately, it has relatively little visibility or political muscle, despite the tremendous challenges ahead.[87]

There is also room for action within the Framework Convention on Climate Change, as there is language that obliges signatories to cooperate in adapting to the impacts of climate change, including land use and water resource planning as well as disaster mitigation.[88]

Many have concluded that the time has come for a profound shift in how we approach disasters. As Kunda Dixit and Inam Ahmed put it, when writing about floods in the vast Himalayan watershed: "Complete flood control...is impossible. Even partial control is...problematic....So the question arises: Should we try to prevent floods at all? Or should we be looking at what it is we do that makes floods worse? Is it better to try to live with them, and to minimize the danger to infrastructure while maximizing the advantages that annual floods bring to farmers?" The same questions must be asked about natural hazards everywhere.[89]

If we continue on a course of undermining the health and resilience of nature, putting ourselves in harm's way, and delaying mitigation measures, we set ourselves up for more unnatural disasters, more suffering, more economic losses, and more delayed development. If instead we choose to work with nature and each other, we can reduce the waves of unnatural disasters that have been washing over the shores of humanity with increasing regularity and ferocity.

Chapter 8

Ending the Debt Crisis

David Malin Roodman

In January 1949, President Harry Truman delivered the first U.S. inaugural address of the cold war. "Each period of our national history has had its special challenges," he declared. "Those that confront us now are as momentous as any in the past. Today marks the beginning...of a period that will be eventful, perhaps decisive, for us and for the world."[1]

Truman did not exaggerate. Already, he had presided over the first use of atomic power in anger and the founding of the United Nations, the World Bank, and the International Monetary Fund (IMF). He had witnessed the descent of the Iron Curtain around Eastern Europe, and had begun aiding reconstruction of Western Europe with the Marshall Plan. Now Truman vowed to press further in what he cast as a biblical battle between freedom and oppression. He urged the formation of what would become the North Atlantic Treaty Organization. And he called for the first-ever peacetime foreign aid program for poor nations.[2]

Thus was born the modern project of foreign aid—the giving of advice, grants, and loans to help poor nations. From the start, its motives were complex—geopolitics, missionary zeal for democracy and capitalism, and a genuine desire to alleviate suffering. Lenin had rightly condemned the West for exploiting poor nations for their resources and enslaving their people. To contain communism, Truman apparently believed, the West had to meet that criticism head-on by aiding, not just exploiting, poorer nations. With a helping hand, he now argued, those nations could become wealthier, democratic, more like the United States. In short, they could "develop."[3]

Half a century later, in April 2000, thousands of protesters from an international movement called Jubilee 2000 locked arms and ringed the very spot where Truman spoke—indeed, ringed the entire U.S. Capitol—to dramatize how lending programs he fathered had gone badly awry. The human chain, they explained, symbolized the "chains of debt" that financially enslave poor nations to rich ones.[4]

The governments of poor nations now owe so much to those of rich nations (and to lenders such as the World Bank and IMF) that many are spending more on foreign debt payments than on basic social services for their desperately poor citizens. Zambia devoted 40 percent of its national budget to foreign debt payments in 1997 and only 7 percent to basic health and education, clean water, sanitation, family planning, and nutrition. Yet the death rate among children there is rising, partly because a third of them are not fully vaccinated. Meanwhile, the number of Zambian children not in school rose to 665,000 in 1998. In Mozambique, the government spends $7 per person on debt service, compared with $3 on health, even though 160,000 children under five are dying each year partly from lack of basic drugs and health services.[5]

The Third World debt crisis is not new. Wealthier "middle-income" countries such as Brazil, Mexico, Poland, and the Philippines borrowed heavily from commercial banks in the 1970s, ran into debt trouble in the 1980s, and pulled free in the early 1990s after banks finally compromised on repayment terms. The poorest nations, in contrast, borrowed mainly from official agencies (both aid agencies and export credit agencies that offer loans to finance imports from the lending country). Their debt trouble grew more slowly and steadily and has seen little resolution. If anything, it took a turn for the worse after Truman's key rationale for aid—the cold war—disappeared and aid flows fell 25 percent between 1991 and 1998.[6]

At the heart of the debt crisis in the poorest countries lies a basic problem: official bodies lent—and borrowers accepted—billions of dollars in the hope that the projects and policy changes they funded, from roads to food imports to schools to electric company privatization, would stimulate the borrowers' economies and exports enough to make full repayment economically and politically possible. In many nations, that did not happen.

There is plenty of blame to go around—corruption and counterproductive economic policies among borrowers, arrogance and bad planning among lending agencies, contradictory pressures on those agencies from western politicians, and more. It seems common sense that all parties that contributed to the failure should bear the consequences. But to date, those on the lending side have worked hard to shield themselves from blame and shift the costs of failure onto the borrowers. Many borrowers, meanwhile, are scarcely more than dictatorships that care little for their poorest citizens. Thanks in no small part to pressure from nonprofit groups, western governments have made strides in recent years toward reducing the debt burden, under the Debt Initiative for Heavily Indebted Poor Countries (HIPC initiative). But the program does not go far enough in holding all the institutions involved accountable for the mistakes that led to the crisis in the first place—far enough, that is, to prevent debt trouble from continuing and recurring.

Costs of Crisis

The present debt trouble began in 1982, when Mexico, Brazil, and other developing nations announced that they could no longer service debts they had run up since the early 1970s. Within years, the crisis involved dozens of debtors on every continent. Its effects varied from debtor to debtor and were in all cases complex. On balance, however, the experience of the last two decades strongly suggests that continuing debt problems in the poorest nations are taking a toll on poor people and the environment. Thus the crisis is undermining the very goals for development in whose name most of the debts were contracted.[7]

The crisis has affected debtor economies in several ways. From the start, it placed heavy payment pressure on the debtors. The world's most powerful banks, governments, and multilateral lenders presented a solid front. They gave debtors a choice: repudiate some of their debts, alienate great powers such as the United States, and risk losing all ability to borrow for the foreseeable future—or accept the debt burden and try, as economists say, to "adjust."[8]

To squeeze debt payments out of their economies, countries had to spend less or earn more. Spending less was more realistic in the short run. This rebalancing had to occur within the budget of the government, which owed most of the money. It also had to occur within the nation's trade flow, since the excess of exports over imports earned the foreign exchange to service the debts. Each government could choose from a handful of unpleasant options for its "adjustment," "stabilization," or "austerity" package. It could cut government spending (aside from debt payments). It could raise taxes. It could print money out of thin air, which would enrich the government but cause inflation and make people poorer. To prevent inflation from feeding on itself, the government could attempt to cap monthly wage increases, or hike domestic interest rates, or limit lending by domestic banks. It could tax or forcibly block imports. Or it could devalue its currency in order to make the country's exports cheaper and more attractive to the rest of the world and also raise the prices of imports for locals.

Whatever the details, austerity needed to quickly extract national wealth such as food and tropical hardwood for sale abroad. This it had to do even though rich industrial nations were erecting import barriers against shirts and sugar and other exports of developing countries—and indeed were subsidizing their own exports of those products. Worse, each debtor had to compete with dozens of others pursuing the same strategy. As gluts developed, the IMF index of dollar prices for commodities fell 56 percent between 1980 and 1999, accounting for inflation. Rice exports from Thailand rose 31 percent in volume in the first half of 1985 but fell 8 percent in total value.[9]

Zambia devoted 40 percent of its national budget to foreign debt payments in 1997 and only 7 percent to basic social services.

Commodity gluts, an investor-spooking atmosphere of economic uncertainty, and government cutbacks for important investments such as roads and rails combined to extinguish economic growth. On average between 1950 and 1980, gross domestic product (GDP) per person had risen 2.4 percent a year in Latin America and 1.8 percent a year in Africa. But between 1980 and 1999, GDP per person fell in Latin America and then gradually recovered for a net increase of only 4.3 percent. (See Figure 8–1.) It fell 6.2 percent per person in Africa over the same period. Statistical analysis by economist Daniel Cohen at the Ecole Normale Supérieure in Paris supports the conclusion that debt trouble played an important role: the more prone a country was to reschedule its debt payments in the 1980s—a good indicator of debt trouble—the less it grew, other factors being equal.[10]

The costs of crisis have not spread even-

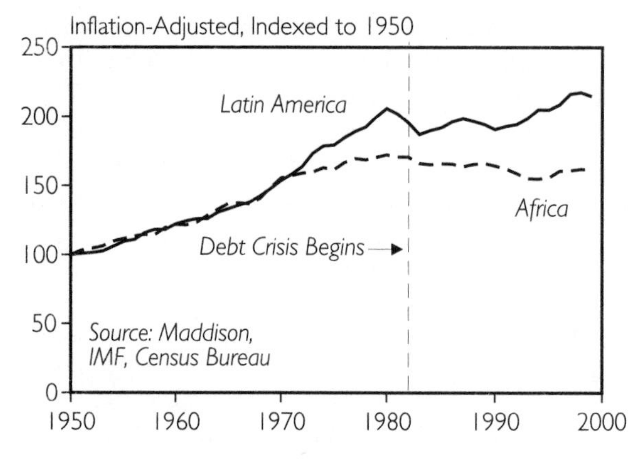

Figure 8–1. Gross Domestic Product Per Person, Latin America and Africa, 1950–99

ly within debtor nations. For instance, not all poor people suffered. In fact, the poorest of the poor—such as Indians living in the Sierra Madre mountains of Guatemala beyond paved roads and power lines—were often too isolated from the commercial economy to notice its ups and downs. Some small, land-owning farmers plugged into global markets, on the other hand, benefited from adjustment's stimulus to exports. But they were the exceptions. In general, constituencies that supported the government and gained from the original loans, such as the urban middle class and civil servants, used their influence to shift costs of repayment onto those less fortunate. UNICEF concluded that "it is hardly too brutal an oversimplification to say that the rich got the loans but the poor got the debts."[11]

As noted earlier, some poor countries now spend more servicing debts to rich countries—debts mostly run up in the name of development—than on providing basic social services. (See Table 8–1.)

Comparisons between what governments spend on debt service and what they spend on basic social services, however, deserve cautious interpretation. It is naive to assume, for example, that if governments had no obligations to repay debts they would pour the savings into social services or other enlightened investment programs, and that the programs would all work. It is naive, in other words, to assume that every dollar in debt payment costs the poor. Yet that is what Jubilee campaigners assumed when they asserted that the debt crisis is killing thousands of children a day. Responsive, effective governments are relatively unlikely to get into debt trouble in the first place.[12]

But it is equally mistaken to abandon all hope for change by assuming that debtors would use none of the savings to fight poverty. One chronic problem debtor, Uganda, hit bottom after its civil war in 1986 and emerged with a strong political consensus for reforms to restore economic growth and reduce poverty. Since 1995, the government has consulted with poor communities on a comprehensive strategy to fight poverty—an effort that won praise from the World Bank. Yet the country's high debt obligations continue to sap its efforts to aid the poor.[13]

A 1997 study by the World Bank's internal auditors analyzed the impact of debt on social spending systematically, and found that the truth does indeed lie between these extremes. Government social spending fell between 1980 and 1993 in countries adjusting to debt problems—but more slowly than government spending as a whole (excluding debt service). On the other hand, within education budgets, primary education—including the one-room schools in dirt-poor hamlets—took larger cuts than universities serving urban elites. Tanzania, a country once known for its nearly universal primary schooling system, introduced

Table 8–1. Share of Government Spending Covered by Foreign Borrowing and Devoted to Foreign Debt Service and Basic Social Services, Selected Countries, 1996–97[1]

Country	Share Covered by Foreign Borrowing	Share Devoted to Foreign Debt Service	Share Devoted to Basic Social Services
		(percent)	
Belize[2]	8	6	20
Benin[3]	29	11	10
Bolivia[3]	16	10	17
Burkina Faso[3]	20	10	20
Cameroon[4]	36	36	4
Costa Rica[2]	7	13	13
El Salvador[2]	47	27	13
Jamaica[2]	14	31	10
Nepal[3]	33	15	14
Nicaragua[2]	14	14	9
Peru[3]	36	30	19
Sri Lanka[2]	21	22	13
Zambia[3]	13	40	7

[1]Basic social services include basic health and education, clean water, sanitation, family planning, and nutrition. [2]Data for 1996. [3]Data for 1997. [4]Data for 1996–97.
SOURCE: U.N. Development Programme, Management Development and Governance Division, *Debt and Sustainable Human Development*, Technical Advisory Paper No. 4 (New York: 1999), 14–15; shares covered by foreign borrowing are Worldwatch estimates, based on ibid., and on World Bank, *Global Development Finance 2000*, electronic database, Washington, DC, 2000, and include net flows on short-term debt.

school fees in 1986 that shut out the very poorest. Zaire fired a fifth of its teachers in 1984.[14]

In Brazil, some of society's weakest members paid for adjustment with their lives, as mothers struggled to nourish themselves and their families amidst economic recession and government cutbacks. In the northeast of the country, the incidence of low birth weight among babies, which that had held stable around 10.2 percent in the five years through 1982, jumped to 15.3 percent in 1984. Infant mortality climbed too. In 1982, 9 out of every 100 babies died in their first year, down from 16 in 1977. But by 1984, the figure had climbed back to 12. Meanwhile, in the southeastern state of São Paulo, similar economic trends combined with cutbacks in government immunization to open the way for a measles epidemic that killed thousands of babies.[15]

In many countries, budget cuts combined with economic recession and rising prices to raise unemployment and reduce the buying power of those still working. All this led World Bank Chief Economist Stanley Fischer to conclude in 1989 that "most of the burden has been borne by wage earners in the debtor countries." In Mexico, inflation-adjusted wages halved between 1982 and 1988. Turkey's 1980 adjustment program led to price increases for goods sold by government companies, ranging from 45 percent for gasoline to 400 percent for fertilizer. This helped cut the buying power of workers' wages by 45 percent between 1979 and 1985 and the real value of crop prices for farmers by 33 percent.[16]

The debt crisis has also taken a toll on the environment, but again the story is complex. In Thailand, the government lowered tariffs to stimulate exports of rice and rubber. This encouraged farmers to shift toward those crops and away from more soil-damaging cassava farming. But it also encouraged them to clear rainforests for new rubber plantations. Similarly, economic recession slowed consumerism in Latin America and cut pollution by shutting down factories—but it sent poor Jamaicans up steep, erosion-prone hillsides to grow coffee and Venezuelans up tributaries of the Amazon to blast rocks for bits of gold.[17]

The current debt initiative demands too much from debtors in terms of policy reform and offers too little in terms of debt reduction.

Less ambiguously, debt burdens reduced funding for environmental protection agencies. Nine of 12 case studies of adjusting countries commissioned by the World Wide Fund for Nature found such cuts. "In Venezuela," wrote project director David Reed, "the environmental ministry lost one-half its professional staff. In El Salvador, environmental institutions were systematically gutted over the years." Such cuts may hurt more in the long term, since young environmental agencies in developing countries often wield little effective power over such activities as industrial pollution and logging, but can be expected to gain strength with time if properly supported.[18]

The debt-generated need for foreign exchange has spurred export-oriented mining, logging, and agriculture in developing countries. Several studies have found a statistical link between high debt burdens and deforestation. Debt pressure is one reason many developing countries made their laws more inviting to foreign logging and mining companies in the 1980s. Venezuela attracted firms prospecting for oil, bauxite, iron, and gold in, among other places, the country's relatively pristine eastern jungles.[19]

The Year of Jubilee

The World Bank currently classifies 41 nations as heavily indebted and poor, including 33 in sub-Saharan Africa, plus Bolivia, Guyana, Honduras, Laos, Myanmar, Nicaragua, Viet Nam, and Yemen. Twenty-nine of these HIPCs rank in the U.N. Development Programme's lowest category of human development. This means that babies born today can be expected to live 40–50 years, that fewer than half the eligible children and young adults attend school or college, and that barely half the adults can read.[20]

Governments of rich industrial nations first acknowledged the plight of the poorest debtors in the late 1980s. In Venice, in June 1987, the Group of Seven leading industrial nations (the G–7) agreed to allow debtors to stretch out payments on outstanding debts—though not to cancel any outright. They required countries seeking payment delays to implement economic policy reforms, known as structural adjustment, under the auspices of the IMF. At succeeding summits through the early 1990s, the G–7 gradually softened their stance, allowing modest debt reductions.[21]

Still, HIPCs fell further behind on their debts, which by 1998 reached $214 billion in face value—a huge sum for them (but equal to only 4.5 months of western military spending). The deepening problem goaded Christian aid groups into launching a movement called Jubilee 2000 in the United Kingdom in 1996. They took the name from the Bible's Book of Leviticus, which recorded God's laws for ancient Israel, among them that every fiftieth year was to be a year of Jubilee, in which debtors were forgiven and farmers reclaimed land they had once sold off to survive a bad year. Jubilee 2000's members, ranging from small nongovernmental organizations (NGOs) in Uganda and Bolivia to heavyweights such as Oxfam, were not the first to assail the debt, but they thrust the issue onto the international stage as never before.[22]

Creditor governments responded to NGO pressure with the HIPC initiative, announced at a G-7 meeting in Lyon in 1997 and "enhanced" in Cologne in 1999. In some respects, the initiative represented a major step forward in grappling with the debt problem. For the first time, it put debt owed to the IMF, World Bank, and other development banks on the negotiating table alongside that owed directly to governments. It created a HIPC Trust Fund that uses creditor government contributions to compensate these institutions for writing down debt. And the initiative committed the creditors to meeting goals defined from the debtors' point of view and specified in terms of the ratio of a country's foreign debt to its exports (150 percentmaximum). The goals use the "present value" of a nation's debt, a measure that recognizes that a $1 million loan at 10 percent interest costs a country much more than a $1 million loan at 1 percent. In all, the initiative aims to fight poverty while bringing the total debt reduction to 50 percent.[23]

Despite these innovations, the current HIPC initiative is unlikely to end the debt crisis or make major strides against poverty. It demands too much from debtors in terms of policy reform and offers too little in terms of debt reduction. More fundamentally, by casting debt reduction as a one-time act of charity, it insulates both lenders and borrowers from criticism about their role in creating the problem and discourages a deeper examination of the causes of the crisis. In the end, the HIPC initiative invites new debt problems by relieving almost everyone involved from studying and applying the lessons of history.

Debt History

In a review of international lending since 1820, Swiss sociologist Christian Suter identified four major waves of borrowing by governments of what would now be called developing countries. They occurred roughly every 50 years. (See Figure 8–2.) Each terminated with an outbreak of "debt service incapacity"—a widespread inability to keep up with interest and principal payments—causing banks and bond investors to beat a hasty retreat.[24]

These cyclical shifts of capital into and out of developing countries are not fully understood. Suter suggests that the long-term rhythm of technological revolutions may play a role. The commercialization of the automobile and, later, the computer, for instance, created major investment booms within rich countries, which offered lenders there plenty of lucrative opportunities at home. Between these investment peaks, however, lenders searched abroad for higher returns. The manic-depressive psychology of financial markets amplified any such

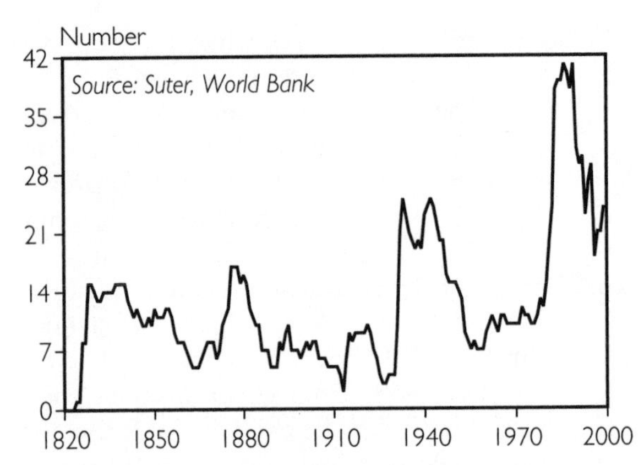

Figure 8–2. Number of Countries Not Servicing All Their Foreign Debts, 1820–1999

underlying dynamics. Just as in the recent boom and bust of East Asian stocks, bonds of foreign governments would go in and out of favor with western investors swiftly and unpredictably. During the boom phase, enthusiasm made fertile ground for waste and fraud.[25]

Remarkably, the same nations often held the roles of debtor or creditor from cycle to cycle. Since 1820, Latin American and East European nations have frequently run into debt trouble. (See Table 8–2.) (Most of Africa was off the lending map until the 1960s.) Of the countries that ran into trouble between 1820 and 1929, 64 percent did so again in the 1980s. Meanwhile, a small handful of countries—Britain, France, Germany, the United States, and Japan—accounted for most of the overlending.[26]

This regularity suggests that the tendency to default or overlend emerges from deep within a nation's character and history. One generalization that seems useful is that the most internally divided nations make the least dependable debtors. Almost all the Latin American states, as well as the Philippines, inherited from their colonial rulers Spain and Portugal a highly unequal distribution of landownership, which has perpetuated itself down through the centuries. In Bolivia, where in the mid-1980s the wealthiest fifth of the people earned 15 times as much as the poorest fifth, the result was a "degeneration of politics into . . . fierce battles of the 'ins' versus the 'outs'," according to economists Juan Antonio Morales and Jeffrey D. Sachs. When politicians find it nearly impossible to forge societal consensus, loans from outsiders are sorely tempting as a way to aid one group, say, by financing coal plants to generate jobs and cheap power, without taxing another group—or at least to forestall the day of reckoning.[27]

Nevertheless, international credit has generally benefited both lending and borrowing nations. Despite some profligacy and defaults, foreign lending since 1820 has profited investors who diversified and persevered—more so than lending at home. International bonds, with their higher interest rates to reflect risk, were honored enough to cover any losses. Thus while some of the borrowings were wasted in war, corruption, white elephants, and extravagant court life, most were productively invested.[28]

When that investment was in banana plantations in Central America or gold mines in South Africa that exploited workers in slave or near-slave conditions, it did most people in the receiving nation little good. But within Europe and North America, cross-border lending sped economic development that lifted millions out of poverty. Countries that began industrializing earliest, most of all Great Britain, financed factories and railroads in places that began industrializing later, including the United States and the Italian peninsula.

Table 8–2. International Debt Crises Since 1820

Period	Selected Major Debtors	Selected Major Creditors	Approximate Debt with Payment Problems
			(billion 1999 dollars)
1826–30	Greece, much of Latin America, Spain	British bond investors	3
1840–45	Mexico, Portugal, Spain, nine state governments in United States	British and French bond investors	6
1875–82	Egypt, Greece, much of Latin America, Ottoman Empire,[1] Spain, Tunisia, 10 southern state governments in United States	British bond investors; German banks; French government	40
1890–1900	Greece, Latin America, Portugal, Serbia	German and British bond investors; Barings Bank	20
1911–18	Bulgaria, Mexico, Ottoman Empire,[1] Russia[2]	French bond investors	200
1931–40	China, France, Germany, Greece, Hungary, much of Latin America, Poland, Turkey, United Kingdom, Yugoslavia	World War I victors (owed by Germany); U.S. government (owed by other victors); Western bond investors	80
1982–	Much of Latin America, Morocco, Philippines, Poland, much of sub-Saharan Africa, Turkey, Yugoslavia	French, West German, Japanese, United Kingdom, U.S. banks; G–7 governments; IMF, World Bank, African Development Bank	430

[1]Predecessor to modern Turkey. [2]Accounted for as much as 85 percent of defaulted amount in this period.
SOURCE: Christian Suter, *Debt Cycles in the World-Economy: Foreign Loans, Financial Crises, and Debt Settlements, 1820–1990* (Boulder, CO: Westview Press, 1992), 69.

More recently, foreign borrowing helped South Korea industrialize rapidly.[29]

The lending upswing that led to the current crisis began in the late 1960s and terminated abruptly in August 1982, when Mexico admitted that it could no longer meet its obligations to U.S. banks. The immediate causes of the crisis included high interest rates triggered by U.S. Federal Reserve Chairman Paul Volcker's effort to stamp out U.S. inflation, the ensuing global recession, and resulting drops in prices for exports of many borrowing nations.[30]

The new crisis broke all records. Developing and Eastern bloc nations owed $800 billion ($1.3 trillion in 1999 dollars). (See Figure 8–3.) About $300 billion of that ($430 billion in today's terms) became problem debt, on which countries had to defer or cancel promised payments. During 1983–89, about 35 governments were in hastily negotiated "consolidation periods" at any given time, during which they made no loan payments.[31]

As in previous crises, the debt trouble concentrated in a few regions. (See Table 8–3.) Latin America and the Caribbean accounted for 44 percent of all foreign debt

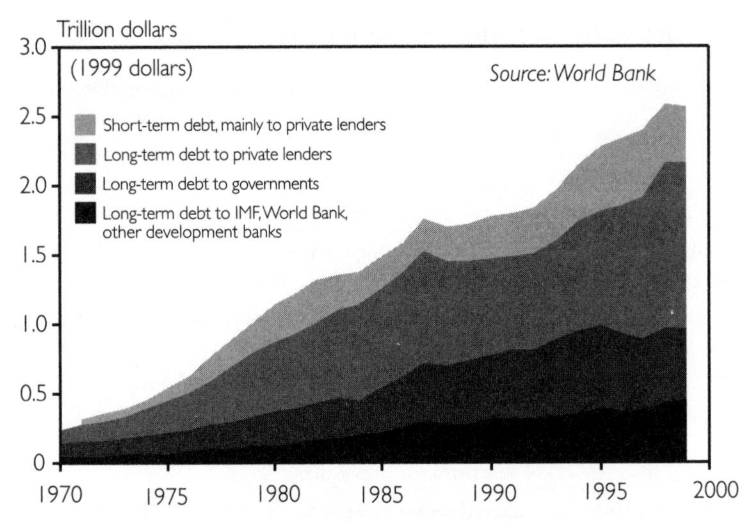

Figure 8–3. Foreign Debt of Developing and Former Eastern Bloc Nations, 1970–99

of developing and Eastern bloc countries at the end of 1982 (and more than half of the problem debt). The region owed $939 per person to foreign creditors, just under half its gross national product (GNP) of $2,000 per capita. The debt-GNP ratio was also high in Sub-Saharan Africa, at 33 percent. Major Eastern bloc and Asian nations, not so financially tied to the democratic industrial powers, were in much better shape, though notable exceptions included the Philippines, Poland, and South Korea.[32]

Total debt was a poor measure of debt trouble, however. A $1-million bank loan due for full repayment in one year to Citibank along with 20 percent interest burdened a government far more than a $1-million loan due for repayment to the World Bank over 40 years with only 0.75 percent annual interest. Yet both showed up in the statistics as $1 million in debt. Thus a better indicator of the immediate crisis was the ratio of a country's debt service (interest and principal payments) to its export earnings—the funds from which foreign debts had to be paid. In Latin America and the Caribbean, this debt-service ratio reached 47 percent in 1982, compared with an average 13 percent for other developing countries. It reached an insupportable 51 percent in Mexico and 82 percent in Brazil. (Like height and weight, debt-GNP and debt service–export ratios are different measures of size and should not be compared with each other.)[33]

This crisis is distinctive in being the first involving governments as important lenders as well as borrowers. Indeed, the crisis can be seen as two smaller disasters. One mainly involved South American and East European governments and the commercial banks. It peaked in the 1980s and then dissipated in the early 1990s as the two sides compromised on more realistic payment terms.

The other crisis principally involves official lenders and the ultrapoor nations of Central America and sub-Saharan Africa, which commercial banks generally avoided. The official lenders in turn are of two kinds. The bilateral lending agencies of individual governments include aid agencies, as President Truman had envisioned, and export credit agencies that lend to foreign governments and companies to help them buy food, weapons, and other exports of the lending country. In addition, there are the jointly financed multilateral lenders, notably the IMF, the World Bank, and the Inter-American, African, and Asian Development Banks. The development banks (but not the

Table 8–3. Foreign Debt Burdens of Developing and Eastern Bloc Countries, 1982[1]

Region/Country	Debt	Debt-GNP Ratio	Debt Service[2]-Exports Ratio	Debt per Person
	(billion dollars)		(percent)	(dollars)
Eastern Bloc	87	11	8	224
South Asia	49	21	15	52
Middle East and North Africa	111	25	9	547
East Asia and Pacific	129	27	18	78
Sub-Saharan Africa	77	33	12	191
Latin America and Caribbean	355	48	47	939
All	808	27	18	204

[1]Figures are for gross debt: foreign assets are not subtracted from foreign debts. Middle Eastern nations in particular had large assets abroad. [2]Principal and interest payments paid during the year, excluding short-term debt balances. Often paid for at least in part with new borrowing from same creditors.
SOURCE: World Bank, *Global Development Finance*, electronic database, Washington, DC, 1998 and 2000; U.S. Bureau of the Census, *International Data Base*, electronic database, Suitland, MD, 18 August 2000.

IMF) are allowed to use their government backing to borrow cheaply on international capital markets and relend the funds to developing nations. Some agencies also make grants as well as loans, and offer something in between: "concessional" loans with terms far easier than commercial loans. Since 1960, for instance, the World Bank has offered the poorest countries loans at 0.75 percent interest, with a 10-year grace period and 35–40 years to repay.[34]

Whereas private lending followed the usual boom-bust arc in middle-income countries, official lending climbed more slowly and dominated only in the poorest countries. As this stopped growing in the 1990s, however, and as grace periods expired and interest compounded, the cost of servicing old official loans rapidly caught up with and surpassed receipts from new ones. Between 1992 and 1999, developing and former Eastern bloc nations spent $53 billion more (in 1999 dollars) servicing debts to official creditors than they received from the creditors in new loans. For many in developing countries, this was a bitter reminder of the colonial past. The lowest-income developing countries as a group still received more than they paid in the late 1990s—but only barely, and some individual countries did not. (See Figure 8–4.)[35]

Sources of the Official Debt Crisis

President Truman's vision for development aid was striking, in retrospect, in the contrast between its ambitious aims and narrow means. For Truman, development meant nothing less than freeing a people from want, war, and tyranny, a definition it is hard to improve on even today. Yet the tools he offered—technical knowledge and investment from the West—seem inadequate to the task of erasing the effects of

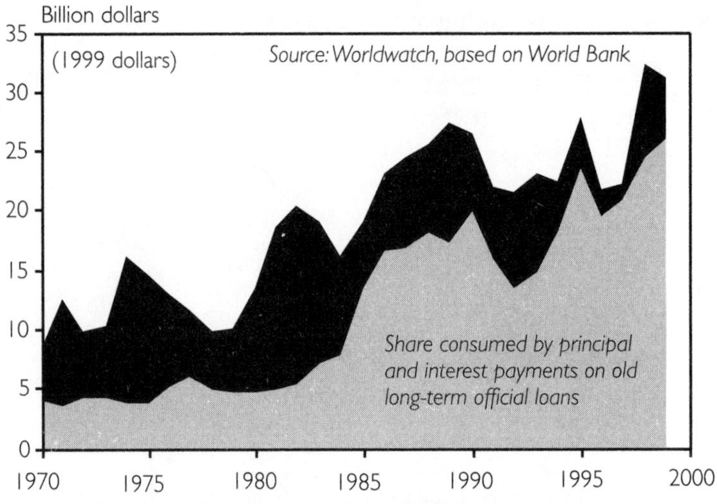

Figure 8–4. New, Long-term Official Lending to Low-income Developing Countries, 1970–99

centuries of colonialism and engineering radical transformations of whole societies.[36]

Truman's conception of development as a matter of injecting technology and finance into a country through discrete projects such as irrigation networks, roads, rail links, power plants, and dams was superficially understandable. It would soon work amazingly well in reconstructing Western Europe. But it fixated on the most visible aspects of development. In fact, effective use of such physical assets, which the World Bank and other aid agencies would enthusiastically finance, required less-visible "assets" that take decades to develop, including the rule of law, a reasonably competent government, skilled workers, a fabric of small and medium-sized businesses, and civic organizations. But these were far less present in most developing countries than they were in postwar Europe. After Mozambique gained independence from Portugal in the 1970s, for example, its Ministry of Education had five people, including a 23-year-old Minister. To graft the surface features of a heavily industrialized society onto a much-less-industrialized society was to build a skyscraper on quicksand. Many well-intended loan-financed projects failed to pay for themselves.[37]

As a result, a top World Bank official concluded by the late 1960s that many borrowers were as deep in debt as they could safely go: "The fact is...that the debt service...on development loans we made during these past twenty years...are more than many, perhaps most, of the countries we have made the loans to can carry....This means that we simply made a major misjudgment—a mistake—when we assumed that conventional banking terms and conditions could be applied to such countries—India, Pakistan, countries in Africa."[38]

The history of development aid since Truman has seen aid agencies reluctantly, but repeatedly, widen their understanding of what development requires, each time broadening their agendas to match. The seeming lack of curiosity about the bigger picture springs from many sources. For geopolitical reasons, western politicians pressured agencies to channel funds to favored nations regardless of how they were ultimately used. At other times, they demanded proof that the lenders were using tax dollars responsibly, and were reassured by dams completed and roads paved—again, regardless of the ultimate benefit.

Among agencies, belief in the superiority

of western economic systems and cultures—sometimes almost a missionary belief in democracy and capitalism—blinded officials to the histories and complexities of borrowing nations and the difficulty of change. Yet the urgency of the problems they worked on—poverty, disease, illiteracy—made criticism unseemly. Implicit in many development projects was the assumption that engineers and economists in Washington, DC—or in Bonn or Tokyo—could easily analyze societies far different from their own well enough to give governments reliable and realistic advice. In fact, project designers frequently tripped over barriers that were obvious to people with more experience in the borrowing countries or with expertise in political science, anthropology, or ecology. In 1970, the World Bank financed the drilling of what were to be communal irrigation wells in Bangladesh (then East Pakistan). Each $12,000 well was meant to be shared by all the farmers in a village, but following traditional power lines, rich farmers tended to take control of the wells and charge poor ones for water they were supposed to get free.[39]

A final hindrance to learning was that aid agencies, like almost all institutions, were concerned above all with their own survival and growth. Each time an agency admitted that development is more complex and slow and difficult to assist than its policies previously suggested, it implicitly criticized its past work even while casting doubt on its future usefulness.

By the same token, when aid agencies did come to recognize that their approaches were floundering, that economic development was more complex than they had thought, they almost always responded with more ambition rather than humility. Agencies that lent money drove forward with particular force. Paul Collier, director of the World Bank's Development Research Group, explained that "an aid cutoff is very likely to trigger a suspension of debt service payments by the government. While the net financial impact on the IFIs [international financial institutions] may well be the same whether they lend money with which they are repaid or suspend lending and are not repaid, the impact on staff careers is liable to be radically different. People do not build careers in financial institutions on default and a lack of loan disbursement." Increasing lending deferred any reckoning with debt burdens that were outstripping countries' abilities to repay—by allowing the debts to grow more.[40]

For example, World Bank President Robert McNamara, who came to the Bank in 1968, responded to debt worries of that time by expanding lending. Infrastructure and city-based industry alone could not help a country grow, the Bank now officially recognized, without parallel investment in health, education, and agriculture. By funding these new areas, the Bank would foster a supportive environment for its traditional projects, speed economic growth, and bolster countries' abilities to service their debts. It was not obvious, however, that investing in the education and health of poor people would generate enough exports to pay back loans. It was not obvious, in other words, that lending was a good way to support these worthy causes. At any rate, 90 percent of the Bank's lending during McNamara's 13 years at the Bank still went for the big-iron industrial projects.[41]

Toward the end of his time in office, McNamara appeared to harbor even deeper doubts about the Bank's approach to development. Perhaps the nearly exclusive focus on projects, whether dam building or school building, was missing the mark.

Many failed projects seemed to have been foredoomed by the economic environment in which they operated. Governments that kept the value of their currencies artificially high against foreign currencies, for example, effectively subsidized imports and hurt their own Bank-financed factories.[42]

The development effort of the last 50 years has spawned economic, environmental, and human rights dissasters—and run up a lot of unpayable debt.

Once again, the dominant conception of development widened, this time to encompass economic policy. As the 1980s began, the World Bank, along with the IMF and U.S. government lending agencies, increasingly lent money to countries on the condition that they enact major economic policy reforms, such as privatizing state-run enterprises or cutting spending. These reforms, eventually called structural adjustment, soon played a major role in the debt crisis.

In late 2000, the World Bank released its latest *World Development Report*, which "urges a broader approach to reducing poverty" that embraces not only economic growth but "opportunity, empowerment, and security." Implicit in this message is a self-criticism, an admission of past mistakes, but the very implicitness indicates how reluctant the institution still is to examine those mistakes openly. The learning process has been welcome, but it is fair to ask why it took 50 years to reach this point.[43]

The problems of lack of curiosity about local circumstances, limited accountability, and the drive to push projects through were even more serious among borrowing governments. One Bank anthropologist, for example, recalled a "village meeting with very upset farmers, watching the government officials fall asleep one after the other as the farmers made their desperate petitions to stop the evictions from their forest homes."[44]

In the worst cases, central banks seemed to channel foreign loans right to the dictator's Swiss bank account. President Mobuto Sese Seko of Zaire personally accumulated $5 billion by the early 1980s, almost matching his country's external debt. On a smaller scale, many government officials in developing countries siphoned off shares of foreign loans for private use and sent the wealth abroad for safekeeping. During 1976–84, the flow of capital out of Latin America roughly matched the increase in external debt, according to the World Bank. The 15 HIPCs included in Transparency International's *2000 Corruption Perceptions Index* scored an average 2.8 out of 10 (with a lower number indicating more corruption).[45]

With too much drive and too little guidance, the development effort of the last 50 years has spawned enough economic, environmental, and human rights disasters to fill many books—and run up a lot of unpayable debt. To take one extreme example, the World Bank lent $72 million to Guatemala to help finance a dam on the Chixoy River in 1978. In 1980, local people whose land would be flooded by the reservoir, mostly indigenous Achi Indians, began protesting against the project. In response, soldiers massacred at least 294 villagers, after raping or torturing many of them. But the World Bank, "did not consider it to be appropriate to suspend disbursements." Five months after the reservoir filled, the dam stopped working because of poor design. The Bank lent another $47 million for repairs. Such problems, along with corruption, eventually raised the dam's total cost from $340 million to $1 billion, accounting for a substantial fraction of Guatemala's current $4.6-billion foreign debt.[46]

Adjusting to Reality

Recent protests against globalization have attacked international financial institutions not only for the projects they have funded but for the policy reforms they have demanded from countries in debt trouble. Indeed, the debt crisis turned structural adjustment lending into a major line of business for the World Bank and IMF. But the attacks on adjustment lending oversimplify. In fact, the two international financial institutions influence economic policies of debtor nations less than they, and their fiercest critics, often appear to recognize. In dozens of case studies, independent economists and political scientists have documented how most of the reforms IFIs have pushed for either never occurred or probably would have occurred anyway.[47]

Western governments, led by the United States, demanded two sorts of adjustment in exchange for loans from international financial institutions. The first type, focusing on stabilization and austerity, dominated in the early 1980s. Many of the measures, however, including devaluations and spending cuts, were essentially inevitable as long as governments shouldered their debt burdens (as western governments pressured them to do). Indeed, some countries pursued adjustment independent of the IFIs. The domestic and international political dynamics that led countries to run up debts and then service them were the driving force—not the IFIs.[48]

As it became clear in the mid-1980s that austerity was prolonging the debt crisis rather than solving it, western lenders began to argue that debtors needed to make still deeper reforms—to make *structural* adjustments. If governments privatized state enterprises, ended subsidies, removed barriers to foreign trade and investment, and generally got out of the way of business, the argument went, then investors would take risks again, economies would grow, and tax revenues and foreign exchange would flow into government coffers.

In contrast with austerity programs, which had a certain inexorable logic, structural adjustment stood on shaky economic ground. True, countless heavy-handed government interventions were hampering economic development in many countries in the 1970s. Bloated civil services drained away tax dollars. Complex, constantly changing import controls bewildered traders. But it did not follow that paring government down wholesale would bring economic rebirth. The economies that have grown fastest in the last 200 years—including China, Japan, South Korea, and the United States—saw extensive government involvement.[49]

In a statistical survey of developing countries, Harvard economist Dani Rodrik found that two factors best predicted the growth rate of an economy after 1975: how much the country invested, and whether the government kept the economy on an even keel despite turbulence in the global economy. Keeping inflation moderate and the exchange rate realistic, major goals of basic austerity plans, mattered. But the openness to trade and investment that structural adjustment aimed for helped in only a minority of countries. "The lesson of history," Rodrik concluded, "is that ultimately all successful countries develop their own brand of national capitalism."[50]

Further muffling the impact of structural adjustment agreements with the IFIs is that borrowers have routinely sidestepped the provisions. One sign of slippage is that IFIs often cancel or delay second installments on their loans. (They usually disburse loans in installments to extend their leverage over borrowers.) In the 1980s and early 1990s, for example, the World Bank delayed 75 percent of the second installments on its adjustment loans because borrowers were slipping on agreed reforms. Similarly, between 1979 and 1993 the IMF abandoned half its adjustment loans after the first installments.[51]

The reasons for this surprising record are many, and relate closely to the problems that have plagued development lending since 1949. For example, pressure from western politicians on the IFIs to lend to favored nations, as well as pressure within IFIs on individual employees to increase lending, greatly weakened the institutions' leverage over debtors. Tellingly, although the World Bank delayed 75 percent of those second installments, it eventually granted at least 92 percent, usually after the borrower made a face-saving gesture. Similarly, the IMF frequently negotiated a new program with a borrower after abandoning an old one. Other forces undermining the IFIs' leverage included domestic politics and natural and human-made disasters. (See Table 8–4.) The relative inability of outside lenders to influence economic policy has now become conventional wisdom at the World Bank. A new Bank report admits that although almost every African nation has taken out structural adjustment loans since the early 1980s, only Ghana and Uganda have made much progress on growth and poverty.[52]

Within each borrowing nation, then, an official lender such as the IMF or World Bank is best seen as one more political actor. Like every such actor, a lender has particular goals, strategies, and tactics for changing the country—and particular strengths and weaknesses. With skill and luck, and under the right circumstances, lenders can tip the political balance toward policies they favor. In 1996, for example, the World Bank refused to release the second installment of an adjustment loan to Papua New Guinea because of that government's failure to make agreed anti-corruption reforms in the logging industry. The government protested bitterly, but gave in after a year out of desperation for the money and appeared for the moment to be genuinely reforming. A World Resources Institute case study credits this success to the Bank's desire to be seen fighting for the environment, to commitment and intimate understanding of local circumstances among local Bank staff, and to a strong domestic constituency for reform. Such a constellation of circumstances, however, is rare—certainly too rare for the IFIs to have affected the debt crisis much through structural adjustment loans.[53]

Yet by common agreement, the debt crisis is now mostly over in most middle-income nations. In Latin America, net transfers on debt (new loan receipts less payments on old debts) became positive in 1993. A larger measure of Latin American capital flow—including investment in stock markets, factories, and office buildings—went from $3.5 billion outward in 1990 to a $38 billion inward in 1993. Suddenly, foreign exchange was plentiful and servicing debt was much easier.[54]

The beginning of the end had come when Nicholas Brady, the Treasury Secretary in the administration of President George Bush, called for a new approach to the debt crisis in 1989. Without naming names, he described the previous administration's insistence on full repayment through adjustment as a failure. The key to resolution, he said, was pragmatism and flexibility—just as it had been in every debt crisis before. The low road of messy compromise led more surely to resolution than the high road of structural adjustment.[55]

Under the Brady Plan, a debtor nation gave its commercial lenders a menu of choices. One option was to lend more

Table 8–4. Reasons IMF and World Bank Pressure for Structural Adjustment Is Weaker Than It Seems

Reason	Examples
Geopolitics	In the Philippines, the U.S. government, which viewed the country as an important ally against communism, apparently pressed the IMF and World Bank to keep lending despite low compliance with conditions until the final years of the Ferdinand Marcos regime (1984–86).
Finance	Adjustment lending to Argentina continued through most of the 1980s with little effect except to bolster forces for the status quo with additional funds. Withdrawal after 1988 deepened the economic crisis that soon forced reform.
Pressure to lend and maintain appearance of success	In 1990, the World Bank promised a $100-million loan to support Kenya in ending its rationing of foreign exchange for purchasing imports. The government then resumed rationing. But local Bank officials did not alert the Bank's board for fear that it would not approve the loan.
Multiple conditions	Among a sample of 13 sub-Saharan African nations, adjustment agreements in 1999–2000 had an average of 114 policy reform conditions each. It is unrealistic for lenders to demand complete compliance with so many conditions. Other surveys have found that countries formally comply with a typical 50–60 percent of conditions.
Ease of undermining commitments formally met	Starting in 1984, the government of Turkey levied special taxes on imports to support Extra-Budgetary Funds that in turn financed housing and other popular programs. This undermined but did not technically violate loan agreements to reduce conventional import tariffs and cut public investment's share of the budget.
Disasters	In Malawi, in 1986, drought in southern half of the country and an influx of 700,000 refugees from civil war in Mozambique necessitated large corn imports and contributed to the government's decision to reverse commitments to end fertilizer and crop subsidies.
Debt trouble	In Bolivia, debt crisis in the early 1980s led the government to print money to pay bills, sparking hyperinflation and chaos. In 1985–86, a new administration adopted a harsh stabilization program that raised taxes, laid off state workers, and stopped hyper-inflation. IMF and the World Bank adjustment loans came later and so contributed little to the policy shift.
Domestic politics	In Zambia, in 1986–87, strikes by government workers and riots over a 120-percent price increase for staple foods forced the government to restore food subsidies and abandon its agreed adjustment program.

SOURCE: See endnote 52.

money to cover payments coming due. Another was for the bank to tear up some of its IOUs in exchange for new ones. The new IOUs could have the same principal as the old ones but a lower interest rate. Or they could have the same interest rate and a lower principal. Or banks could accept shares of state-owned companies. In 1990, for example, Argentina initiated several "debt-for-equity" swaps to retire $5 billion of its debt in exchange for its national airline and phone company.[56]

One option that was not on the banks' menus was doing nothing. If it had been, then Citibank, for example, might have waited for Deutsche Bank to strike a deal with Brazil, accept a loss, and help Brazil regain creditworthiness—at which point Citibank would demand full repayment. But Deutsche Bank and other competitors, of course, would have waited too. In the ban on doing nothing lay the genius of the Brady Plan. Debtors including Argentina, Brazil, Mexico, Nigeria, and the Philippines struck Brady agreements between 1990 and 1994, in each case ending years of repeated, crisis-atmosphere negotiations. Nearly forgotten amid the deal-making, though, were the low-income nations that were heavily indebted not to banks but to other governments.[57]

Even before Brady announced his plan, charitable groups had begun experimenting with debt-reduction transactions in the same family. One kind, debt-for-nature swaps, worked like this: In August 1987, the U.S. group Conservation International (CI) bought Bolivian IOUs with a face value of $650,000 from commercial banks. Because the banks viewed the IOUs as nearly worthless, they agreed to a sale price of only $100,000. CI then cancelled $400,000 of the debt and accepted the remaining $250,000 in Bolivian currency rather than dollars, which were scarce in Bolivia, and donated it for management of the 15,000-square-kilometer Beni Biosphere Reserve. In effect, Bolivia paid off loans by "exporting" the service of protecting its forests to the rest of the world, rather than exporting the trees.[58]

Soon other environmental groups joined in, making deals with Costa Rica, Madagascar, the Philippines, and two dozen other nations. Over time, the national aid agencies that already "owned" billions in Third World bilateral debt became the major players. To date, donors have retired some $7 billion in debt and freed up more than $1 billion for environmental projects, half of that in Poland, where an "EcoFund" is cleaning up some of the worst environmental legacies of the communist era.[59]

In the same vein, UNICEF bought or received donations of debt with a face value of $200 million between 1989 and 1995, which it used to finance debt-for-development swaps. The goals of projects funded ranged from providing clean water in Sudan to helping street children in Jamaica.[60]

Lending, Development, and Accountability

Spanish-American philosopher George Santayana famously observed that "Progress, far from consisting in change, depends on retentiveness.... Those who cannot remember the past are condemned to repeat it." History offers three principal lessons for closing the last, sad chapter of the Third World debt crisis, the continuing impasse between official creditors and the poorest nations on Earth.[61]

First, there is value in compromise. In the long history of international debt trouble, full repayment after crisis is the exception, not the rule. In the current crisis, realistic compromises have brought relief much more than demands that debtors shoulder every dollar of debt burden and overhaul policies in order to make payments.

Second, there is value in humility. Development projects, including adjustment programs, fail more than they succeed. Processes of economic development are extremely complex and the barriers to progress many. Rarely can a single individual or institution, whether native or foreign, shape the course of a nation's development—its history.

And third, there is value in speed. Continuing crisis can harm the poor and the environment, undermining goals for which the debts were avowedly run up.

Underlying all these lessons is a more basic one: there is value in accountability. We cannot expect perfection from the agencies of our governments. But we should expect them to learn from their mistakes and from the past generally. Yet the parties to the debt crisis do not have a strong history of learning from history. In the 1980s, for instance, the U.S. government, the IMF, and the World Bank seemed unaware of how historically unrealistic it was to expect that they could elicit radical economic reform, rekindle economic growth, and extract full repayment in debtor nations. It is rarely easy or pleasant for institutions, as for people, to engage in the sort of self-criticism that is needed, so they will usually only do it in response to strong pressure—whether from street demonstrations or a loss of capital.

Viewed through this lens, the Debt Initiative for Heavily Indebted Poor Countries, which aims to end the debt crisis between official lenders and the poorest nations, represents a half-step forward. The initiative, jointly administered by the World Bank and the IMF, embodies a major shift in official creditors' approach to the crisis, recognizing the value of compromise, humility, and speed in principle. But it falls short in its particulars and will almost certainly not end the crisis. Most problematically, it will likely shield official lenders such as the World Bank from the costs of past mistakes.

The initiative accepts the importance of compromise in principle by offering many countries deep debt reductions. Tanzania's debt, for example, should shrink two thirds—from an amount equivalent to 480 percent of its annual export earnings to the target level of 150 percent. But in fact the initiative hardly compromises in practice. Countries that are not quite poor or indebted enough to qualify, including Indonesia and Nigeria, will receive no additional relief. For countries that do qualify, the definition of "sustainable debt"—the level to which a country's debt would be reduced—is high and, World Bank and IMF staff have admitted, rather arbitrary.[62]

For example, the initiative aims to cap a country's annual debt service (as distinct from total debt) at 20–25 percent of export earnings. In contrast, the victors in World War I demanded that Germany pay reparations worth 13–15 percent of its exports, a demand so excessive it spurred nationalist anger and helped bring Hitler to power. After World War II, West Germany's creditors settled for 3.5 percent of exports, so as not to slow the nation's recovery. If those creditors were as solicitous of the need of HIPCs to develop, they would offer far deeper relief.[63]

Using mathematical models of HIPC economies, the U.S. General Accounting Office has concluded that many HIPCs' debts will probably stay unrealistically high. Its model for Tanzania, for example, suggests that although the initiative would cut the country's debt from $5.7 billion to roughly $2 billion, new borrowing would lift it to $6.4–7.3 billion by 2018—an unpayable amount unless the country's economy and exports grow improbably fast. One reason: if Tanzania shifts spending from debt service to social services, it will need just as much money as before—just as much tax revenue, grant income, and new loan receipts—barring major changes. The 30-percent reduction in the country's annual debt payments under the initiative would therefore be short-lived.[64]

Uganda shifted funding toward providing safe drinking water in response to comments from poor communities.

Drawing on the experience of middle-income debtors in the 1980s and early 1990s, economist Daniel Cohen has estimated that HIPCs are economically and politically capable of paying back roughly 28 percent of their current debts. (Many HIPCs are now maintaining the appearance of fully servicing their loans only by borrowing more.) Thus, demanding 50-percent repayment instead of 100 percent will have little meaning. Debtors will be pressed almost as hard as before to service their debts, and will spend almost as much over the long run doing so. Cohen's figures suggest that only 5 percent of the billions of dollars in pledged debt relief will reach HIPCs through lower payments in coming decades. Ninety-five percent will stay with the lenders.[65]

The largest immediate winners under the HIPC initiative, then, will be the World Bank and other multilateral development banks. In effect, the HIPC Trust Fund, financed by creditor government contributions, will recapitalize these banks to clean up some of their bad loans. (The IMF, however, will pay for its debt write-downs out of its large gold reserves.) In a strange way, the HIPC initiative has co-opted Jubilee 2000 lobbyists who pressed legislatures to fund the program fully. Without HIPC, the only option for development banks wanting to cover problem loans was to grant still more loans. With HIPC, creditor governments will reimburse banks for forgiving debts—all in the name of charity.

The "enhanced" version of the HIPC initiative, announced in 1999, shows some humility by conditioning debt relief on a country developing a Poverty Reduction Strategy Paper (PRSP) in consultation with civil society—including domestic NGOs, churches, or village representatives—and implementing the plan with savings from debt relief. (The requirement also applies now to countries seeking new structural adjustment loans.) These rules tacitly admit that lenders and governments do not hold a monopoly on information and wisdom relevant to fighting poverty. Uganda, which began such a consultation in 1995, offers an example: it shifted funding toward providing safe drinking water in response to comments from poor communities.[66]

The PRSP process could become an important new approach to planning aid and development in poor countries, allowing greater consultation and coordination among civil society groups, government agencies, and donors. Already, PRSP requirements have raised the profile of the poverty problem among government officials and led them to work somewhat more with nonprofit groups. In Kenya, for instance, the government held seminars with aid donors, local businesses, women's groups, and opposition movements to present its poverty strategy.[67]

But again, important details undermine the appearance of change. First, the new requirement joins rather than replaces old ones: debtors must also stick to a standard structural adjustment plan for three to six years to obtain permanent debt reduction. Indeed, according to the NGOs polled by the Brussels-based European Network on Debt and Development, most local IMF representatives still seem to care more about structural adjustment than poverty reduction strategies. Debtor governments are taking the cue, emphasizing measures such as privatizing state companies in their economic plans.[68]

Second, PRSP requirements, like structural adjustment requirements, seem based on a naive understanding of internal dynamics of lending institutions, domestic politics in debtor nations, and the problem of poverty itself. In reality, debtors' resistance to outside pressure will probably combine with creditors' eagerness to cancel debt and thus weaken their influence over debtors' policies. Debtors less committed to poverty reduction—or less responsive to pressure from civil society—may put on insincere PRSP shows in order to obtain debt relief, as NGOs in Kenya fear is happening there.[69]

Indeed, a host of thorny questions confronts lenders attempting to gauge the quality of consultation. Who should participate in designing the poverty reduction strategy, asks Joan Nelson, a political scientist at the Overseas Development Council in Washington, DC: Churches? NGOs based in the capital? Elected leaders of certain villages? And what constitutes a legitimate process for selecting the participants? Once selected, should participants merely air their views or be given a veto over final plans? What role should experts play? There can be no clear answers to these questions, which is one reason that IFIs generally have not enforced conditions for which good performance is hard to define.[70]

Poverty reduction, moreover, is a complex and uncertain undertaking. Aid agencies have lent and spent billions to fight poverty, yet indigenously driven reforms, such as the shifts toward the market in China and Viet Nam, have generated most of the successes. Were ending poverty quick and easy, there would probably be no official debt crisis. It takes time, especially for a cash-strapped government, to assess the extent and nature of poverty, consult with the groups affected—from slum dwellers to landless farmers—then assign priorities to potential responses and draw up plans. Uganda and Ghana, two leaders in consultative development of anti-poverty strategies, had been working on their plans for five years as of mid-2000.[71]

Herein lies one of the central tensions of the HIPC initiative, between the need for speed in debt relief and the need for time to develop anti-poverty programs. The PRSP requirements could delay debt reduction for years so countries can assemble high-quality plans. Or governments could ram hastily designed poverty reduction strategies through quick, ineffective consultations in order to win quick debt relief. The initiative's designers grapple with this tension by allowing debtors to submit preliminary, interim poverty plans in return for prompt but provisional debt relief—lower annual debt payments. Permanent reductions in the debt stock are to come several years later, upon approval of a full PRSP.[72]

Unfortunately, this approach does not eliminate the tension over tempo. In 2000, for example, the government of Honduras decided to shoehorn initial consultations with civil society into a few months so that it could submit an interim PRSP to the IMF and World Bank by March 2001. And once interim poverty reduction plans receive the stamp of approval from the IMF and the World Bank, they can take on a life of their own, however hastily drawn up. Burkina Faso's final PRSP closely follows the interim one, for example.[73]

Overall, the current HIPC plan is seriously biased toward placing costs of past mistakes, and risks of future ones, on poor countries. Most of the money will go toward reimbursing lenders for writing off debts, not to the struggling debtors. Yet it is the debtors, not the lenders, who must jump through hoops for their money. Insulating lenders from the costs of past failures invites a repetition of the lending patterns that led to those failures, and a renewal of debt crisis.

Official lenders should estimate how much debt will never be repaid—and cancel it, without imposing conditions.

This critique points to ways to improve international efforts to solve and prevent official debt crises. Broadly, the HIPC process should operate more like a bankruptcy court. It should serve as a permanent forum for quick resolution of debt problems as they arise. All countries, not just the 41 HIPCs designated by the World Bank, should be eligible. As a bankruptcy court would do, official lenders should estimate how much debt will never be repaid—and cancel it, without imposing conditions for structural adjustment or poverty reduction. Just as in a commercial bankruptcy, this would be an act of realism, not charity. Indeed, some debt should stay on the books so as not to reward debtors for waste any more than necessary.

Attacking the roots of official debt trouble is much less straightforward than treating the symptoms. In the ideal, politicians in creditor nations would not make unrealistic demands on development lenders for quick results, nor on debtors to jump through hoops to preserve the appearance that creditors are receiving something in exchange for financially meaningless debt relief.

Similarly, aid agencies would rarely push money through for its own sake, lending more selectively. They would favor countries that show commitment to fighting poverty, such as Guyana and Uganda—but not favor them exclusively. Within each developing country, they would see themselves as part of the political fabric rather than above it, and recognize the practical limits on their influence. Like other political actors, they would develop a deep understanding of the nation's history, institutions, and culture, then build alliances and watch for opportunities to make a difference, as the World Bank did in Papua New Guinea when it held back the second installment of a loan. Borrower governments would operate with much more efficiency and responsiveness to the needs of the poor, and much less corruption.

It is an impossible ideal. So the practical challenge for concerned citizens and policymakers is not so much to develop a complete reform agenda but to pressure the parties to evolve in the needed directions. Yet just as it is hard for lending agencies to influence policies in borrowing countries, so is it hard for any individual to influence the policies of creditor governments and lending agencies. Barber Conable, who entered the presidency of the World Bank with an ambitious reform agenda in 1986, left five years later feeling that he had hardly changed the institution. There are no pat solutions.[74]

Nevertheless, if future official debt crises are to be prevented or lessened, the current system—in which the costs of failure fall exclusively on borrowers—must be reformed. It is possible here to give only a few examples of how accountability might be more equitably and appropriately shared.

First, donor governments can require that for each $1 million that multilateral development banks accept from the HIPC Trust Fund, they cancel, say, $1.5 million in debt. Depending on their internal accounting systems, governments can apply similar conditions to their own bilateral lending agencies. Lending agencies would then suffer from the debt crisis through a loss of capital or budget, which might also lower their credit ratings and raise the interest rate they had to pay on funds they borrowed on capital markets. The lack of such a rule is, ironically, creating what economists at development banks call a "moral hazard": by insulating banks from the costs of bad loans it actually encourages imprudent lending.

Second, governments can delay contributing new capital to lending agencies until they are satisfied that the agencies are reining in pressure on individual employees to lend, becoming more strategic in their lending, and coming to terms with their limited influence in developing countries.

Third, governments need to rethink the very notion of lending for development. Development lending is an extremely risky business: failure is frequent. Aid can still be worthwhile, since even occasional success can help millions of deserving people. But loans, by definition, force borrowing nations with poor, fragile economies to bear most of the risk. One alternative is for creditor governments to increase foreign aid so that agencies could grant more and lend less. Another is to tie loan repayments to the increase in the borrower's GDP or exports. Lenders would then become more like stock market investors, standing to lose in the event of failure—or to reap high returns in the event of success. Strange as it may seem to link debt payments to GDP or exports, the HIPC initiative does exactly that by capping a country's debt-exports ratio at 150 percent. Applying the practice to new loans would merely make it routine rather than an ad hoc response to failure.

Finally, there is no substitute for activism. Pressure from NGOs has been the most powerful force for holding creditor governments accountable for their role in the debt crisis, and for bringing them as far as they have come in confronting it. NGOs spotlighted the Year of Jubilee. The jubilee has not occurred yet, but if NGOs fight hard enough, it will.

Chapter 9

Controlling International Environmental Crime

Hilary French and Lisa Mastny

In early May 2000, the Bering Sea was the scene of events that could have been out of a Tom Clancy thriller—but with a distinctly post–cold war twist. For most of a week, a U.S. Coast Guard cutter called the *Sherman* pursued a Korean-owned, Russian-crewed, Honduran-registered vessel named the *Arctic Wind* across some 1,500 kilometers of water off the coast of Alaska. A Coast Guard surveillance airplane had earlier spotted the *Arctic Wind* fishing for salmon on the high seas with kilometers-long driftnets that are banned under a 1991 U.N. General Assembly resolution and several other international agreements.[1]

Things came to a head when the captain of the *Sherman* ordered his crew to uncover the ship's machine guns and prepare its gun deck for action. He had received authorization from Coast Guard headquarters to fire on the *Arctic Wind* if necessary to bring it to a stop. Faced with this imminent threat, the captain of the *Arctic Wind* backed down and allowed the Coast Guard aboard to inspect the ship for illegal catch. In addition to finding a ton of salmon already in the vessel's hold, the inspectors watched as some 14 kilometers of nets were pulled up that had collectively ensnared 700 salmon, 8 sharks, 50 puffins, 12 albatross, and a porpoise.[2]

Just a month or so earlier, it was Russian officials who were cracking down on international environmental crime. On March 31st, in the port city of Vladivostok, undercover Russian police officers trailed two investigators from environmental groups as they posed as eager purchasers of a stockpile of Siberian tiger skins from a corrupt official. The Russian officers arrested the wildlife trader on the spot, and oversaw two similar sting operations over the next few days. Earlier in the year, Russian investigators had infiltrated the wildlife trade crime ring and determined that it was raking in some $5 million a year from smuggling such items as wild ginseng, tiger skins, and bear paws and gallbladders across the Russian border. Trade in these wildlife products is restricted under the terms of the 1973

Convention on International Trade in Endangered Species of Wild Fauna and Flora (CITES).[3]

These stories are indicative of broader trends worldwide. The last few decades have seen an explosion in international environmental agreements as countries have awakened to the seriousness of transboundary and global ecological threats and the need to address them.

But reaching agreement on an international accord is only the first step. The larger challenge is seeing that paper agreements are translated into on-the-ground reality. While some environmental treaties have generated clear and measurable results, many others are plagued by weak commitments and lax enforcement.

Under international law, countries that have ratified treaties are responsible for carrying them out. But governments cannot do this alone; they must also mobilize a broad range of domestic constituencies, including private businesses and ordinary consumers and citizens. Ensuring that environmental treaties make a difference on the ground requires compliance by countries with specific legal commitments, enforcement by national governments of the domestic laws passed to implement them, and attention to broader questions about the overall effectiveness of the accord.

The general public is often skeptical of international treaties, viewing them as paper tigers that lack enforcement teeth. Yet some analysts arguably go too far in the opposite direction, maintaining that countries almost always comply with their treaty commitments. To the extent that this is true for environmental treaties, it may have more to do with the lack of specific requirements in most treaties than the high level of motivation of treaty members to put existing accords into practice.[4]

Growing evidence suggests that many national laws and policies passed by governments to implement their international commitments are inadequately enforced. These deficiencies have resulted in an explosion of international trafficking in restricted substances such as wildlife, fish, waste, and ozone-depleting chlorofluorocarbons (CFCs).

The growth of illegal trade in environmentally sensitive products stems from a range of factors, including overall increases in international trade, lax border controls, uncoordinated national enforcement efforts, and the rising strength of transnational organized crime. Ironically, international environmental crime has also resulted from the proliferation of environmental treaties, which by creating rules for international commerce also create opportunities for people to evade them.[5]

The last few years have brought rising international attention to the range of enforcement and compliance challenges posed by environmental treaties, which has in turn sparked various innovative responses from diverse quarters. These initiatives augur well for broader efforts to build environmental governance structures that are up to the task of reversing global ecological decline.[6]

The Treaty Landscape

The flurry of international environmental diplomacy in recent decades has caused the number of accords to soar from just a handful in 1920 to nearly 240 today. (See Figure 9–1.) More than two thirds of these pacts have been reached since the first U.N. conference on the environment was held in Stockholm in 1972. These agreements cover a broad range of issues, including biological diversity, international fisheries, ocean pollution, trade in hazardous wastes and substances, and atmospheric pollution. (See Table 9–1.)[7]

The task of overseeing the implementation of treaties at the international level falls to many different institutions and actors. Small offices known as secretariats are a linchpin of these efforts. Secretariats are staffed by international civil servants, and are often managed by U.N. agencies, including the U.N. Environment Programme (UNEP) and the International Maritime Organization (IMO). Secretariats have diverse functions, including gathering and analyzing data, advising governments on implementation measures, maintaining convention records, and servicing the Conference of the Parties (COP)—the regular meeting of treaty members that is the accord's governing body.[8]

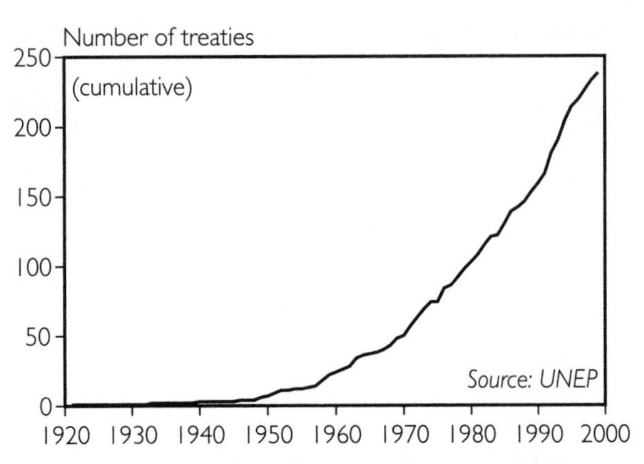

Figure 9–1. International Environmental Treaties, 1921–99

At Conferences of the Parties, treaty members review implementation, consider new information from diverse sources, and revise the treaty if necessary. They may also decide how to respond to any noncompliance by treaty members. Many COPs create sub-bodies responsible for reviewing implementation of the accord on an ongoing basis. Environmental conventions also commonly include scientific bodies, which provide relevant scientific and technological information.[9]

One important tool that most environmental treaties rely on to encourage compliance is transparency of information. A treaty typically requires member governments to supply secretariats with detailed reports that generally include relevant data as well as information about national policies to put the agreement into practice. This information is then made available to the Conference of the Parties, where it can be used to assess compliance by individual countries and to address broader questions about the overall effectiveness of the agreement.[10]

Although reporting is a key requirement of most environmental treaties, experience with it to date has been uneven. One problem is that countries do not always provide timely information. For example, only 39 percent of the parties to the London Convention on ocean dumping submitted the required reports for their 1997 deadline, while just 48 percent of the parties of the Bonn Convention on Migratory Species did so for a 1999 deadline. And the documents that are submitted may be inconsistent, outdated, or incomplete.[11]

But recent attention to these problems has begun to generate somewhat more timely reporting. Parties to CITES, for example, agreed in 1991 to consider the failure of members to submit annual trade data reports on time to be an infraction of the accord that would be publicized to other countries. This peer pressure has borne fruit: by 1998, reporting under CITES had risen to nearly 60 percent of members, almost double the share in 1990. The penalties for

Table 9–1. A Timeline of Selected International Environmental Agreements

1971 Ramsar Convention on Wetlands of International Importance, Especially as Waterfowl Habitat (123 parties)

A framework for national action and international cooperation for the conservation and sustainable use of wetlands and their resources.

1972 Convention Concerning the Protection of the World Cultural and Natural Heritage (161 parties)

Encourages countries to protect areas with outstanding physical, biological, and geological formations, habitats of threatened animal and plant species, and areas with scientific, conservation, or aesthetic value.

1972 London Convention on the Prevention of Marine Pollution by Dumping of Wastes and Other Matter (78 parties)

With amendments, bans dumping at sea of high- and low-level radioactive and highly toxic wastes as well as incineration and dumping at sea of all forms of industrial waste.

1973 Washington Convention on International Trade in Endangered Species of Wild Fauna and Flora (CITES) (152 parties)

Restricts trade in species that are either threatened with extinction or that may become endangered if their trade is unregulated.

1973 (updated 1978) International Convention for the Prevention of Pollution from Ships (MARPOL) (112 parties)

Restricts intentional discharges of oil, sewage, garbage, noxious liquids, packaged toxins, and airborne emissions at sea. Sets standards for ship construction and operation.

1979 Bonn Convention on the Conservation of Migratory Species of Wild Animals (CMS) (70 parties)

Outlines strict protection for more than 70 endangered terrestrial, marine, and avian migratory species throughout their range. Establishes conservation and management rules for more than 170 additional species under potential threat.

1982 U.N. Convention on the Law of the Sea (UNCLOS) (135 parties)

Establishes a broad framework governing ocean use that encompasses other oceans agreements. Designates 200-mile Exclusive Economic Zones (EEZs). Includes provisions on conservation of living marine resources, maintenance and restoration of marine populations, and protection of the sea from pollution.

1987 Montreal Protocol on Substances That Deplete the Ozone Layer (175 parties)

With amendments, requires the gradual phaseout of chlorofluorocarbons, halons, and other synthetic chemicals that damage the stratospheric ozone layer that protects life on Earth from harmful ultraviolet radiation.

(continued)

Table 9–1. A Timeline of Selected International Environmental Agreements
(continued)

1989	Basel Convention on the Control of Transboundary Movements of Hazardous Wastes and Their Disposal (141 parties)
	Restricts the international export of hazardous wastes from industrial to developing countries, unless the importing country agrees to accept them. Far-reaching 1994 amendment completely bans all hazardous waste exports to developing countries for final disposal and recovery operations, but is not yet in force.
1992	Convention on Biological Diversity (CBD) (178 parties)
	Establishes a broad framework for the conservation of biological diversity, sustainable use of its components, and fair and equitable sharing of the benefits arising from the use of genetic resources. Biosafety protocol of 2000 addresses the effects of transboundary shipment and use of genetically modified organisms.
1995	U.N. Agreement Relating to the Conservation and Management of Straddling Fish Stocks and Highly Migratory Fish Stocks (28 parties; not yet in force)
	Expands the scope of UNCLOS, prescribing a precautionary approach to the management and conservation of fisheries that straddle EEZ boundaries and migrate across the high seas. Grants parties the right to board and inspect vessels of other parties and obligates parties to collect and share data.
1997	Kyoto Protocol to the 1992 U.N. Framework Convention on Climate Change (30 parties; not yet in force)
	Requires industrial countries to reduce emissions of carbon dioxide by 6–8 percent by 2008–12, in order to meet the overall goal of stabilizing the atmospheric concentrations of gases that contribute to warming of Earth's climate.
1998	Rotterdam Convention on the Prior Informed Consent Procedure for Certain Hazardous Chemicals and Pesticides in International Trade (PIC Convention) (11 parties; not yet in force)
	Restricts the international export of 27 harmful pesticides and industrial chemicals that have been banned or severely restricted in many countries, unless the importing country agrees to accept them.

SOURCE: See endnote 7.

failing to submit were boosted further still in early 2000: CITES parties failing to submit annual reports in three consecutive years now face trade restrictions in species controlled under the treaty.[12]

Governments are also increasingly granting secretariats powers to actively oversee compliance and implementation. The CITES secretariat, for example, can request information from countries about alleged lapses, and ask for explanations from any country it believes is falling short of meeting treaty obligations. The secretariat of the Montreal Protocol on ozone layer depletion is also empowered to investigate suspected cases of noncompliance.[13]

Some secretariats even occasionally undertake on-the-ground inspections. Under the Ramsar Convention on the protection of wetlands, for instance, on-site inspections by a team including secretariat staff as well as independent experts can be launched when national reports or independent sources indicate that a protected wetland may be threatened. A report based on the visit provides recommendations to the government for responding to the situation. More than 20 inspections have occurred since the procedure was established in 1989, in countries as diverse as Austria, Iran, Pakistan, and the United Kingdom.[14]

Despite the importance of these secretariats, governments typically give them limited authority and resources. They often lack the wherewithal or authority to verify the information on implementation efforts contained in government reports. A typical secretariat has fewer than 30 staff and an annual budget of less than $5 million (see Table 9–2)—a drop in the bucket compared with the budgets of the agencies charged with implementing environmental laws in large industrial countries.[15]

The efforts of secretariats to monitor treaty implementation and compliance are often supplemented by those of nongovernmental organizations (NGOs). In one particularly successful example, a network of national organizations called TRAFFIC plays an important role in monitoring international wildlife trade. Created in 1976,

Table 9–2. Selected Treaty Secretariats—A Comparison of Resources

Treaty	Budget[1] (million dollars)	Staff Size	Secretariat Location	Web Site
Climate Change Convention	12.0	100	Bonn	www.unfccc.de
Biodiversity Convention	9.0	50	Montreal	www.biodiv.org
CITES	5.0	27	Geneva	www.cites.org
Ramsar Wetlands Convention	3.8	19	Gland, Switzerland	www.ramsar.org
Montreal Protocol	3.0	16	Nairobi	www.unep.org/ozone
Basel Convention	1.7	17	Geneva	www.basel.int
Migratory Species Convention	1.1	11	Bonn	www.wcmc.org.uk/cms
London Dumping Convention	0.23	3	London	www.londonconvention.org

[1]Budget figures represent the most recent year available.
SOURCE: See endnote 15.

TRAFFIC tracks national customs enforcement efforts, documents areas of unsustainable wildlife trade, identifies trade routes for wildlife commodities, and investigates smuggling allegations. It channels this information to CITES members through regular contact with the CITES secretariat and the publication of a bimonthly bulletin, and provides recommendations to governments on proposed species listings. NGOs are also active in monitoring other forms of international environmental crime, including the hazardous waste trade, the smuggling of CFCs and other ozone-depleting substances, and illegal fishing and logging.[16]

NGOs as well as governmental institutions are using a range of new technologies in their efforts to uncover international environmental crime. Satellites are tracking illicit fishing and logging, and DNA tracing is being used to monitor both fishing and wildlife trade. The latter technology enables researchers to link seafood items and wildlife parts back to the species or even the animal of origin. In one recent application, scientists from the University of Auckland in New Zealand confirmed that a wide range of whale species protected under the International Whaling Commission's moratorium on commercial whaling were nonetheless being sold in fish markets in Japan.[17]

Species at Risk

One burgeoning form of international environmental crime is wildlife trading that violates the terms of CITES. This treaty was finalized in 1973 in response to widespread public concern over the growing extinction crisis. CITES bans international trade in some 900 animal and plant species in danger of extinction, including all tigers, great apes, and sea turtles, and many species of elephants, orchids, and crocodiles. Through a requirement for export permits, it also restricts trade in some 29,000 other species that are directly or indirectly threatened by commerce, among them birdwing butterflies, parrots, black and stony corals, and some hummingbirds.[18]

CITES is credited with reducing trade in many threatened animals, including cheetahs, chimpanzees, crocodiles, and elephants. But despite these notable successes, unchecked trade in many endangered species that are theoretically protected under CITES continues, earning smugglers high profits on black markets worldwide. Illegal traders are estimated to earn some $5 billion annually through such commerce, representing roughly a quarter of the value of global wildlife trade. Among the most coveted black market species are tigers and other large cats, rhinos, reptiles, rare birds, and botanical specimens. (See Table 9–3.)[19]

The bulk of illegally traded wildlife originates in developing countries, home to most of the world's biological diversity. In Latin America, for example, dealers in Brazil secret out some 12 million animals a year, generating annual revenues of $1.5 billion—while those in Colombia smuggle some 7 million animals, earning an estimated $40 million. And in Central Asia, locals in Kazakhstan, Mongolia, and elsewhere sell growing numbers of endangered falcons and pelts from bears and snow leopards to foreigners in exchange for desperately needed cash.[20]

The demand for illegal wildlife, meanwhile, comes primarily from wealthy collectors and other consumers in Europe, North America, Asia, and the Middle East. In 1999, customs officials in the United Kingdom alone confiscated some 1,600 live ani-

Table 9–3. Selected Species Threatened by Illegal Wildlife Trade

Species	Threats/Population Status
Tiger	Live tigers can fetch up to $15,000 each in the pet trade, while pelts used in the fashion and curio industries can cost $10,000 or more. Ground bones and other parts are sought for use in traditional medicines. The global tiger population has fallen 95 percent since 1900, to some 5,000–7,500 animals. Three of eight sub-species have already gone extinct.
Giant panda	A single panda pelt, valued in the fashion and curio industries, can sell for up to $100,000 on the black market. Only 1,000 pandas are thought to remain in the wild.
Elephant	Ivory from tusks is valued for use in sculpture and jewelry, especially in Japan. A recent discovery of 78 smuggled tusks in Egypt had a street value of more than $200,000. African populations fell by half between the early 1970s and late 1980s—to some 600,000 animals—and are only now beginning to recover. Fewer than 20,000 elephants remain in India.
Horned parakeet	Coveted in the exotic pet trade. Because of heavy poaching, fewer than 1,700 individuals now remain in the wild, primarily in New Caledonia.
Musk deer	Raw musk is extracted for use in perfumes and traditional medicines, particularly in Europe and Asia. Up to 100,000 deer are killed each year to satisfy Chinese demand alone. Populations are declining throughout their range in Asia and parts of eastern Russia.
Ginseng root	Used as a general health tonic in Asia and elsewhere. The Asian root can fetch more than $10,000 per kilogram; each year, up to 600 kilograms are smuggled from Russia. The oldest American roots can sell for up to $30,000 in China. Ginseng now grows only in small areas of Asia, Russia, and North America, and could disappear from the wild within decades.
Caspian Sea sturgeon	Smuggled caviar can sell for as much as $4,400 a kilogram in Europe and the United States. More than half of all shipments from Russia to the United States, the leading importer, are thought to be illegal. Adult sturgeon population has fallen 80 percent since 1986.

SOURCE: See endnote 19.

mal and birds, 1,800 plants, 52,000 parts and derivatives of endangered species, and 388,000 grams of smuggled caviar.[21]

Among other uses, wildlife is in demand for food and medicine, as clothing and ornamentation, for display in zoo collections and horticulture, and as pets. In the United States—the world's largest market for reptile trafficking—exotic pets such as the Komodo dragon of Indonesia, the plowshare tortoise of northeast Madagascar, and the tuatara (a small lizard-like reptile from New Zealand) reportedly sell for as much as $30,000 each on the black market.[22]

The multibillion-dollar traditional Asian medicine industry has been a large source of the demand for illegal wildlife, with adherents from Beijing to New York purchasing potions made from ground tiger bone, rhino horn, and other wildlife derivatives for their alleged effect on ailments from impotence to asthma. In parts of Asia, bile from bear gall bladders—used to treat cancers, asthma, eye disease, and other afflictions—can be worth more than narcotics.[23]

Wildlife smuggling is an attractive proposition to many traders. It generally requires minimal investment and carries low risk, with domestic penalties often small or nonexistent. And it can be extremely lucrative, with prices climbing substantially as the items change hands from local merchants to international dealers—transforming creatures into valuable commodities. A Senegalese wholesaler, for instance, may buy a grey parrot in Gabon for $16–20 and sell it to a European wholesaler for $300–360, who then gets $600–1,200 for it.[24]

In early 2000, the Tanzanian government estimated that more than 70 vessels were fishing illegally in its waters.

Large-scale illegal wildlife trade reflects serious obstacles to enforcement under the CITES regime. Smugglers have devised numerous ways to evade the treaty's controls, from using unauthorized crossing points to shipping illegal species along with legal consignments. Some conceal the items on their person or in vehicles, baggage, or postal and courier shipments—typically resulting in a high fatality rate for many live species. Another common fraud is to alter the required CITES permits to indicate a different quantity, type, or destination of species, or to alter the appearance of items so they appear common. In February 2000, the U.S. Fish and Wildlife Service arrested someone from Côte d'Ivoire for smuggling 72 elephant ivory carvings, valued at $200,000, through New York's John F. Kennedy airport—many of them had been painted to resemble stone.[25]

As it grows in sophistication, wildlife smuggling accounts for a rapidly rising share of international criminal activity. Wildlife traders commonly rely on the same international trafficking networks as smugglers of other contraband goods, such as gems and drugs. In the United States, consignments of snakes have been found stuffed with cocaine, and illegally traded turtles have entered on the same boats as marijuana. The proceeds from black market ivory sales and other wildlife trafficking are sometimes laundered or funneled to other illicit activities, such as the purchase of drugs or weapons.[26]

Adequate enforcement of CITES requires strong, coordinated efforts within governments. As with most international treaties, the CITES secretariat has little centralized authority and no official agents or inspectors of its own. Each country must enforce CITES through its national legislation, designating permitting authorities and training customs inspectors and police to detect and penalize illegal activity. But a 1993 study by the World Conservation Union–IUCN reported that some 85 percent of CITES parties had incomplete implementing legislation, with one of the most common deficiencies being a lack of appropriate penalties to deter treaty violations.[27]

Developing countries, in particular, have had difficulty implementing CITES. Many lack the money, equipment, or political will to monitor their own wildlife populations effectively, much less wildlife trade. Customs offices are chronically understaffed and possess limited resources, while agents are often either untrained to identify illegal activity or overwhelmed by the sheer numbers of species they must track. Local authorities sometimes issue permits for species they do not know are restricted, or let consignments slip through because they cannot identify a fake or altered permit. In extreme cases, corrupt local officials facilitate the illegal trade, deliberately issuing false permits or overlooking illegal consignments in order to obtain bribes or other kickbacks.[28]

As awareness of the scope of illegal wildlife trade grows, worldwide enforcement of CITES is gradually improving. The United Kingdom, for instance, recently established a new national police unit to combat wildlife crime, and announced tougher penalties for persistent offenders. And in South Korea, a team at the Seoul airport now uses specially trained dogs to detect tiger bone, musk, bear gall bladders, and other illegal wildlife derivatives smuggled in luggage and freight—the first program of its kind in East Asia.[29]

Some progress against illegal wildlife trade has also been made at the regional level. Under the 1996 Lusaka Agreement, for example, Kenya, Lesotho, the Democratic Republic of Congo, Tanzania, Uganda, and Zambia have conducted joint operations that have led to the seizure of hundreds of tons of illegal elephant ivory and the arrests of several poachers in the region.[30]

Although it has typically focused mainly on terrestrial animals and plants, CITES offers limited protection to several highly endangered fish species as well. The world's fisheries are also ostensibly protected by a wide range of other agreements at the global and regional levels. But many of these face growing enforcement hurdles, resulting in illegal fishing in all types of fisheries and in all the world's oceans, including national waters, regionally managed fisheries, and the high seas. As with other forms of wildlife trade, booming consumer demand is an important driving force. In Japan, for instance, species such as the threatened Southern bluefin tuna now fetch up to $50,000 per fish.[31]

One major form of illegal fishing occurs when foreign vessels fish without authorization in other national waters, often those of developing countries that are unable to patrol their shores adequately. In early 2000, the Tanzanian government estimated that more than 70 vessels were fishing illegally in its waters—mainly from Mediterranean countries and the Far East. Tanzania has few police boats or enforcement equipment of its own, and has had to rely on assistance from France and the United Kingdom to crack down on offenders.[32]

Commercial fishers are also turning to the largely unmonitored high seas—including the Mediterranean Sea and the Indian, South Atlantic, and Southern Oceans. These vessels may illegally straddle the borders of regional fishing grounds that are restricted under international laws, deliberately hiding their flags or other required identification markings to avoid recognition. They may then offload their illicit catch to other vessels to further disguise its origin and to minimize the penalties if discovered. Often these fishers do not report their activity—or if they do, they falsify the species caught, the amount caught, or the fishing area frequented.[33]

One of the most serious challenges to adequate fisheries enforcement is the rapid rise in so-called flag-of-convenience (FOC) fishing. Increasingly, commercial fishing companies register their ships in countries that are known to be lax in enforcement of international fisheries laws or that are not members of major maritime agreements. By transferring their allegiance to these new "flags," the companies can easily enter the waters of their adopted countries or operate undercover on the high seas where agreements currently in force permit only the country of registry (the flag state) to make an official arrest. Some vessels change their flags frequently in order to hide their origins or identities, making it difficult for enforcement officials to track them down. This practice not only leads to unfair competition among vessels, it also undermines the overall effectiveness of international fisheries agreements.[34]

An estimated 5–10 percent of the world fishing fleet now flies flags-of-convenience. This includes more than 1,300 large industrial fishing vessels registered in such countries as Belize, Honduras, Panama, St. Vincent, and Equatorial Guinea. In total, some 27 countries are considered flag-of-convenience states. Under the 1982 U.N. Convention on the Law of the Sea and other maritime laws, all flag states are expected to exercise control over their vessels and ensure that minimum safety and pollution standards and other international obligations are being met. But FOC countries typically ignore these commitments and open their registries to outside vessels in order to earn substantial tax revenue.[35]

The countries and companies that own the re-flagged vessels are also partially to blame for the problem. Fishing companies from Europe and Taiwan own the highest numbers of flag-of-convenience vessels. Many of these firms receive special government subsidies to register their ships abroad. In an effort to relieve fishing pressures in European waters, for example, the European Union announced a subsidy in 1999 that rewards companies that permanently re-flag their vessels elsewhere.[36]

With the scope of the illegal fishing problem becoming increasingly evident, momentum is building for stronger enforcement measures. The U.N. Food and Agriculture Organization (FAO), which has played a large role in global fisheries discussions, took a significant step in 1999: it developed a new international plan of action to combat illegal fishing, including by FOC vessels. The plan calls for improved oversight of vessels by flag states, better coordination and sharing of information among countries, and stronger efforts to ratify, implement, and enforce existing fisheries accords. It is expected to work in conjunction with other FAO efforts, including a 1993 pact known as the "Compliance Agreement." This requires member countries to keep detailed records on all vessels flying their flags on the high seas and to ensure that the vessels do not engage in activities that would undermine the effectiveness of international conservation and management measures. But it does not yet have legal force, as only 18 of the required 25 countries have ratified it.[37]

Enforcement on the largely unregulated high seas is expected to improve significantly when a pathbreaking new treaty, the U.N. Agreement Relating to the Conservation and Management of Straddling Fish Stocks and Highly Migratory Fish Stocks, enters into force. The agreement, adopted in 1995 as an implementing agreement of the Law of the Sea Convention, specifically addresses the management and conservation of fish stocks that straddle national and international waters or migrate long distances, including in international waters. It requires member countries to issue licenses, permits, or other authorizations to all their vessels registered to fish on the high seas, and to ensure that the fishers report on the locations of the vessels, the quantity of their catch, and other relevant scientific and technical data. The agreement also includes an unprecedented measure allowing authorized officials to board and inspect suspect vessels in high seas areas covered by regional fisheries agreements, even if the ships are registered in other countries. It will come into force as soon as 30 countries ratify it.

As of November 2000 some 28 nations had taken this step, but several major fishing nations—including China, Indonesia, Japan, and the Philippines—had not yet ratified the treaty.[38]

Many countries are already stepping up efforts to enforce international fisheries laws, though much more needs to be done. These activities include exchanging information about illicit activity, making vessel registries more transparent, increasing vessel inspections at sea and in ports, denying landing and transshipment rights for illegally caught fish and improving monitoring and surveillance of fleets using satellite-linked Vessel Monitoring Systems and other technologies. These steps appear to be working, at least in some places: in the North Pacific, Canadian officials attribute the recent drop in the number of vessels using illegal driftnets to stepped-up air patrols by Canada and its partners in regional enforcement—the United States, Russia, and China.[39]

In another promising step, growing numbers of fisheries agreements are targeting fish traders for their role in fostering illegal fishing. Many regional bodies, such as the 28-member International Commission for the Conservation of Atlantic Tunas, now require exporters to provide evidence of the origin of their catch, along with other important data. These measures are designed to make it harder for illegal traders, and particularly flag-of-convenience fishers, to market their goods.[40]

Several other treaties are also aimed at reversing today's unprecedented biological losses, most notably the June 1992 U.N. Convention on Biological Diversity. This agreement currently contains few concrete commitments, and thus has faced few major enforcement challenges. But as it is strengthened over time, greater attention will need to be devoted to ensuring that countries and private actors adhere to its terms.[41]

Cracking Down on Dumping

Illegal dumping, both at sea and on land, is another form of international environmental crime that poses growing challenges for international treaty bodies and national-level enforcement officials.

In the middle of the twentieth century, an upsurge in ocean-borne shipping led to a growing awareness of the pollution risks from marine transport. International cooperation on oceanic management began in earnest in 1948, when the International Maritime Organization was established under the United Nations as a forum where countries could address common concerns over maritime safety and pollution. The 158-member body has adopted more than 40 international conventions and agreements, many aimed exclusively at protecting marine ecosystems from pollution and other toxic threats. These were all incorporated in 1982 into the U.N. Convention on the Law of the Sea, which sets minimum rules and standards for global maritime activities.[42]

Two IMO conventions forged in the 1970s have proved particularly successful in reducing marine pollution—the London Convention regulating ocean dumping and the MARPOL convention restricting releases from ships. Largely as a result of the first, for instance, the amount of industrial waste dumped at sea dropped from 17 million tons in 1979 to 6 million tons in 1987, and dumping of dredged material and sewage sludge also declined. The MARPOL convention led to a 60-percent decline in the amount of oil spilled into the oceans between 1981 and 1989.[43]

Despite their accomplishments, these agreements are showing signs of strain. The United Nations reports increasing violations of MARPOL's strict restrictions on the disposal of plastic wastes and derelict fishing gear, particularly in the North Pacific. Moreover, despite regulations on ship operation and construction—such as requirements calling for conversion to double-hulled tankers—serious oil spills and other accidents still occur, in part because these rules are ignored or inadequately implemented or enforced. Indeed, since the notorious *Exxon Valdez* incident in Alaska's Prince William Sound in 1989, at least 1.1 million tons of oil have spilled from tankers worldwide—the equivalent of some 30 *Valdez* accidents.[44]

Cruise ships are the culprits in many recent violations of international pollution laws. The San Francisco–based watchdog group the Bluewater Network reports that these "floating cities" account for an estimated 77 percent of all waste generated by ships, which includes such items as human waste, plastic dinnerware, and other garbage, as well as oil and chemical releases from general ship operations. The companies running these ships may decide to discard their waste at sea either because they lack the storage or incineration capacity onboard or because they wish to avoid the high costs of eventual disposal on land. Often this at-sea discharge occurs in the waters of countries where monitoring and enforcement are minimal, and the activity can go undetected. But this is not always the case: in 1999, officials in the United States fined Royal Caribbean Cruises Ltd. a record $18 million for 21 separate violations of U.S. marine pollution laws, including some passed to uphold U.S. commitments to MARPOL. The charges included routine disposal of oil waste into the sea, falsification of vessel logbooks, and deliberate efforts to cover their tracks.[45]

As with illegal fishing, flag-of-convenience vessels pose a serious problem in the cruise and shipping industries. The International Transport Federation reports that shippers in industrial countries control nearly 70 percent of the world's shipping fleet by weight, and that 67 percent of this tonnage is carried by vessels registered in flag-of-convenience countries. Shipping and cruise companies may re-flag their vessels in order to avoid higher taxes, labor, and operating costs at home, or to take advantage of the lower safety standards and weaker pollution laws in countries like Liberia, Panama, and Malta. The re-flagged ships account for a disproportionate share of pollution and accidents at sea, as well as detentions in ports for violations of maritime laws.[46]

Stronger enforcement is needed to address these problems. But like the CITES secretariat, the IMO has no enforcement power of its own, and must rely on member governments to ensure compliance with its various accords. Although there have been efforts to build stronger enforcement into both MARPOL and the London Convention, neither treaty contains sufficient measures to ensure effective domestic enforcement of its provisions. Most flag states remain reluctant to pursue MARPOL violators, in part because the governments are then also required to pay the legal expenses of the prosecution. Possible solutions to these problems include granting the IMO greater central authority to review and investigate specific cases, centralizing and strengthening the monitoring of maritime activity, creating a fund to help more developing countries come into compliance, and establishing a tribunal that would allow citizens to bring lawsuits against companies directly when pollution or other violations are discovered.[47]

On a positive note, efforts to address the flag-of-convenience issue in shipping have gained momentum in recent years. The IMO has encouraged flag-state governments to exercise greater control over their ships. Another significant development has been the shifting of power to countries other than the flag states, in particular to countries with ports. Under an accord known as the 1982 Paris Memorandum of Understanding, for example, some 19 port countries in Europe and North America have agreed to systematically inspect and detain any docked foreign vessels if they suspect violations of international maritime laws.[48]

The international community has been somewhat slower to address the dumping of hazardous wastes on land than it was to take on dumping at sea. In 1989, countries adopted the Basel Convention on the Control of Transboundary Movements of Hazardous Wastes. This treaty aims to curb the uncontrolled dumping of hazardous wastes from the industrial world in developing countries by requiring exporting countries to notify importing nations in advance of all waste shipments, and to obtain permission before proceeding. Nearly all industrial countries have ratified the treaty except the United States.[49]

Under the terms of the convention, roughly one tenth of the 300–500 million tons of hazardous wastes produced annually is now shipped legally across international borders. Industrial countries produce up to 90 percent of all hazardous waste, which includes industrial sludges, used batteries, toxic incinerator ash, and contaminated medical and military equipment. The bulk of the waste is transferred among industrial countries, but an estimated 20 percent heads to the developing world and countries in transition.[50]

In addition to this ongoing legal trade, the illegal waste trade is believed to be flourishing, although it is impossible to quantify, given that most such activity never comes to light. Asia—and in particular South Asia—is one of the biggest destinations for illegal hazardous waste. Greenpeace India reports that more than 100,000 tons of illicit wastes entered India in 1998–99—including toxic zinc ash and residues, lead waste and scrap, used batteries, and scraps of chromium, cadmium, thallium, and other heavy metals. The unauthorized imports, originating in Australia, Belgium, Germany, Norway, the United States, and several other industrial countries, violated the Basel Convention's notification rules as well as a 1997 Indian government ban on waste imports.[51]

Since the *Exxon Valdez* incident in Alaska in 1989, at least 1.1 million tons of oil have spilled worldwide—equivalent to some 30 *Valdez* accidents.

Developing countries are particularly vulnerable to the health and environmental effects associated with the illegal waste trade. Many governments do not have the infrastructure or equipment needed to store waste safely, clean up dumpsites, or monitor waste movements. And a recent survey by the Basel secretariat revealed that many countries lack adequate—or any—legislation on preventing and punishing illegal waste traffic. On a global scale, the absence of uniform definitions of hazardous waste and of coordinated enforcement efforts among ministers, customs officers, and port authorities has contributed to the spread of illegal waste trading. Even for countries that do have the resources, the lack of hard data on the extent or geographic flow of this trade makes it hard for officials to know how to allocate the resources properly.[52]

Like other criminals, hazardous waste smugglers go to great lengths to conceal their activities. Many rely on false permits or other documents to export the material, mislabeling wastes as raw materials, less severe substances, or other products. Others use unauthorized routes in order to ship wastes without prior notification or without the consent of the importing country. In many cases, organized crime is thought to be behind large-scale waste trading, which can be closely linked with money laundering, the illegal arms trade, and other criminal activities.[53]

Illicit waste dealers increasingly pass off hazardous shipments as recycled material, abusing a provision in the Basel Convention that permits trade in waste destined for reuse or recycling. Indeed, while the Basel controls have significantly reduced the quantity of waste exported for final disposal, exports of "secondary" material have skyrocketed. Greenpeace estimates that 80–90 percent of waste shipped to the developing world, particularly plastics and heavy metals, is now labeled as destined for recycling. Although most of this waste is legitimate, some is illegal and is simply dumped after being imported. In 1999, for example, a Japanese waste disposal company was caught attempting to dump some 2,700 tons of used hypodermic needles and other infectious medical waste in the Philippines in containers marked "paper for recycling."[54]

The recycling loophole is expected to close when a far-reaching 1995 amendment to the Basel Convention, known as the Basel Ban, enters into force. This significantly strengthens the treaty by prohibiting all waste exports from industrial to developing countries, for recycling as well as for final disposal. Many countries are already respecting the terms of the ban, though 41 more ratifications are needed before it comes into effect. But a few key governments—including Australia, Canada, and the United States—continue to object to many of the ban's provisions, in part because the measure could impede the lucrative trade in recyclable materials.[55]

Members of the Basel Convention have taken other positive steps to address illicit waste smuggling. The treaty requires countries to implement domestic legislation to curb and punish illegal waste traffic and outlines how to handle illegally traded waste once it is discovered. In 1994, the Basel secretariat strengthened information sharing among member countries by agreeing to set up a centralized system for reporting suspect activity.[56]

A pathbreaking new liability protocol to the Basel Convention, adopted in December 1999, could also curb some of the disastrous effects of the hazardous waste trade, including illegal traffic. The first of its kind in any multilateral environmental accord, the measure makes exporters and disposers of hazardous waste liable for any harm that might occur during transport, both legal and illegal. It requires the dealers to be insured against the damage and to provide financial compensation to those affected. These requirements are expected to discourage illegal activity by increasing waste traffickers' costs. But the protocol will not enter into force until 20 ratifications are received. Moreover, it still contains significant loopholes: it only covers harm that occurs in transit—not after disposal—and only applies to damage suffered in the jurisdiction of a treaty member.[57]

Lessons being learned in implementing the Basel Convention will be put to good use in other international toxics treaties recently finalized or still under development, such as a 1998 agreement regulating international commerce in hazardous chemicals and a treaty on persistent organic pollutants that is expected to be completed by early 2001.

Atmospheric Assaults

In the mid-1980s, the world community came face to face with the first environmental problem of truly global proportions: damage to the stratospheric ozone layer that protects life on Earth from harmful ultra-violet radiation. Negotiators were spurred to action by concern over the "ozone hole" over the Antarctic and mounting scientific evidence that chlorofluorocarbons and other synthetic chemicals were to blame. In September 1987 they finalized a landmark in international environmental law: the Montreal Protocol on Substances That Deplete the Ozone Layer. Experience to date with the accord suggests that making it work will require innovative approaches to implementation and enforcement.

The 175-member Protocol, which has been strengthened several times since 1987, calls for the gradual phaseout and elimination of some 95 ozone-depleting chemicals, including CFCs, halons, HCFCs, and methyl bromide. Industrial countries were required under the treaty to phase out CFC production for nearly all domestic uses by the beginning of 1996. Developing countries, however, were given until 1999 to freeze their CFC consumption and production, and must completely phase out the chemicals by 2010. Similar phaseout timetables exist for the other ozone-depleting substances covered under the treaty.[58]

Scientists estimate that if all countries comply fully with the Protocol, the ozone shield will gradually begin to heal within the next few years, with a full recovery to pre-1980 levels expected by about 2050. So far, however, little progress is visible, as the ozone layer continues to be damaged by chemicals that have built up in the stratosphere over decades. Some of the worst ozone holes ever recorded still appear annually over the northern and southern poles as a result, posing a threat to human and wildlife populations in these areas.[59]

Overall, most countries have met their obligations to the Montreal Protocol in a timely fashion. By 1997, global CFC production was down 85 percent from its 1986 level, as the treaty began to spur the development of alternatives for the chemicals. But the picture has not been uniformly bright. The most serious cases of noncompliance occurred when growing economic instability prevented Russia and several other countries in transition—including Belarus, Bulgaria, Poland, and Ukraine—from meeting their 1996 deadline. Under pressure from the treaty's secretariat, most of these countries were able to catch up fairly easily—except Russia, the only major CFC producer of the group. By 1998, Russia still housed roughly half of the global CFC manufacturing capacity and produced some 9 percent of the world's CFCs.[60]

When it was clear that Russia would be unable to make the transition alone, the international community agreed to provide funding to compensate Russian producers for shutting down production and converting equipment at key manufacturing facilities. After two years of negotiations, the World Bank and several other international donors stepped in with $26 million in October 2000. In exchange, Russia promised to adhere to a detailed schedule to phase out production at its seven main manufacturing sites. Similar financial incentives are being used to promote compliance in Azerbaijan, Bulgaria, Latvia, and Turkmenistan.[61]

The next major challenge will be ensuring that developing countries meet their 2010 CFC phaseout targets, though it is not yet clear whether a significant noncompliance problem will emerge. Indeed, a recent UNEP analysis found that at least two thirds of the developing countries surveyed were well on track to meeting their commitments.[62]

Illicit CFC trade violates the spirit of the Montreal Protocol, undermining the effectiveness of the global phaseout.

The different phaseout scheduling for industrial and developing countries under the Montreal Protocol has also contributed to the emergence of a flourishing illegal trade in CFCs and halons. Chemicals that are legally produced in the developing world make their way to lucrative black markets in the United States and Europe, where substances like Freon (the trade name for CFC-12) are still in demand for use in auto air conditioners, refrigerator chillers, and other equipment.[63]

Estimates of the amount of illegal trade are bound to be uncertain, but government and industry reports suggest that at the height of the problem in the mid-1990s, as much as 15 percent of global production of the chemicals—or some 16,000–38,000 tons—entered industrial countries annually. In the case of smuggled CFCs, the bulk of the supply is thought to originate in China and India, which together produce roughly half of the world's remaining CFCs. The illicit CFC trade violates domestic import laws that many industrial countries have implemented for the phaseout, which require importers to obtain official permits for any shipments of new and recycled CFCs. More significantly, it violates the spirit of the Montreal Protocol, undermining the overall effectiveness of the global phaseout.[64]

CFC importers resort to fraud and other evasive tactics to smuggle the banned chemicals into industrial countries. As in the illegal hazardous waste trade, traders abuse existing loopholes in the Montreal Protocol and domestic laws to pass off shipments of new CFCs as "recycled" material or as CFC replacements—neither of which are yet restricted in most countries. The chemicals are typically colorless and odorless, making them easy to disguise and virtually impossible to differentiate without chemical analysis. Even small-scale CFC smuggling can be lucrative: in 1997, a 12-ounce (340 gram) can of Freon bought in Mexico for $2 could be resold in California for $23.[65]

In the United States, the illegal CFC trade reached its peak in the mid-1990s, when an estimated 10,000–20,000 tons of the chemicals—valued at $150–300 million—entered the country each year. Most of these early shipments originated in Russia and were smuggled through Europe, en route to southern Florida. By 1995, CFCs were considered the most valuable contraband entering Miami after cocaine. Following a subsequent crackdown on large consignments through East Coast ports, much of the illegal trade shifted to the Canadian and Mexican borders. Between April 1998 and March 1999, the U.S. Customs Office in Houston, Texas, reported 619 seizures of Freon, totaling nearly 20 tons—all but two of which originated in Mexico.[66]

But the illegal CFC trade in the United

States now appears to be on the decline. A recent study by the consulting firm ICF Incorporated reports that large shipments of contraband Freon are becoming less frequent. This is in large part because interest in the chemical is fading as older, CFC-dependent equipment is gradually replaced or retrofitted. The number of cars with older-model air conditioners, which account for some 82 percent of the remaining U.S. demand for Freon, is expected to fall from 55 million in 1999 to only 7 million by 2005. Conversion and replacement of the tens of thousands of CFC-dependent refrigerator chillers is proceeding at a slower pace, however, and black market demand in this sector is predicted to jump as the remaining legal CFC supplies shrink.[67]

Europe has been another significant market for illegal CFCs. In the mid 1990s, researchers with the London-based Environmental Investigation Agency (EIA) reported on a thriving regional trade in smuggled CFCs, amounting to between 6,000 and 20,000 tons of chemicals annually. Well after the phaseout deadline, supplies were still abundant and prices disproportionately low—suggesting that the market was being swamped with illegal imports. Meanwhile, regional sales of CFC replacements were growing more slowly than expected, amounting to only about three fifths their expected volume in late 1997.[68]

The European black market has thrived in part because regional refrigerant management programs have been poorly organized, and because consumers perceive alternatives to CFCs to be too costly and less efficient. In Central and Eastern Europe, where the illegal CFC trade is thought to be increasing, a major problem is inadequate detection capacity: border officials are typically untrained in identifying the chemicals and have difficulty deciphering their often vague customs codes. Poland, Romania, and the Czech Republic have all reported recent attempts by importers to smuggle CFCs under falsified customs codes.[69]

Worldwide, the illegal trade in halons is also believed to be on the rise. Once widely used for fire suppression in military, aerospace, and oil exploration activities, halons contain bromine and are an estimated 40–60 times more efficient than chlorine-based CFCs at depleting ozone. While total atmospheric chlorine concentrations have started to decline, those of bromine are still growing. Scientists estimate that continued release of halons and other bromine-containing chemicals could eventually offset the gains from reduced CFC use, delaying the projected recovery of the ozone layer.[70]

Although the Montreal Protocol required industrial countries to phase out production and sales of halons by 1994, developing countries are allowed to manufacture the chemicals until 2010. China is now the predominant source of the illegal halon trade, according to the Environmental Investigation Agency. Halon production in China more than doubled between 1994 and 1996 alone, and now accounts for nearly 90 percent of global production. As with CFCs, smugglers pass halon shipments off as "recycled" in order to supply them to industry and military users in the United States and Europe.[71]

In the United States, domestic enforcement efforts have played an important role in curbing the smuggling of ozone-depleting chemicals, though the quantities seized still only account for a fragment of the total illegal trade. The 1994 establishment of an inter-agency task force, code-named Operation Cool Breeze, has led to more than 17 convictions and the seizure of some 450 tons of contraband CFCs in southern Florida. And Operation Frio Tejas, along the U.S.-Mexican border, resulted in the confiscation of more than $2 million worth of Freon within its first two years. By September 1999, the U.S. Department of Justice had convicted more than 80 individuals and businesses of illegal CFC smuggling, resulting in jail sentences and nearly $64 million in fines and restitution.[72]

In general, European enforcement efforts have lagged behind. A regional body organized in 1994 to assess the illegal CFC trade has only met a few times, in part due to lack of data on violations and weak interest in catching smugglers. Differences among member states regarding fines and other enforcement procedures have also hampered coordinated enforcement efforts.[73]

But there are signs that European enforcement is improving. In 1997, authorities in Belgium, Germany, the Netherlands, and the United Kingdom jointly nabbed a multimillion-dollar crime ring that had illegally imported more than 1,000 tons of Chinese-made CFCs for redistribution in Europe and the United States. And in an unprecedented move, in September 2000 the European Union adopted a regional ban on CFC sales and use, which will further simplify policing and enforcement activities.[74]

Significant effort has also been made at the global level to crack down on smuggling. In 1999, an amendment to the Montreal Protocol entered into force, requiring all signatory countries to issue licenses or permits for the import and export of new, used, and recycled ozone-depleting substances and to exchange information about these activities on a regular basis. This system will help discourage the illegal trade by making it easier for police and customs officials to track the movement of the substances worldwide.[75]

Illegal trade in these chemicals will likely fizzle once controls limiting their production and consumption worldwide are in place. But as global supplies dwindle, black markets in CFCs in particular could also explode in developing countries like China and India, where there are still millions of users of CFC-based equipment. This problem is exacerbated by the fact that industrial countries are exporting much of their used CFC-dependent equipment, such as old cars and refrigerators, to the developing world, further fueling demand in these countries.[76]

The obstacles to addressing the ozone threat may foreshadow difficulties to be faced in enforcing other agreements aimed at protecting the atmosphere, such as the 1997 Kyoto Protocol to the U.N. Framework Convention on Climate Change. While many components of the treaty—including its compliance and enforcement mechanisms—are still being worked out, it is clear that effective enforcement will be a key determinant of success.

From Words to Action

As the magnitude of the international environmental crime problem becomes increasingly apparent, the world community has begun to sound the alarm. In April 1998, environment ministers from the leading economic powers issued a Communiqué that expressed "grave concern about the ever-growing evidence of violations of international environmental agreements," and called for a range of cooperative actions aimed at stepping up enforcement. UNEP has joined in this effort by spearheading the development of guidelines aimed at promoting compliance with multilateral environmental agreements and preventing cross-border environmental crime.[77]

Responding to the range of challenges that fall under the broad rubric of international environmental crime will require a multipronged approach. One priority is to address the shortages of financial, technological, and administrative capacity that often frustrate countries' best efforts to implement environmental treaties.

Governments created an important precedent for this approach with the Montreal Protocol on ozone depletion, which created a sizable fund to assist developing countries with implementation. Industrial countries have so far contributed more than $1 billion to the fund, financing some 3,300 projects in 121 countries that have, among other things, helped refrigerator manufacturers switch to CFC-free models and farmers find substitutes for the soil fumigant methyl bromide. Governments built on the ozone fund model in 1991 when they created the Global Environment Facility to finance investments aimed at ameliorating other global problems, including the loss of biological diversity, climate change, and the pollution of international waterways. The facility has so far allocated a total of $3.4 billion to projects in these areas.[78]

Several other international institutions are actively engaged in helping countries to control international environmental crime. The World Customs Organization is working with governments to harmonize classification systems for waste and other environmental contraband. The international police organization INTERPOL is training national enforcement officers and customs agents to identify illicitly traded goods more easily, and is also working with national police forces to bring international environmental criminals to justice. Both institutions have established close working relations with UNEP and with the CITES and Basel Convention secretariats.[79]

Although positive incentives such as financial and technical assistance have a key role to play in promoting better treaty compliance, sometimes a more punitive approach is called for. A range of sanctions can be used to promote adherence to international environmental accords, including applying diplomatic pressure; suspending treaty privileges, such as the right to trade in a particular substance; withholding funds that would normally be available to a country for implementation; and imposing financial penalties.[80]

INTERPOL is training national enforcement officers and customs agents to identify illicitly traded goods more easily.

Several multilateral environmental agreements rely on trade restrictions to encourage countries to abide by their terms. The Basel Convention on hazardous waste export, for instance, restricts signatories from trading in wastes with countries that have not joined the accord, and the Montreal Protocol restricts trade in CFCs and products containing them with countries that are not members. CITES is empowered to recommend that members suspend wildlife trade with countries identified as out of compliance with its terms. It did just that regarding China, Italy, Taiwan, and Thailand in the early 1990s, and Greece in 1998. When the sanctions have been imposed, they have generally prompted stronger government enforcement.[81]

International fisheries agreements are also increasingly using trade restrictions as a means of bringing offending countries into line. The International Commission for the

Conservation of Atlantic Tunas, for example, recently adopted a measure permitting members to ban tuna and swordfish imports from countries that are undermining its rules, whether or not they are Commission members.[82]

Sometimes countries impose environmental trade restrictions unilaterally but on behalf of stronger multilateral accords. Under a U.S. law known as the Pelly Amendment, the government is authorized to level sanctions against countries whose nationals are known to be diminishing the effectiveness of international wildlife or fishery agreements. Though the sanctions have only rarely been invoked, the threat that they might be has helped strengthen a number of accords. In 1999, for instance, the United States threatened to impose a trade embargo on some $1 billion in Italian fish imports and other products in protest over the continued use of driftnets by Italian vessels. In response, Italy agreed to step up enforcement, seize illegal driftnets, and impose stronger penalties for illegal driftnet fishing by Italian vessels in the Mediterranean.[83]

Although trade levers can be a powerful tool to ensure compliance, their use is controversial. One danger is that sanctions will be used as a tool of the economically powerful to impose their will on weaker nations. In some cases, they may also be counterproductive. When the CITES secretariat considered halting aid for tiger protection to India in early 2000 for the government's failure to curb the tiger trade adequately, Indian officials pointed to the folly of further crippling a country that already lacks the equipment, personnel, or funding to address the problem properly.[84]

An added problem is the possible collision between the trade provisions in multinational environmental agreements and the rules of the World Trade Organization (WTO). Although no country has lodged a formal WTO protest against an environmental treaty, the potential for such challenges clearly exists. Even without one, worries about inconsistencies with WTO rules undoubtedly enter into the calculus of treaty negotiators, dissuading them from using tools that could give teeth to environmental accords.[85]

One solution to this problem would be to amend the environmental exceptions to the WTO to protect trade measures taken pursuant to environmental treaties from being challenged at the trade body. A precedent for this approach is provided by the North American Free Trade Agreement, which stipulates that when its provisions conflict with three earlier environmental treaties (the Basel Convention, CITES, and the Montreal Protocol), these agreements shall prevail in any formal dispute. But some recent environmental treaties take just the opposite approach, stipulating that nothing in the accord should be construed as superseding pre-existing international obligations, including WTO rules.[86]

A major difference between the WTO and most international environmental agreements is that WTO rules are enforced through a binding dispute resolution procedure, which means that if a national law or policy is found to violate WTO rules, a dispute resolution panel can require the offending country to either amend its legislation or pay a financial penalty equal to the value of the foregone trade. In short, the WTO has teeth. Environmental treaties, in contrast, generally provide for voluntary and nonbinding dispute resolution procedures.[87]

One exception is the Law of the Sea convention, which includes compulsory and binding dispute resolution procedures to enforce most of its rules. Parties are given a choice of several venues for resolving disputes, one of which is a newly created U.N. International Tribunal for the Law of the Sea. The jury is still out on the tribunal's effectiveness, as it has so far ruled only a few times. In one notable decision in February 2000, the Tribunal fined the *Camouco*—a Spanish-owned ship flying a Panamanian flag—for illegally catching the threatened Patagonian toothfish in French sub-Antarctic waters. Although the fine was minimal, the ruling was expected to send a powerful message to illegal fishers that they may face penalties.[88]

The tribunal has created an important precedent for more enforceable international environmental commitments. Italian supreme court justice Amadeo Postiglione has proposed building on this model to create an International Court of the Environment roughly analogous to the recently created International Criminal Court. The new court would be empowered to enforce environmental treaties and other forms of international environmental law. Unlike at today's World Court, NGOs and private citizens would have standing to bring charges against transgressors, and member countries would have to accept the Court's jurisdiction. For the time being, though, few governments are willing to subject themselves to such a body. More support exists for proposals to transform UNEP into a World Environment Organization that could serve as an umbrella for the current scattered collection of treaty bodies and promote more coordinated implementation and enforcement.[89]

Another priority is to improve public participation in governance and access to justice at the national level, as this is where international environmental treaties are translated into on-the-ground action. In the Rio Declaration on Environment and Development that emerged from the June 1992 Earth Summit, governments agreed that for nations to respond effectively to environmental problems, their citizens must have access to environmental information, the chance to participate in decision-making processes, and access to judicial and administrative proceedings. These principles were enshrined in legally binding form in the June 1998 Aarhus Convention, negotiated under the auspices of the U.N. Economic Commission for Europe. Work is also now under way in Latin America and Africa to develop regional agreements on procedural rights, although it is not yet clear how closely these efforts will follow the Aarhus model.[90]

NGOs are actively working to control international environmental crime within countries and at the global level.

Despite the constraints NGOs face, they are nonetheless actively working to control international environmental crime both within individual countries and at the global level. In Brazil, for example, the new nongovernmental National Network Against Wild Animal Traffic, known as RENCTAS, is cooperating with the police and the federal environment ministry to train officers in wildlife inspection, and is investigating anonymous tips about wildlife smuggling that are left on its Web site. Internationally, NGO networks such as the Basel Action Network and the Climate Action Network monitor treaty negotiations, pushing for tougher commitments and stricter enforcement provisions.[91]

Targeted industries have bowed to the combined might of NGO and consumer pressure in several recent cases. In 1999, pressure from the World Wide Fund for Nature and increased public awareness of the threats that traditional medicine usage poses to wildlife caused leading practitioners and retailers in China to pledge not to prescribe or promote medicines containing parts from tigers, rhinos, bears, and other endangered species. Later that same year, Mitsubishi—a major supplier of tuna to the Japanese market—responded to aggressive lobbying by Greenpeace and other groups by promising to stop trading in fish caught by flag-of-convenience vessels.[92]

The global scope of today's economy coupled with the crossborder reach of Earth's natural systems portend a growing role for environmental treaties in the years ahead. But these accords will only be able to reverse deteriorating ecological trends if they become both more specific and more enforceable. A broad-ranging effort involving governments, international institutions, businesses, NGOs, and ordinary citizens is needed to rein in international environmental crime and restore the ecological integrity of the planet.

Chapter 10

Accelerating the Shift to Sustainability

Gary Gardner

In 1601, an English sea captain named James Lancaster conducted an important experiment. Commanding four ships on a voyage from England to India, he served lemon juice every day to the crew on one of the ships. Most remained healthy. But on the other three ships, 110 of the 278 sailors died of scurvy by the journey's midpoint. The experiment was of immense import to seventeenth-century seafarers, since scurvy claimed more lives than any other single cause, including warfare and accidents.[1]

Surprisingly, however, this vital information had little impact on the British Navy. The Navy did not conduct its own experiments until 1747, nearly 150 years later, and did not stock citrus fruit on its ships until 1795. And the British merchant marine followed suit only in 1865, some two-and-a-half centuries after the first experiment with lemon juice was carried out. Despite the magnitude of the scurvy problem, and despite the availability of a simple solution, people were slow to respond.[2]

Fast forward to New Year's Eve, 1999. At the stroke of midnight, the entire world seemed frozen in nervous anticipation. Calamitous economic disruptions were widely feared from a programming flaw thought to be lurking in computers worldwide, the binary time bomb known as Y2K. The bug could jam computer systems and potentially stall the economies that depended on them. Governments and businesses had worked for years to find and root out the bug around the globe, yet the degree of success could not be known. But as the clock struck midnight in one country after another, virtually nothing happened. The bomb, it turned out, had been successfully defused, the fruit of worldwide recognition of a problem and effective planning to avert it.

The two cases, extreme examples of the human capacity to respond to a critical challenge, raise important questions for a world facing momentous environmental and social stresses and in need of rapid change. Unfortunately, our response to the global environmental crisis looks more like the

British Navy's of the seventeenth century than the global community's of the 1990s.

Despite abundant information about our environmental impact, human activities continue to scalp whole forests, drain rivers dry, prune the Tree of Evolution, raise the level of the seven seas, and reshape climate patterns. And the toll on people and the natural environment is growing as stressed environmental and social systems feed on each other. The death toll in Hurricane Mitch in Central America, for example, in which some 10,000 people perished, was elevated by the interplay of climate change, population pressures that led to deforestation, and soil erosion. (See Chapter 7.) As environmental systems continue to weaken, and as human demands on them increase, the case for a shift to sustainable economies becomes increasingly urgent.[3]

Indeed, this threshold moment is virtually unprecedented in world history. Only the Agricultural Revolution that started 10,000 years ago and the Industrial Revolution of the past two centuries—which brought unparalleled prosperity as well as environmental pathologies to a large share of humankind—rival the current era as moments of wholesale change in human societies. But those global transformations unfolded much more slowly, and began in different regions at different times. The changes under way today are compressed to just a few decades and are global in scope. The question facing this generation is whether the human community will take charge of its own cultural evolution and implement a rational shift to sustainable economies, or will instead stand by watching nature impose change as environmental systems break down.

Orchestrating change is not easy, but neither is it impossible. After millennia in which human servitude was commonplace, for example, freedom has been increasingly secured in the last 150 years. The abolition of slavery in the United States, the nonviolent movement for India's independence, the end of apartheid in South Africa, the peaceful transition away from communist rule in Eastern Europe and the former Soviet Union—in each case leaders, organizations, and citizens demonstrated the flexibility and courage needed to respond to the moral imperatives of their day and to work for change. This generation will need to summon the same courage and commitment—and then some, given the daunting challenge facing the human family today.

Civil society, business, and government each have an important and distinctive role to play in what is arguably the most exciting moment in human history. But each sector needs to become sophisticated in the language and mechanics of change. Each needs to learn to think strategically—a great challenge for a species that is biased toward the immediate and the local. And each will need to develop strategies that promote change quickly without increasing human suffering. The more we understand how change can be leveraged rapidly and responsibly, the more likely it is that we can safely navigate the choppy waters that lie between us and sustainability.[4]

Anatomy of Change

Human change is distinguished from natural change—biological evolution, for example, or the birth and death of stars—by its willfulness and purposefulness. As the only creatures known to plan change, we boldly dip into our own history and alter the course of our own development. The "cultural evolution" driven by human interventions is now a more powerful determinant of our fate than biological evolution is.[5]

Unlike biological evolution, which passes change from one generation to the next, cultural evolution also facilitates change within a generation and to people we are not related to. As a consequence, cultural change now outpaces biological evolution, and is rapidly accelerating.[6]

The Agricultural "Revolution," for example, unfolded over more than 5,000 years, with each generation inching its way closer to full dependence on farming. Most early agriculturalists would have been astounded to learn that they were part of an economic revolution. But since the Industrial Revolution, and especially in the past 100 years, the pace of change has quickened dramatically. In the United States, it took 46 years for a quarter of the population to adopt electricity early in the twentieth century. But as the century unfolded, 35 years were needed for a similar proportion to adopt the telephone, 26 years were needed for television, 16 years for the computer, 13 years for the mobile phone, and only 7 years for the Internet. At this dizzying pace, societies have little opportunity to comprehend the full consequences of their own activities.[7]

Cultural change is also increasingly complex, the result of a loosening of the centuries-old constraints on economic expansion—limited energy, materials, and knowledge. In the late 1600s, for example, nearly 60 percent of all energy used in Europe came from horses and oxen and 25 percent from wood—a far more limited source than the coal, oil, and gas that fueled twentieth-century machinery. And while earlier societies were built using wood, ceramics, and metals, today's economies use composites and synthetics that draw from all 92 naturally occurring elements in the periodic table. Perhaps most impressive, the stock of specialized human knowledge has expanded dramatically: some 20,000 scientific journals are published today, compared with only 10 in the mid-1700s. As the boundaries on energy, materials, and knowledge have expanded, the combination of development options has increased in number and complexity.[8]

Human activities continue to scalp whole forests, drain rivers dry, and raise the level of the seven seas.

Organizing the complex use of energy, materials, and knowledge became possible in the twentieth century with a vastly expanded human capacity for administration. Management became a science, as Max Weber and Frederick Taylor rationalized organizational structure, as Henry Ford popularized mass production techniques, and as universities began to grant MBAs and degrees in public administration. The many different forms of administration—government, business, and nonprofits, for example—have different primary drivers. Businesses are strongly motivated by pursuit of profit, governments by the need to serve constituents, and grassroots nongovernmental organizations (NGOs) by the energy and ideals of their constituents. The drivers behind each organizational form produce their own particular forms of cultural change.

Consciousness allows humans to study and to some degree influence the diffusion of change. Like natural changes whose growth generates further growth—witness the spread of algae across a pond—cultural change often spreads in a predictable pattern. Communication as simple as word of mouth or as sophisticated as advertising can lead to changes that build on themselves, spreading slowly at first, then gradually picking up speed as people become familiar with them, finally slowing again as the population becomes saturated with change. Computer companies in the United States, for instance, are beginning to see slowing of sales to first-time buyers, as the market for computers becomes increasingly saturated. Another example of this comes from a 1998 European Wind Energy Association scenario for meeting 10 percent of global electricity demand in 2020 with wind power. (See Figure 10–1.) This S-curve pattern of reinforcing change holds for the spread of cultural changes as dissimilar as the use of slang and the adoption of fashion.[9]

For advocates of change, the importance of this pattern is that particular personalities and demographic profiles are associated with various stages of adoption. At the leading edge of change, for example, are what Everett Rogers of the University of New Mexico calls innovators, who tend to be well educated and socially connected. Change then spreads to early adopters, the early majority, and the late majority, each accounting for a particular share of the population. The last to adopt are the laggards, traditionalists who are often socially isolated. If opinion leaders—the innovators and early adopters—can be identified for a particular issue, change efforts can be focused on this group in the expectation that they will help the innovation or idea to spread.[10]

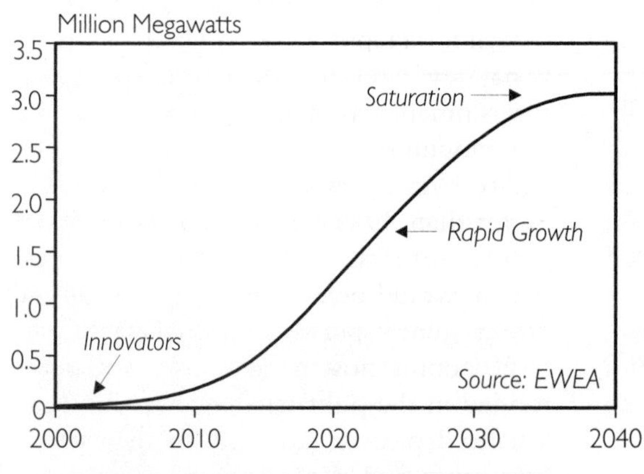

Figure 10–1. Hypothetical Global Wind Generation, 2000–40

The intelligence that allows humanity to intervene in its own development, however, is not advanced enough to foresee all the disruptive secondary effects of human activities. This unassimilated feedback is thus another common characteristic of cultural change. When chlorofluorocarbons (CFCs) were introduced in the 1930s, for instance, they were viewed as a godsend to industry. Nonflammable, nontoxic, and noncorrosive, they were widely used for decades in refrigerants and aerosol propellants. Scientists 70 years ago had no idea that their miracle contribution to science and industry would soon threaten environmental and human health worldwide: CFCs are now much better known for eating away the planet's protective ozone layer. This sort of feedback is the reason ecologists remind us that "you can never do just one thing." It teaches advocates of change to look at whole systems, not just single causes, in assessing problems and prescribing solutions.[11]

Cultural change surrounding environmental issues often has its own characteristic features. Environmental quality is a public good, which is often governed by a "tragedy of the commons" dynamic: individuals have a natural incentive to overuse natural resources, and little incentive to conserve them. This dynamic is one form of an age-old human dilemma—how to get self-interested people to act in the common good. Over time, societies have developed four strategies to persuade people to change by redefining their self-interest using a larger and longer-term perspective.[12]

The most common approach is for societal authorities to use the force of law, or officially sponsored incentives, to determine how resources are to be used and by whom. Ancient Mesopotamia's famous Code of Hammurabi, for example, had several laws governing irrigation, a critical common resource for that society. It required landholders to keep their irrigation ditches in good repair and mandated that negligent farmers compensate those whose production suffered because of the neglect. Today, of course, laws and regulations, along with taxes and subsidies, are important tools used to protect water, forests, minerals, the atmosphere, and other common resources.[13]

Education is also used to convince people to act in the common interest. As an important institution of socialization, schools influence the course of any society. But some critics see formal education as strongly biased toward status quo societal values. Mary Clark, author of *Ariadne's Thread*, a book about historical change, cites only two occasions in western history in which education has brought a sweeping change in people's worldview: during the Golden Age of Greece and during the European Enlightenment. In both periods, philosophers who deliberately critiqued the societal worldview were esteemed by society. But since the Enlightenment, education has rarely challenged the modern worldview. That perspective, which places nature at the service of humankind and splinters knowledge into disconnected disciplines, has contributed greatly to today's global environmental crises. Whether formal education today is capable of standing outside of society and critiquing it in a way that creates a worldview grounded in sustainability is unclear.[14]

Appealing to people's values, especially through religious beliefs, is another tool for persuading people to act in the common good. Political leaders have long had difficulty governing without the support of religious authorities; witness the Islamic revolution that brought down the Shah of Iran in 1979 or the Catholic-supported Solidarity movement that unhinged the Polish government in the 1980s. Religion's role in the formation of values is not as great today as in the past in many societies, however, so appeals to values may need to take multiple forms, only some of which are faith-based.

Finally, many cultures have used small group management to protect common resources. Examples are innumerable, and are typified by the village of Torbel, in the Swiss Alps, which has managed its forest and grazing lands on a communal basis since the thirteenth century. Each village family can send only as many cows to graze the Alpine meadows each summer as they can feed in the winter. And trees are cut only once a year. The village forester determines which trees will fall, villagers cut and stack the logs, and each family receives a share of the harvest. Often overlooked, this kind of small group administration remains a viable option for getting people to act in the common good.[15]

Understanding the anatomy of change is critical for a rapid shift of economies to sustainability. Even today's most encouraging environmental trends are barely unfolding fast enough for a rapid shift toward sustainability. Wind power, for example, grew at better than 30 percent annually in the late 1990s. But this phenomenal growth, if sustained, would boost wind's share of global power generation to only 10 percent by 2020. For a host of other issues not yet headed in the right direction—species loss, aquifer depletion, deforestation, and materials reuse and recycling, among others—the need for rapid change is still more urgent. Only by understanding the drivers of change can civil society, business, and government hope to play a part in producing rapid change.[16]

Empowering the Base: The Role of Civil Society

The diffuse set of actors known as civil society—individuals, civic associations, and churches, among many others—has little formal power compared with government or business. But civil society is crucial in a campaign for sustainability. As consumers and voters, citizens carry immense influence. And individuals who can flex their collective muscle may need to lead the shift to sustainability when government and business initiatives are stymied by special interests. Moreover, citizen support is usually necessary to sustain whatever initiatives are undertaken by government or business. For all these reasons, individuals and the NGOs they belong to are important and powerful allies in the change process.

Civil society appears to be fertile ground for sparking environmental change. A 1998 survey for the International Environmental Monitor found that majorities polled in 28 of the world's 30 most populous countries believe that environmental laws need to be strengthened. In 26 of the countries, majorities or near majorities said that strong action is needed immediately to reduce human impacts on climate. And in most countries, concern over the environment is on the rise, sometimes dramatically so. In India in 1992, only 6 percent of respondents named the environment as the most important problem facing their country. By 1998, 27 percent did.[17]

More broadly, a socio-psychological study of the United States published in 2000 reports that a sizable chunk of American adults—some 26 percent—have adopted a new worldview in the past 40 years that is largely consistent with the values of sustainability. Dubbed "cultural creatives" because of their potential to create a new culture, these individuals are characterized by a concern for the environment, desire for meaningful personal relationships, commitment to spirituality and psychological development, disaffection with the large institutions of life, and rejection of materialism and status. Members of this group are likely to be active in their communities, to choose work consistent with their values, and to value healthy living. Although this demographic segment is not organized—the authors report that most cultural creatives are unaware that a sizable bloc of kindred spirits is out there—it represents a potential base from which to launch a change movement.[18]

These encouraging trends stand in contrast to actual behavior, however. Americans, for example, are among those who consistently express support for an improved environment, yet a growing share drive gas-guzzling sport utility vehicles and live in ever-larger homes. Environmental psychology, a relatively new subdiscipline of psychology, helps to explain why people interested in creating a sustainable society may be slow to change their behavior.

A key and consistent finding of environmental psychologists is that people do not recycle more, drive less, conserve energy, or make other environmental behavior changes simply as a result of hearing about the adverse impact of their current behavior. As an example, a 1981 study of an energy conservation workshop found that participants demonstrated greater awareness of energy issues and conservation measures following the meeting, and said they were willing to adopt what they had learned. But few actually did. Follow-up home visits showed that only one of the 40 participants had lowered the hot water thermostat, and none had added an insulating blanket to the hot water heater. Eight had installed low-flow showerheads—a rather modest rate of adoption, since each workshop participant had been given a free showerhead. The findings are consistent with those from studies on a range of different environmental behaviors, from recycling to water conservation to anti-littering campaigns. Raising awareness, it turns out, is important but often not sufficient in persuading people to change behaviors.[19]

Nor do people necessarily respond to appeals based on fear. Psychologists note that fear prompts behavior change only if the consequences of a behavior are truly frightening, if those consequences are likely to be faced, and if alternative behaviors are available. In some cases these conditions hold: the rapid abandonment of British beef following the outbreak of mad cow disease in 1996, and the subsequent rush to organic foods in the United Kingdom, appears to be a case in point. But even when fear changes behavior, it does so only as long as fear remains active. And the use of fear can backfire in a change campaign: in the absence of credible alternatives, a fear appeal simply leads to despair or denial, attitudes that may prevent people from discussing important topics. For this reason, environmental messages that stress "doom and gloom" alone are less likely to be effective than messages that also offer solutions.[20]

What people do respond to, say environmental psychologists, are the actions of their peers, direct appeals, effective communication, and enticing incentives. And when these influences help people sidestep the psychological obstacles that block them from sustainable behavior, actions can quickly become consistent with beliefs.[21]

Peers are especially powerful influences. Clear evidence of this comes from a study at the athletic center of the University of California at Santa Cruz, where a sign was posted in the shower encouraging users to turn off the water while soaping up. The sign was heeded by only 6 percent of users. But when a researcher introduced to the situation followed this practice, 49 percent of other students turned off the water while soaping up. And with two role models present, the participation rate shot up to 67 percent. The encouraging lesson is that individuals who model sustainable behavior might have more influence over others than they realize.[22]

Some 26 percent of American adults have adopted a new worldview in the past 40 years that is largely consistent with the values of sustainability.

People also have a strong internal need to behave consistently, in order to be seen as reliable and trustworthy. We feel internal pressure to follow through on our commitments, especially if they are made publicly or as part of a group. So strong is our need to be true to our word that small commitments have been shown to make people more receptive to larger commitments. People who were asked to wear a lapel pin publicizing the Canadian Cancer Society, for example, were nearly twice as likely to donate in a subsequent appeal as people who were not asked to wear the pin. Taking the first step in living our principles of sustainability not only models the behavior for others, it gets us closer to even greater commitments.[23]

Through modeling and through commitments, then, individuals can begin the process of building a sustainable world, no matter what the rest of the world is doing. Indeed, the psychological power of these behaviors infuses the inspirational exhortation of Mahatma Gandhi, "be the change you wish to see in the world," with a social dimension that may not be widely appreciated.[24]

The International Campaign to Ban Landmines mobilized more than 1,000 organizations in 60 countries to work for the landmine treaty.

But psychological insights are also useful to NGOs working to help people overcome barriers to sustainable behavior. The field of community-based social marketing—the application of marketing principles to social goods—is based on these insights. The approach starts with community-level research, typically surveys and focus groups, to determine what obstacles stand between a person and a particular behavior change. These obstacles can be psychological or external to the individual; infrequent bus service, for example, may prevent someone from using mass transit. Then social marketing uses tools like modeling and commitment to address common concerns.[25]

Another behavioral tool important to social marketing is effective communication. Information that is vividly presented, crafted for the intended audience, delivered credibly, specific, easy to remember, and designed to spread socially is more likely to affect behavior than other information. As a social species with a strongly developed visual system, people-centered communications and pictures can be especially effective. The photo of open waters at the North Pole published in newspapers in the summer of 2000, for example, may bring home the reality of a melting ice cap and of climate change more effectively than dozens of scientific reports ever could.[26]

Finally, social marketers use incentives, financial or otherwise, that help overcome the behavioral challenge posed by public goods. When people bike instead of driving or buy organic rather than conventional vegetables, the sacrifice they bear yields benefits largely for society and only marginally for themselves. Well-crafted incentives and disincentives help change the calculus by providing an additional benefit for the individual. When a San Jose, California, garbage collection program began charging people by the amount collected rather than a flat fee, while collecting recyclables at no charge, waste flows to landfills fell by 46 percent, and recyclable collections jumped by 156 percent.[27]

NGOs, of course, direct their advocacy efforts not just to citizens, but to governments and businesses as well. This nonprofit sector is large, growing rapidly, and becoming an effective voice for sustainability. A 1998 Johns Hopkins University study found that the nonprofit sector is a $1.1-trillion industry—larger than the gross domestic product of all but seven countries in the world—that employs 19 million people in 22 countries. Much of the sector's strength is new: half of all NGOs in Europe were founded in just the last decade, and 70 percent of those in the United States are less than 30 years old. Meanwhile, the ranks of international NGOs have expanded steadily throughout the century, from 176 in 1909 to more than 23,000 in 1998.[28]

The democratization of previously closed societies in Eastern Europe and the former Soviet Union and the move away from dictatorships in Latin America, Asia, and Africa account for some of the growing strength of civil society. But the sector is prospering in many other regions as well, thanks to a growing middle class, an increase in corporate power (which has sparked countervailing activism from citizens), and the advent of new communications technologies, especially the Internet and e-mail.[29]

NGO strength was demonstrated in a number of high-profile international cases in the 1990s, including the defeat of the Multilateral Agreement on Investment (MAI) and the signing of a treaty to ban landmines. The MAI was an international investment protocol that would have established rules for foreign investment that were largely favorable to investors. Negotiated in secret by a small group within the Organisation for Economic Co-operation and Development (OECD) over the course of 16 months, a draft agreement was leaked in January 1997 to Public Citizen, a watchdog group in Washington, which immediately published it on the Internet. Within less than two years, the agreement was dead, largely because of an Internet-centered campaign that mobilized more than 600 organizations in 70 countries, including environmental, labor, and human rights groups.[30]

NGOs were perhaps even more decisive in the landmine treaty. A network of groups known collectively as the International Campaign to Ban Landmines (ICBL) had developed by 1993 to promote a treaty that would outlaw these weapons. With the help of e-mail and the Internet, and without a staff or central office, ICBL mobilized more than 1,000 organizations in 60 countries to work for the ban. By March 1999, a treaty banning the use, stockpiling, production, and transfer of landmines had been signed by 131 nations.[31]

Beyond the Internet, other elements played important roles in these NGO successes. The groups were unusually successful at working with governments in both cases. By persuading landmine nonproducers and nonusers—such as Austria, Canada, Denmark, and the Netherlands—that a ban would be consistent with their foreign policies, the coalition was able to get key states on board. Treaty proponents also eventually persuaded France and the United Kingdom—both producers and users, and both members of the U.N. Security Council—to join as well. In the MAI case, NGOs exploited divisions among the states that had negotiated the draft agreement—some of whom had serious reservations about it. By mobilizing public opinion, NGOs were able to raise the stakes with governments. One observer noted that protests generated by NGOs "elevated the question of the MAI from the 'level of civil servants' to the 'ministerial level'."[32]

In these cases, NGOs were also particularly adept at framing issues in ways that could win public support. The landmine ban was promoted as a humanitarian issue rather than a military one, making it more comprehensible to the general public. Noting that 80 percent of victims are civilians, and enlisting landmine victims as part of the campaign, organizers were able to humanize and personalize the issue. Similarly, NGOs characterized the MAI not as a financial instrument, as the OECD had, but as an agreement with adverse impacts on labor, human rights, and the environment—issues the public cared about and understood.[33]

Greening Commerce: The Role of Business

Most societies organize their economies around private business, giving this sector a potentially powerful role in shifting economies toward sustainability. Many firms see their profit-making responsibilities as a serious constraint on their capacity to make environmental change, but this is not necessarily the case. Combined with visionary leadership, the profit motive can be channeled to launch businesses toward sustainability. Indeed, motivated firms have used several common business strategies to lighten their impact on the natural world—even as they boosted profits.

One of these strategies is product differentiation based on a product's lighter environmental impact. Consumers are increasingly interested in green products, and in some cases are willing to pay more for them. For example, consumers often pay premiums of 50–100 percent for produce grown without chemical fertilizers and pesticides. As a result, the organic industry has blossomed globally in the 1990s, growing by more than 20 percent annually in the United States, with comparable growth rates in other industrial countries. The approach does have its limits, however. StarKist, in an early effort to market more costly dolphin-safe tuna as an alternative to conventional tuna, found that consumers were unwilling to pay a premium. People may be most willing to pay more when the value added includes a private benefit—such as better health and taste, which organic food is perceived to convey—in addition to the public good of improved environmental quality.[34]

Waste reduction is another common greening business strategy that is consistent with increased profits. Companies increasingly view pollution and waste as proof of inefficiencies in the production process rather than as inevitable byproducts of production. The 3M Company recognized this as far back as 1975 when it established its Pollution Prevention Pays program. The initiative gave employees incentives to look for pollution prevention opportunities early in the production process, rather than continuing the traditional but expensive approach of supplying filters or scrubbers on the end of smokestacks and waste discharge pipes. The 3,000 projects initiated by the company between 1975 and 1992 prevented more than a billion pounds of polluting emissions and saved the company more than half a billion dollars. More recently, Interface, the world's leading manufacturer of commercial carpet, has eliminated more than $90 million in waste since 1994.[35]

More and more businesses are taking pollution prevention strategies to their maximum by becoming "zero-waste" companies. By recycling the byproducts of production, companies avoid sending waste to landfills—and often generate new revenues in the process. Asahi Breweries of Japan, for example, sends the dregs of its beer production to cattle raisers as feed, its plastic packing bands to be recycled into carpet, its bottle tops for use in construction fill, and its cardboard for reprocessing into paper. Indeed, zero-waste production is growing quickly in Japan, largely because landfill space is at a premium—the cost of landfilling more than doubled between 1991 and 1997—and because waste is a hot political issue. Sensing cost savings and marketing advantages in achieving zero-waste production, companies as diverse as Asahi Breweries, Sanyo, Canon, and Toyota report having achieved zero-waste production. The newspaper *Nikkei Shimbun* reported in December 1999 that 27 companies—3.1 percent of those surveyed—claimed to be zero-waste.[36]

Companies can also lobby governments to influence the regulatory climate, or work with competitors to set environmental standards for the industry. These strategies, of course, have often been used to avoid or lessen responsibility for a firm's environmental impact. But they can work the other way as well. By persuading governments to constrain competitors' unsustainable behavior, a firm can position itself to be a green leader and possibly gain market advantage. DuPont did this when it abandoned the manufacture of chlorofluorocarbons, which it was the world leader in producing. With growing scientific evidence that CFCs degrade stratospheric ozone, the company developed substitutes—but at a cost several times greater than CFCs. Undaunted, the company came out in favor of an international ban on CFCs, worked with governments to enact the ban, and—with patents in hand—soon dominated the market for the substitute chemicals.[37]

A different version of this strategy occurs when firms band together to "green" practices industry-wide, either to forestall expensive government mandates, improve public image, or both. One example is the product take-back program of battery makers in the United States and Canada. Prompted by inconsistent recycling laws, some 300 manufacturers and marketers of rechargeable batteries now sponsor battery recycling through the non-profit Rechargeable Battery Recycling Corporation (RBRC). This group licenses its Battery Recycling Seal for a fee to manufacturers of batteries and battery-powered products. Consumers can take spent batteries with the RBRC Seal to more than 29,000 U.S. and Canadian retail outlets that serve as collection sites. The RBRC then collects the batteries and pays a recycling company to process them. Under the initiative, the recycling rate of batteries increased from 2 percent in 1993 to 25 percent in 1998. If the industry's recycling goal of 80 percent is achieved by 2005, and if that momentum is sustained, nearly all batteries in the United States could be recycled by 2010—a rapid shift toward sustainability. (See Figure 10–2.)[38]

Finally, companies can rethink the reason they exist in a way that lightens their impact. Businesses that shift from supplying goods to providing services—furnishing mobility services rather than cars, for example—are a case in point. IBM, once the leading producer of mainframe computers, now sees itself as a provider of business services. Xerox, known for decades for its copiers, now promotes itself as "The Document Company," a reflection of its new emphasis on information management. Herman Miller, a large manufacturer of office furniture with sales in 60 countries, has established a subsidiary to provide office furnishing services, and expects it to become a strong engine of the company's growth. To the extent that providing a service eliminates the manufacture of a product, it lowers the material and energy demands of a company, with corresponding savings in environmental wear and tear. Motivations for the shifts range from anticipation of product "take-back" legislation (as in Europe, where producers are increasingly responsible for products over their

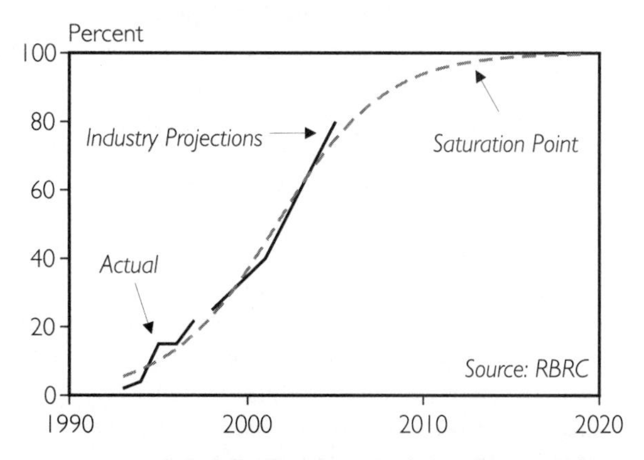

Figure 10–2. Battery Recycling in the United States, 1993–97, with Projections to 2005 and Hypothetical Saturation Point

entire lifetime) to the opportunity to enhance market share through cost cutting or an enhanced corporate image.[39]

Several of these business strategies were used when British Petroleum (BP) recently reversed its thinking on climate change. In the mid-1990s, the company had no intention of saying anything about climate change—but then it witnessed Shell's public relations fiasco surrounding the Brent Spar, an offshore loading buoy that Shell planned to sink off the U.K. coast in 1995. Consumer boycotts of Shell and pressure from Greenpeace brought home to BP the intensity of public feeling around environmental issues. This, together with speculation that Shell might shift its position on climate change, led BP CEO John Browne to look more closely at climate change. He decided to set a new company policy that would set BP apart from the competition—the product differentiation strategy.[40]

Browne made a landmark speech at Stanford University in 1997 acknowledging the risks of climate change. His remarks, and the subsequent activities of BP, demonstrate how the company maneuvered to advance its interests while reducing its environmental impact. Browne was careful to preserve the company's flexibility. Citing the "effective consensus" of leading scientists about the role of human activities in climate change, he nevertheless noted that "there remain large elements of uncertainty" surrounding the issue. And Browne set no deadline for abandoning the oil business.[41]

BP also advanced its own interests when it set out to reduce company-generated carbon emissions by 10 percent below 1990 levels by 2010—a waste reduction strategy whose goals were roughly in line with those agreed to by industrial nations at the Kyoto conference on climate change. (The move is actually quite ambitious, amounting to a 50-percent reduction in projected emissions by 2010.) The company set up an internal emissions trading scheme, which established carbon emission allowances for 12 BP business units. The allowances tighten with each passing year and allow efficient BP units to sell unneeded allowances to other units that are hard pressed to meet the emissions goals. The idea was to give the company experience with emissions trading, a tool the company hoped would become a central part of international carbon reduction agreements. "We have a chance to influence the regulatory apparatus," according to a company consultant. "We influence the emissions trading debate because we have real practical experience to bring to the table."[42]

Finally, the company remade itself—at least rhetorically. It promotes itself as a "green energy company" rather than an oil company, pointing to its increased investments in gas and its ownership of the world's largest solar company, Solarex. But BP's bread and butter is still oil—recently it spent 588 times as much to buy ARCO, an oil and gas company, as it did on Solarex—and it says its transition to renewable energy will take time—decades, perhaps.[43]

How far these business strategies can take a company toward sustainability depends in part on the commitment of corporate leaders. If the strategies are viewed strictly in business terms, the impact is likely to be less than if they are driven by visionary commitments to sustainability. When such a vision is present—as at Interface, where CEO Ray Anderson is working to make the company not just a better environmental citizen, but a "restorative enterprise" that gives back to the environment as much as it extracts—companies can substantially reduce their ecological footprint.[44]

Still, the capacity for business to green

itself would be greatly enhanced through government initiatives that align profit-making opportunities with environmental stewardship. Businesses are unlikely to undertake greening initiatives voluntarily if they are very costly. And voluntary initiatives often have a free-rider problem—the RBRC program, for example, recycles all of the batteries it receives, even those from the 20 percent of companies that have not joined the program. Visionary businesses will need to work with government leaders to reshape the taxes, subsidies, and other structures that bias so much of commerce against sustainability, so that the best environmental practices become profitable and widely diffused.[45]

Building Coalitions: The Role of Government

Although governments have formidable regulatory and fiscal powers and easy access to the media, they usually look to citizens or business for cues about what issues to address, especially in democratic nations. Independent government initiatives do occur, of course, but usually when opposition is minimal, or if the government is prepared to spend precious political capital. Harnessing state power in the cause of change, then, often requires building coalitions that can support key officials in change efforts. Without this political base, governments are constrained in their capacity to generate change.

The importance of a viable political base for effective government action is clear in the quite different responses to HIV/AIDS in South Africa and Thailand. In 1990, less than 1 percent of pregnant women in each country—an indicator of the virus's spread to the general population—was infected with HIV. Today, the infected share of pregnant women in South Africa has grown to near 20 percent, while the figure in Thailand is just over 2 percent. (See Figure 10–3.) The distinct outcomes reflect differences in operating latitude for the two governments, which resulted from quite dissimilar sets of political alignments and cultural histories.[46]

In Thailand, the rapid spread of HIV came to light in 1988, and initial government reaction was to hide the problem, out of concern that exposure would dampen tourism. But because infection occurred primarily through the country's sex trade, and because HIV was widespread among sex workers—44 percent of prostitutes in the northern province of Chiang Mai were HIV-positive in 1989—the country was at risk of being swamped by the disease. As infection rates skyrocketed in the early 1990s, a new Prime Minister, Anand Panyarachun, stepped up the government response, launching extensive public education programs and a 100% Condom Campaign that sought to enforce condom use in commercial sex establishments all the time.

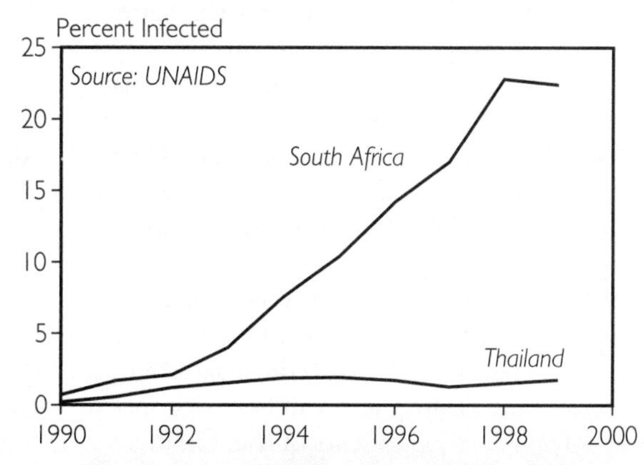

Figure 10–3. HIV Prevalence Among Pregnant Women, South Africa and Thailand, 1990–99

The government commitment was real: the budget for HIV/AIDS prevention and care rose from $180,000 in 1988 to $80 million in 1996.[47]

Behavioral changes were almost immediate: condom use in commercial sex establishments jumped from 15 percent of client visits in 1989 to more than 90 percent in 1994. Between 1990 and 1993, the share of men reporting premarital or extramarital sex fell from 28 percent to 15 percent, while visits to prostitutes fell by more than half. The war is far from won—there are now signs of backsliding—but major progress was made in slowing the spread of the disease. Indeed, the incidence of HIV-1 among Thais was so low in 1999 that researchers could not find a population at risk of infection there—outside of drug addicts—large enough to do efficacy trials of a new AIDS vaccine.[48]

Thailand's success is due in large measure to the intelligent political strategy used by the Prime Minister to fight the epidemic. He approached the crisis not merely as a health problem but as a development issue. Responsibility for formulating the National AIDS Action Plan, for example, was given to the National Economic and Social Development Planning Board, which integrated an AIDS strategy into the country's national development plan. This is a prudent approach for managing a disease with widespread economic, social, and health causes and effects. But it also increases the number of political players with a stake in fighting the disease. A range of government ministries was enlisted in the effort, and each received a share of the huge budget. And domestic and international NGOs were asked for help as well.[49]

Meanwhile, opponents of the initiative—the tourism industry and brothel owners—quieted down as tourism continued to flourish, and as the requirement of condom use at brothels was evenly enforced, leaving no brothels at an advantage.[50]

To be sure, Thailand has several advantages in battling the disease. The country is relatively well off socially, with a high level of literacy, extensive media, and an accessible health care system. Its decades-old network of venereal disease (VD) clinics offered a natural infrastructure for condom distribution. And in contrast to many countries where people are ashamed to admit they are infected, sex workers were accustomed to frequenting the VD clinics and talking openly about sexually transmitted disease. But by giving a stake in the process to a broad coalition of political players, these advantages were used to maximum effect.[51]

This coalition building success was not replicated in South Africa's struggle with HIV/AIDS. There, the legacy of apartheid continued to cast a long shadow over the country, despite the adoption of majority rule. White civil servants, whose jobs were protected under an agreement with the African National Congress, either knew little about HIV/AIDS or did not pursue it aggressively. Black South Africans, recalling the sterilization programs of the apartheid regime, viewed the promotion of condoms by white doctors at government clinics as a ploy to depopulate black communities. And African suspicions of western solutions, along with the possibility of generating an indigenous cure, diverted precious resources to unproductive uses, and led the current President, Thebo Mbeke, to voice doubts publicly about the cause of the disease. These problems have divided the country and allowed the disease to spread freely. HIV prevalence jumped from 12 percent in 1998 to 20 percent in 2000.[52]

Government leadership and coalition

building might be spurred in South Africa through an understanding of the costs of not acting. South Africa's health department already spends an amount equal to 10 percent of its 2002–03 fiscal year budget to treat AIDS-related illnesses—and this is before the expected wave of HIV-positive individuals who have gone on to develop AIDS. Preliminary HIV testing of the South African Defence Force indicates that 60–70 percent of the troops may be infected. Business is also deeply affected. A South African insurance company estimates that one in five South African workers will be infected by 2005, and that employee benefit costs will double by that date and triple by 2010 because of the disease. Such profound adverse impacts on government and industry could potentially create powerful coalitions to combat the disease.[53]

An industrial-country case of effective coalition building comes from Denmark, where the government has promoted wind energy since the adoption of the national energy plan of 1976. Originally written to reduce Denmark's dependence on foreign sources of energy, the plan promoted the development of homegrown renewable energy sources. Successive governments have pursued this goal aggressively, by funding research into renewables, establishing a certification program for wind energy, promoting subsidies for capital investments in wind, and requiring Danish power companies to pay a guaranteed price for purchases of wind energy from private suppliers. As a result of these policies, the wind business grew quickly in Denmark, and now supplies 13 percent of the country's electricity.[54]

Because most of its fossil fuel is imported, Danish reductions in oil and coal do not harm important domestic constituencies. At the same time, the government's pursuit of wind energy has generated a number of domestic supporters. Denmark's turbine manufacturers, for example, are now global suppliers of a cutting-edge technology, accounting for 60 percent of global wind turbine exports. Complementing this activity, the Danish International Development Agency promotes wind turbines in developing countries as a way to advance Denmark's foreign aid and foreign commercial goals. And the wind program can help the country to meet its international commitments to reduce carbon emissions. The country's Energy 21 goal is to reduce carbon emissions to 20 percent below 1988 levels by 2005, which will require Denmark to nearly double its wind generating capacity over 1999 levels. In sum, the wind energy program is rooted strongly in important commercial and bureaucratic constituencies.[55]

In Denmark, the government has promoted wind energy since the adoption of the national energy plan of 1976.

Many Danish citizens also have a stake in the success of wind, thanks to government-established wind cooperatives that allow citizens to invest in the new energy source. The scheme is popular: more than 100,000 Danish families now own wind turbines or shares in wind cooperatives. The cooperatives create a large grassroots constituency for wind, of course, but their structure also helps address the concerns of other Danes who dislike the turbines' noise, their potential harm to wildlife, and their obstruction of landscapes. By requiring members to live within 3 kilometers of the turbine site, the cooperatives ensure that any of the technology's downsides are largely borne by those who benefit from it—and that a potentially damaging opposition movement does not emerge.[56]

An important factor in keeping the pro-wind coalition is the government's steadfast commitment to wind energy development, which lends a stability to the sector that favors its continued expansion. Subsidies for wind and taxes on fossil fuels and carbon emissions are clear and consistently applied, so that developers and purchasers of wind energy know what the financial costs and opportunities are. This is important, since lack of predictability has been a leading cause of the failure of renewable energy programs in other countries.[57]

Accelerating Cultural Evolution

Once individuals, businesses, and governments become advocates for sustainability, a global transformation as great as the Industrial Revolution could be unleashed. Energy will come from the same renewable sources that sustain all of life, rather than from finite and polluting stocks. Production will mimic the natural cycle of birth, death, and rebirth, with waste no longer casting a huge and ugly shadow over cities and factories. And natural resources will be appreciated for their contributions to the planet's living systems, not simply for their value as economic assets. In sum, a sustainable world will be powered by the sun; constructed from materials that circulate repeatedly; made mobile by trains, buses, and bicycles; populated at sustainable levels; and centered around just, equitable, and tight-knit communities.

Such a world may be difficult to imagine, but rapid change is possible. Nobody foresaw, for example, the astonishing unraveling of the Soviet Union and Eastern Europe that began in 1989. That shocking set of events reminds us that change is often not gradual and linear, but sometimes arrives like the floodwaters from a collapsed dam. Biologist Stephen Jay Gould of Harvard has long argued that the natural world evolves in this way, with long periods of drift that are broken by sudden leaps forward. The pattern may describe cultural evolution as well. Only after a critical set of conditions were in place—especially economic stagnation and widespread disaffection with communist ideology—could something as simple as East German refugees trigger the events that would topple governments in Eastern Europe and the Soviet Union. The encouraging consequence of this pattern is that rapid change is indeed possible once critical thresholds have been crossed.[58]

Rapid cultural change is most likely when all sectors of society work together. It is true that sweeping cultural change has often emerged from the grassroots: the Gandhi-led movement for Indian independence, Poland's Solidarity movement, South Africa's struggle against apartheid, and the U.S. civil rights movement—all of these came out of civil society. But change is more likely, and can be accelerated, if all sectors are engaged. During World War II, for example, the United States converted to a war economy virtually overnight by tapping the energy of citizens, businesses, and government. People recycled metal, rubber, and other materials; automobile plants shifted production from cars to tanks and airplanes; households sowed millions of Victory Gardens; and women replaced men in factories. Making a similar effort to build a sustainable society will require that individuals, businesses, and governments become conscious agents of change, acting strategically and building on each other's strengths.

Society-wide change is facilitated if all sectors work from a common base of information. Here, the scientific community plays an important role. The ongoing scientific review by the Intergovernmental Panel on Climate Change (IPCC), for example, has been a credible source of information on this issue, both for the international climate negotiators it officially advises and for the general public. IPCC participants have helped counter the organized propaganda of groups like the Global Climate Coalition, the mouthpiece organization for the fossil fuel and automobile industries. The Panel's progressively stronger assertions that human activities are changing the climate—coupled with growing experience of 100-year storms, melting glaciers, and disappearing islands—make the issue increasingly real for the public. Similar high-profile scientific and educational efforts focused on species loss, population growth, falling water tables, disappearing forests, and widening social inequities would be a valuable boost to a sustainable society.[59]

The media are also crucial to this information effort, given their pivotal position among civil society, business, and government. And their capacity to manipulate images—still, moving, and computer-generated—is a powerful tool to educate a species strongly influenced by visual cues. This image-generating capacity is especially valuable for the many cases of environmental degradation that remain unseen by most people, such as dying coral reefs, depleted aquifers, and the hole in the stratospheric ozone layer.[60]

But in order to have a positive impact, the media will need to understand the issues and develop a long-range outlook. Political elites, because of their power to influence media agendas, are well positioned to foster this understanding. British journalist Ivor Gaber credits two 1988 speeches by Prime Minister Margaret Thatcher, for example, with putting environmental issues on the public agenda, citing a tripling of references to global warming in the British press in the 12 months following the speeches compared with the preceding year. But he also writes that the media are fair-weather environmentalists, covering the issues in times of prosperity but abandoning them during recessions. Developing a big-picture worldview will require sustained media attention to the life-support systems of human culture.[61]

Media inattention is due in part to the extremely short "news cycle"—typically measured in hours—that is poorly matched with the seemingly glacial pace of natural change. Mass extinction of species is arguably the most momentous act ever committed by human beings, but it is unlikely to make the evening news. News bureaus see the issue as little changed from yesterday, and therefore not newsworthy. Only if news organizations expand their time horizons to encompass long-term change will the media realize their potential to educate for sustainability.[62]

With a common base of information, societies can profitably focus change efforts on high-leverage areas. Regions and issues of great environmental impact, for example, such as "biodiversity hotspots" or the driving and eating habits of Americans, are likely high-payoff candidates for change activity. And concentrations of power—in consumer spending, for instance, or through a government's fiscal policy—are an equally important focus of effort. The combination of great environmental impact and powerful leverage is what makes the E–9 nations described in Chapter 1 such a logical strategic focus of change.

In the effort to steer our cultures toward sustainability, civil society, business, and government all have powerful points of leverage available to them. Each also faces particular constraints: fragmentation limits the influence of civil society, the pursuit of profit reduces options for business, and competing interests tie the hands of government. But collectively, the three sectors are capable of wholesale cultural change. By tapping our potential as conscious agents of cultural evolution, we can create a sustainable civilization, one fully worthy of being called human.

Notes

Chapter 1. Rich Planet, Poor Planet

1. Based on author's visit to Bahia, August 2000.

2. James Brooke, "Brazilian Rain Forest Yields Most Diversity for Species of Trees," *New York Times*, 30 March 1993.

3. "Latin America and the Caribbean: Brazil," The Nature Conservancy, <www.tnc.org/brazil/forest.htm>, viewed 12 October 2000.

4. World Bank, *World Development Report 2000/2001* (New York: Oxford University Press, 2000), 21–23.

5. Christopher Flavin, "Wind Power Booms," and idem, "Solar Power Market Jumps," both in Lester R. Brown, Michael Renner, and Brian Halweil, *Vital Signs 2000* (New York: W.W. Norton & Company, 2000), 56–59; Seth Dunn, "Carbon Emissions Fall Again," in ibid., 66–67; Clive Wilkinson, *Status of Coral Reefs of the World, 2000: Executive Summary*, 9th International Coral Reef Symposium, 23–24 October 2000, Bali, Indonesia; National Snow and Ice Data Center, "Mountain Glacier Fluctuations: Changes in Terminus Location and Mass Balance," <www-nsidc.colorado.edu/NASA/SOTC/glacier_balance.html>, viewed 2 February 2000; Alexander Wood, "An Emerging Consensus on Biodiversity Loss," in Alex Wood, Pamela Stedman-Edwards, and Johanna Meng, eds., *The Root Causes of Biodiversity Loss* (London: Earthscan, 2000), 2.

6. Eduardo Athayde, Atlantic Forest Open University, Salvador, Bahia, Brazil, discussion with author, 10 August 2000.

7. The E–9 concept was first introduced as the E–8 in *State of the World 1997*. This chapter adds South Africa to the group and substitutes the European Union (EU) for Germany, which substantially extends its breadth of economic and ecological coverage. The sector-specific tables that follow, however, use statistics for Germany (the EU's most populous member), due to the lack of comparable data for the EU as a whole.

8. "'Love Bug' Suspect Charged," *Associated Press*, 29 June 2000; Mark Landler, "A Filipino Linked to 'Love Bug' Talks About His License to Hack," *New York Times*, 21 October 2000.

9. Casualties from "Payatas Relocation Coordination Ordered," *Manila Bulletin*, <www.mb.com.ph/umain/2000%2D07/mn071805.asp>, viewed 10 September 2000; Typhoon Kai Tak from Roli Ng, "Garbage Slide Kills 46 in Manila's Promised Land," *Planet Ark*, <www.planetark.org/dailynewsstory.cfm?newsid=7412&newsdate=11-Jul-2000>, viewed 10 September 2000; number of residents from "Manila Urges Payatas Residents to Get Out of Dumpsite," *China Daily Information*, <www.chinadaily.net/cover/storydb/2000/07/15/

wnmanilla.715.html>, viewed 10 September 2000.

10. Growth of world economy from Angus Maddison, *Monitoring the World Economy 1820–1992* (Paris: Organisation for Economic Co-operation and Development (OECD), 1995), 227, and from Angus Maddison, *Chinese Economic Performance in the Long Run* (Paris: OECD, 1998), 159, using deflators and recent growth rates from International Monetary Fund (IMF), *World Economic Outlook* (Washington, DC: October 1999); growth estimate for 2000 from IMF, *World Economic Outlook* (advance copy) (Washington, DC: September 2000); International Telecommunications Union (ITU), *World Telecommunication Indicators '98*, Socioeconomic Time- series Access and Retrieval System database, downloaded 24 August 1999, and ITU, *World Telecommunication Development Report 1999* (Geneva: 1999); number of host computers from Internet Software Consortium and Network Wizards, "Internet Domain Surveys," <www.isc.org/ds/>, viewed 20 February 2000.

11. Growth of various economies from World Bank, *World Development Indicators 2000* (Washington, DC: 2000), 182–83; China's economy from ibid., 10–12; consumer products in China from ibid., 300, and from Population Reference Bureau, "2000 World Population Data Sheet," wall chart (Washington, DC: June 2000), with computers from Ye Di Sheng, "The Development and Market of China's Information Industry and its Investment Opportunity," <www.caspa.com/event/augdin2.htm>, viewed 10 November 2000.

12. Wealth disparities from World Bank, op. cit. note 4, 282–83; trends in per capita income from U.N. Development Programme (UNDP), *Human Development Report 1999* (New York: Oxford University Press, 1999), 2–3; income disparities from World Bank, op. cit. note 11, 66, 68; 10 percent based on UNDP, op. cit. this note, 149, 197; inequality groth from ibid., 3.

13. UNDP, op. cit. note 12, 25; number of people living on less than $1 per day from World Bank, op. cit. note 4, 3, 23.

14. Number of people malnourished is a Worldwatch estimate based on U.N. Administrative Committee on Coordination, Sub-Committee on Nutrition in collaboration with International Food Policy Research Institute (IFPRI), *Fourth Report on the World Nutrition Situation* (Geneva: 1999), and on Rafael Flores, research fellow, IFPRI, Washington, DC, e-mail to Brian Halweil, Worldwatch Institute, 5 November 1999, and discussion with Gary Gardner, Worldwatch Institute, 3 February 2000; selected countries with chronic hunger from Gary Gardner and Brian Halweil, *Underfed and Overfed: The Global Epidemic of Malnutrition*, Worldwatch Paper 150 (Washington DC: Worldwatch Institute, March 2000), 17.

15. Number without access to clean water from Peter H. Gleick, *The World's Water 1998–1999* (Washington, DC: Island Press, 1998), 40; percentages by country from World Bank, op. cit. note 11, 14–16; toilets from World Bank op. cit. note 11, 94–96.

16. Joint United Nations Program on HIV/AIDS, *Report on the Global HIV/AIDS Epidemic—June 2000* (Geneva: June 2000), 124; Elizabeth Rosenthal, "In Rural China, a Steep Price of Poverty: Dying of AIDS," *New York Times*, 28 October 2000.

17. John Noble Wilford, "Ages-Old Polar Icecap Is Melting, Scientists Find," *New York Times*, 19 August 2000.

18. D.A. Rothrock, Y. Yu, and G.A. Maykut, "Thinning of the Arctic Sea-Ice Cover," *Geophysical Research Letters*, 1 December 1999, 3469; Ola M. Johannessen, Elena V. Ahalina, and Martin W. Miles, "Satellite Evidence for an Arctic Sea Ice Cover in Transformation," *Science*, 3 December 1999, 1937; Lars H. Smedsrud and Tore Furevik, "Toward an Ice-Free Arctic?" *Cicerone*, February 2000.

19. Paul N. Pearson and Martin R. Palmer, "Atmospheric Carbon Dioxide Concentrations Over the Past 60 Million Years," *Nature*, 17

August 2000, 695; Andrew C. Revkin, "A Shift in Stance on Global Warming Theory," *New York Times*, 26 October 2000.

20. Carsten Rühlemann et al., "Warming of the Tropical Atlantic Ocean and Slowdown of Thermohaline Circulation During the Last Glaciation," *Nature*, 2 December 1999, 511.

21. Percentage of fish species as reef dwellers from Norman Myers, "Synergisms: Joint Effects of Climate Change and Other Forms of Habitat Destruction," in Robert L. Peters and Thomas E. Lovejoy, eds., *Global Warming and Biological Diversity* (New Haven, CT: Yale University Press, 1992), 347; Colin Woodard, "Fall of the Magic Kingdom: A Reporter Goes Underwater in the Belize Barrier Reef," *Tuftonia*, summer 2000, 20.

22. Daniel Cooney, "Coral Reefs Disappearing," *Associated Press*, 23 October 2000; Wilkinson, op. cit. note 5; Ove Hoegh-Guldberg et al., *Pacific in Peril*, available at <www.greenpeace.org>.

23. Hectares of forest per person from Robert Engleman et al., *People in the Balance: Population and Natural Resources at the Turn of the Millennium* (Washington, DC: Population Action International, 2000), 12.

24. Number of people living in water-deficit countries from ibid., 9; Beijing water table from James Kynge, "China Approves Controversial Plan to Shift Water to Drought-Hit Beijing," *Financial Times*, 7 January 2000; oil prices from U.S. Department of Energy, *Monthly Energy Review*, September 2000; near peak production of oil from Colin J. Campbell and Jean H. Laherrere, "The End of Cheap Oil," *Scientific American*, March 1998, 78–83.

25. Pamela Stedman-Edwards, "A Framework for Analysing Biodiversity Loss," in Wood, Stedman-Edwards, and Meng, op. cit. note 5, 15–16.

26. Wood, Stedman-Edwards, and Meng, op. cit. note 5, 283, 231–54.

27. Kanbur resignation from Joseph Kahn, "A Fork in the Road to Riches: Redrawing the Map," *New York Times*, 25 June 2000.

28. World Bank, op. cit. note 4, 3; Russian poverty rate data from Nora Lustig, Director, *World Development Report 2000/2001* team, World Bank press conference, Washington, DC, 12 September 2000.

29. Agricultural revolution from Jared Diamond, *Guns, Germs and Steel: The Fate of Human Societies* (New York: W.W. Norton & Company, 1997), 110–12; Retze Koen, Greenpeace Switzerland, at Seventh International Symposium on Renewable Energy Education, Oslo, 15–18 June 2000.

30. World Bank, op. cit. note 11, 74–76, 82–84; Ceara program from World Bank, *World Development Report 1998/1999* (New York: Oxford University Press, 1998), 122.

31. Lustig, op. cit. note 28.

32. World Bank, op. cit. note 30, 125, 128.

33. Anil Agarwal, "The Poverty of Amartya Sen," *Down to Earth*, 15 December 1998.

34. Area of organic farmland is a Worldwatch estimate based on Helga Willer and Minou Yussefi, *Organic Agriculture World-Wide* (Bad Durkheim: Stiftung Okologie & Landbau, 2000), and on Catherine Greene, Economic Research Service, U.S. Department of Agriculture, discussion with Brian Halweil, Worldwatch Institute, 11 January 2000; Figure 1–1 from Nic Lampkin, University of Wales, Aberystwyth, e-mail to Brian Halweil, Worldwatch Institute, 25 January 2000 (includes area "in conversion" or "transition"); sales figure is Worldwatch estimate based on Willer and Yussefi, op. cit. this note, and on the International Trade Centre, *Organic Food and Beverages: World Supply and Major European Market* (Geneva: U.N. Conference on Trade and Development, 1999).

35. Global Forest and Trade Network, Home Depot, Centex, and 2005 goal from David A.

Ford "Certified Wood: State of the Marketplace," *Environmental Design+Construction*, July/August 2000, 49, 51; number of hectares certified from Forest Stewardship Council, <www.fscoax.org/principal.htm>, viewed 13 October 2000.

36. Andrew Evans, "Buying and Selling Green: Deregulation and Green Power Marketing," *Renewable Energy World*, January 2000; Green-e Renewable Energy Program, "Switch to Green Power," <www.green-e.org/switch/index.html>, viewed 13 October 2000.

37. Figure 1–2 based on BTM Consult ApS, *International Wind Energy Development: World Market Update 1999* (Ringkoping, Denmark: March 2000).

38. Wind power projects from *Wind Power Monthly*, various issues; ABB Group, "ABB Sees Billion-Dollar Growth Opportunity in Alternative and Renewable Energy," press release (London: 8 June 2000).

39. Don Hinrichsen, "The Oceans are Coming Ashore," *World Watch*, November/December 2000.

40. Developing-country share of population from United Nations, *World Population Prospects: The 1998 Revision* (New York: December 1998); Fox proposal from Mary Jordan, "Mexican Touts Open Borders: Visiting President-Elect Pushes N. American Convergence," *Washington Post*, 25 August 2000.

41. E–9 share of energy markets from BP Amoco, *BP Amoco Statistical Review of World Energy* (London: Group Media Publications, June 2000), 9, 25, 33.

42. Herman E. Daly, "Toward a Stationary State Economy," in John Harte and Robert Socolow, eds., *Patient Earth* (New York: Holt, Rinehart and Winston, 1971), 237.

Chapter 2. Uncovering Groundwater Pollution

1. Chalk aquifer from U.N. Environment Programme (UNEP), *Groundwater: A Threatened Resource* (Nairobi: 1996), 6; Museum Victoria, "Ice Age Mammals—Britain and Ireland," <www.mov.vic.gov.au/dinosaurs/mammbrit.stm>, viewed 23 August 2000.

2. UNEP, op. cit. note 1, 4.

3. Groundwater from UNEP, op. cit. note 1, 7; rivers from Igor A. Shiklomanov, *World Water Resources* (Paris: International Hydrological Programme, UNESCO, 1998), 6.

4. Hydrological cycle from Ralph Heath, *Basic Ground-Water Hydrology*, U.S. Geological Survey (USGS), Water-Supply Paper 2220 (Washington, DC: U.S. Government Printing Office, 1998), 5; Broecker quoted in William K. Stevens, "If Climate Changes, It May Change Quickly," *New York Times*, 27 January 1998.

5. Water witches from Francis H. Chapelle, *The Hidden Sea: Ground Water, Springs, and Wells* (Tucson, AZ: Geoscience Press, Inc., 1997), 43.

6. Eastern China and urban dependence from UNEP, op. cit. note 1, 8, 10; U.S. Environmental Protection Agency (EPA), Office of Water, *National Water Quality Inventory: 1998 Report to Congress* (Washington, DC: 2000), 158, at <www.epa.gov/305b/98Report>, viewed 30 June 2000; India from Stephen Foster et al., *Groundwater in Rural Development*, Technical Paper No. 463 (Washington, DC: World Bank, 2000), 2.

7. Current irrigation by aquifers from data supplied by Charlotte De Fraiture, International Water Management Institute, e-mail to author, 21 October 1999; tubewells from R.K. Pachauri and P.V. Sridharan, eds., *Looking Back to Think Ahead* (abridged text) GREEN India 2047 Project (New Delhi: Tata Energy Research Institute, 1998), 12; agricultural product and gross domestic product from World Bank and

Ministry of Water Resources, Government of India, *Groundwater Regulation and Management*, South Asia Rural Development Series (New Delhi: World Bank and Allied Publishers, 1999), 2; U.S. irrigated area today and current contribution of fresh water to irrigation from Sandra Postel, *Pillar of Sand* (New York: W.W. Norton & Company, 1999), 42, 112.

8. Industry-to-agriculture ratio from Lester R. Brown and Brian Halweil, "China's Water Shortage Could Shake World Food Security," *World Watch*, July/August 1998, 13; industrial share of consumption from Sandra Postel, *Dividing the Waters: Food Security, Ecosystem Health, and the New Politics of Scarcity*, Worldwatch Paper 132 (Washington, DC: Worldwatch Institute, September 1996), 14.

9. Shiang-Kueen Hsu, "Plan for a Groundwater Monitoring Network in Taiwan," *Hydrogeology Journal*, vol. 6 (1998), 407; Bangladesh from Peter H. Gleick, *The World's Water 2000–2001* (Washington, DC: Island Press, 2000), 165–66; International Bottled Water Association, Alexandria, VA, *Bottled Water in the U.S.*, 1999 Edition, <www.bottledwater.org/public/gallon_byseg.htm>, viewed 5 April 2000.

10. Postel, op. cit. note 7, 80; overdraft and pollution from Stephen Foster, Adrian Lawrence, and Brian Morris, *Groundwater in Urban Development*, Technical Paper 390 (Washington, DC: World Bank, 1998), 6, 23–25.

11. California from Postel, op. cit. note 8, 19; subsidence from UNEP, op. cit. note 1, 17–18.

12. USGS, *Ground Water and Surface Water: A Single Resource*, USGS Circular 1139 (Denver, CO: 1998), 12; daily addition from aquifers from EPA, *The Quality of Our Nation's Water* (Washington, DC: 1998); the contribution to major rivers from Francis Chapelle, hydrologist, USGS, Columbia, SC, discussion with author, 5 November 1999; wetlands from USGS, op. cit. this note, 19.

13. UNEP, op. cit. note 1, 12–13; Africa from Shiklomanov, op. cit. note 3, 25.

14. Chapelle, op. cit. note 5, 167–80.

15. Ibid.

16. Ibid.

17. Toxicity from D. Pimentel and D. Kahn, "Environmental Aspects of 'Cosmetic Standards' of Foods and Pesticides," in David Pimentel, ed., *Techniques for Reducing Pesticide Use* (New York: John Wiley & Sons, 1997), 415.

18. Postel, op. cit. note 7, 59.

19. Jack E. Barbash, Research Chemist, USGS, Tacoma, WA, discussion with author, 17 November 1999.

20. U.N. Food and Agriculture Organization (FAO), *Fertilizers—A World Report on Production and Consumption* (Rome: 1952), 3; FAO, *Agriculture, Means of Production, Fertilizer*, electronic database, <apps.fao.org>, viewed 10 April 2000; W.L. Zhang et al., "Nitrate Pollution of Groundwater in Northern China," *Agriculture, Ecosystems and Environment*, vol. 59 (1996), 223–31; National Research Council, *Soil and Water Quality: An Agenda for Agriculture* (Washington, DC: National Academy Press, 1993); Paul Faeth, *Fertile Ground* (Washington, DC: World Resources Institute, 2000), 7, 16.

21. Figure of 130 times from Faeth, op. cit. note 20, 11; Brian Halweil, "United States Leads World Meat Stampede," press briefing (Washington, DC: Worldwatch Institute, 2 July 1998).

22. World Health Organization (WHO) limit from UNEP, op. cit. note 1, 22 (the drinking water limit is 10 mg/liter of nitrogen, which is equivalent to 45 mg/liter of nitrate); China from Zhang et al., op. cit. note 20.

23. Table 2–3 from the following: Zhang et al., op. cit. note 20; Geoffrey D. Smith et al., "The Origin and Distribution of Nitrate in Groundwater from Village Wells in Kotagede, Yogyakarta, Indonesia," *Hydrogeology Journal*, vol. 7 (1999), 576–89; Canary Islands and East Anglia from Gordon Conway and Jules Pretty, *Unwelcome Harvest* (London: Earthscan Publications, 1991), 183–84, 196; Nigeria from Foster et al., op. cit. note 6, 84; Romania from European Environmental Agency (EEA), *Groundwater Quality and Quantity in Europe* (Copenhagen: 1999), 53–60; Julia A. Pacheco and Armando S. Cabrera, "Groundwater Contamination by Nitrates in the Yucatan Peninsula, Mexico," *Hydrogeology Journal*, vol. 5, no. 2 (1997), 47; and Nebraska and Kansas (levels converted by author from nitrogen to nitrate equivalent) from David K. Mueller et al., *Nutrients in Ground Water and Surface Water of the United States—An Analysis of Data through 1992* (Denver, CO: USGS, National Water Quality Assessment Program, 1995), 35. India from R.K. Pachauri and P.V. Sridharan, eds., *Looking Back to Think Ahead* (unabridged version), GREEN India 2047 Project (New Delhi: Tata Energy Research Institute, 1998), 215; United States from USGS, *The Quality of Our Nation's Waters—Nutrients and Pesticides* (Reston, VA: 1999), 8.

24. San Joaquin from USGS, op. cit. note 23, 54; Denmark from Larry W. Canter, *Nitrates in Groundwater* (New York: Lewis Publishers, 1997), 58–59; Linda Nash, "Water Quality and Health," in Peter H. Gleick, ed., *Water in Crisis* (New York: Oxford University Press, 1993), 28.

25. Nash, op. cit. note 24, 27–28; Joseph I. Barzilay, Winkler G. Weinberg, and J. William Eley, *The Water We Drink* (New Brunswick, NJ: Rutgers University Press, 1999), 70–71; links to miscarriages and lymphoma from Bernard T. Nolan and Jeffrey D. Stoner, "Nutrients in Groundwaters of the Conterminous United States, 1992–1995," *Environmental Science & Technology*, vol. 34, no. 7 (2000), 1156; ruminant animals from EEA, op. cit. note 23, 51.

26. Grape vines, overall effects of nitrates, and FAO from R.S. Ayers and D.W. Westcott, *Water Quality for Agriculture*, FAO Irrigation and Drainage Paper 29 (Rome: FAO, 1985), 91; other effects from George N. Agrios, *Plant Pathology*, 4th ed. (San Diego, CA: Academic Press, 1997), 149–50, and from Ian McPharlin, *Nitrogen and Phosphorous Disorders of Vegetable Crops*, Bulletin No. 4175 (South Perth, Australia: Western Australian Department of Agriculture, December 1989), 6.

27. Scott Phillips, Michael Focazio, and L. Joseph Bachman, *Discharge, Nitrate Load, and Residence Time of Ground Water in the Chesapeake Bay Watershed* (Baltimore, MD: USGS, 1998); USGS, "The Bay's Recovery: How Long Will it Take?" (Baltimore, MD: April 1998).

28. Phillips, Focazio, and Bachman, op. cit. note 27; USGS, op. cit. note 27.

29. Midwest from Nolan and Stoner, op. cit. note 25, 1163–64; Mueller et al., op. cit. note 23, 38.

30. Mueller et al., op. cit. note 23, 1, 38.

31. Soil as filter from EPA, op. cit. note 12; Foster et al., op. cit. note 6, 77.

32. USGS, op. cit. note 23, 58; Jack E. Barbash and Elizabeth A. Resek, *Pesticides in Ground Water* (Chelsea, MI: Ann Arbor Press, 1996), 115–17.

33. DDT from Barbash and Resek, op. cit. note 32, 141; DBCP from Joseph L. Domagalski, *Pesticides in Surface and Ground Water of the San Joaquin-Tulare Basins, California* (Washington, DC: USGS, 1997), 42–47; dieldrin from USGS, op. cit. note 23, 74.

34. Pachauri and Sridharan, op. cit. note 23, 216.

35. P.J. Chilton, A.R. Lawrence, and M.E. Stuart, "Pesticides in Groundwater: Some Preliminary Results From Recent Research in Temperate and Tropical Environments," in J. Mather et al., eds., *Groundwater Contaminants and their Migration* (London: Geological Society, 1998), 335.

36. Nash, op. cit. note 24, 31–32; Polly Short and Theo Colburn, "Pesticide Use in the U.S.

and Policy Implications: A Focus on Herbicides," *Toxicology and Industrial Health*, vol. 15, nos. 1–2 (1999), 240–43; Robert Repetto and Sanjay S. Baliga, *Pesticides and the Immune System: The Public Health Risks* (Washington, DC: World Resources Institute, 1996); Ted Schettler et al., *Generations at Risk: Reproductive Health and the Environment* (Cambridge, MA: The MIT Press, 1999), 107.

37. USGS, op. cit. note 23, 76; herbicide degradates from Jack E. Barbash et al., *Distribution of Major Herbicides in Ground Water of the United States* (Sacramento, CA: USGS, 1999), 1, 25, 27; degradate toxicity from Conway and Pretty, op. cit. note 23, 40.

38. EPA, Office of Water, "Current Drinking Water Standards," (Washington, DC: EPA Office of Water, 1995), at <www.epa.gov/safewater/pmcl.html>, viewed 7 August 2000; Warren P. Porter, James W. Jaeger, and Ian H. Carlson, "Endocrine, Immune, and Behavioral Effects of Aldicarb (Carbamate), Atrazine (Triazine) and Nitrate (Fertilizer) Mixtures at Groundwater Concentrations," *Toxicology and Industrial Health*, vol. 15, nos. 1–2 (1999), 135, 142–43.

39. Office of Technology Assessment, *Beneath the Bottom Line: Agricultural Approaches to Reduce Agrichemical Contamination of Groundwater* (Washington, DC: U.S. Government Printing Office, November 1990), 34.

40. Chilton, Lawrence, and Stuart, op. cit. note 35, 336–37.

41. Aaron Sachs, "Virtual Ecology: A Brief Environmental History of Silicon Valley," *World Watch*, January/February 1999, 13; EPA, *Safe Drinking Water Act, Section 1429 Ground Water Report to Congress* (Washington, DC: EPA Office of Water, 1999), 16.

42. EPA, op. cit. note 12; United Kingdom from "Aquifer Pollution Inventory Sets the Scene for Tussles Over Clean-Up," *ENDS Report*, September 1996, 17.

43. David A. Bender et al., *Selection Procedure and Salient Information for Volatile Organic Compounds Emphasized in the National Water-Quality Assessment Program* (Rapid City, SD: USGS, 1999), 1; "42 Million Americans Use Groundwater Vulnerable to Contamination by Volatile Organic Compounds," *ScienceDaily Magazine*, <www.sciencedaily.com>, 29 October 1999.

44. James F. Pankow and John A. Cherry, *Dense Chlorinated Solvents and other DNAPLs in Groundwater* (Waterloo, ON, Canada: Waterloo Press, 1996), 7; production from Anne Platt McGinn, "Phasing Out Persistent Organic Pollutants," in Lester Brown et al., *State of the World 2000* (New York: W.W. Norton & Company, 2000), 81; uses from Bender et al., op. cit. note 43, 1, and from Paul J. Squillace et al., "Volatile Organic Compounds in Untreated Ambient Groundwater of the United States," *Environmental Science & Technology*, vol. 33, no. 23 (1999), 4176–87.

45. UNEP, op. cit. note 1, 26; Pankow and Cherry, op. cit. note 44, 9–15.

46. Schettler et al., op. cit. note 36, 73, 77, 82; kidney and liver damage and childhood cancers from Pankow and Cherry, op. cit. note 44.

47. Pankow and Cherry, op. cit. note 44, 9.

48. EPA, op. cit. note 12.

49. Donald Sutherland, "60 Percent of America's Liquid Toxic Waste Injected Underground," *Environment News Service*, <www.ens-news.com>, 7 July 1999; EPA, "Class I Injection Wells Underground Injection Control Regulations for Florida; Proposed Rule," draft (Tallahassee, FL: 24 April 2000), provided by Suzi Ruhl, Legal Environmental Assistance Foundation, Inc., Tallahassee, FL; India from Manish Tiwari and Richard Mahapatra, "What Goes Down Must Come Up," *Down to Earth*, 31 August 1999, 30–40.

50. Squillace et al., op. cit. note 44, 4176–87; Squillace quoted in "42 Million Americans," op. cit. note 43.

51. Netherlands from European Centre for Ecotoxicology and Toxicology of Chemicals, *Joint Assessment of Commodity Chemicals No. 39—Tetrachloroethylene CAS No. 127-18-4* (Brussels: 1999), 42; England from "Aquifer Pollution Inventory," op. cit. note 42, 15; Japan from British Geological Survey (BGS) et al., *Characterisation and Assessment of Groundwater Quality Concerns in Asia-Pacific Region* (Oxfordshire, U.K.: 1996), 49–51; Semiconductor Industry Association, *Semiconductor Forecast Summary, 2000–2003* (San Jose, CA, June 2000).

52. Biswas quoted in Tiwari and Mahapatra, op. cit. note 49, 39.

53. Arnab Neil Sengupta, "Bangladesh Arsenic an Invitation to Catastrophe," *Environmental News Network*, <www.enn.com>, 23 March 1998; WHO, "Arsenic in Drinking Water," Fact Sheet No. 210 (Geneva: February 1999), 4.

54. Numbers of people affected from BGS and Mott MacDonald (U.K.), "Phase I, Groundwater Studies of Arsenic Contamination in Bangladesh," Executive Summary, Main Report, <bicn.com/acic/infobank/bgs-mmi/risumm.htm>, viewed 9 July 1999; Uttam K. Chowdhury et al., "Groundwater Arsenic Contamination in Bangladesh and West Bengal, India," *Environmental Health Perspectives*, May 2000, 393–94; Nadia S. Halim, "Arsenic Mitigation in Bangladesh," *The Scientist*, 6 March 2000, 14; Sengupta op. cit. note 53, 3. The BGS reports that 20 million Bangladeshis may be drinking arsenic-contaminated water; estimates from scientists and medical workers in Bangladesh are considerably higher, on the order of 70–75 million. Number of wells affected from Ross Nickson et al., "Arsenic Poisoning of Bangladesh Groundwater," *Nature*, 24 September 1998, 338; deaths from Kimberley Masibay, "Drinking Without Harm," *Scientific American*, September 2000, 22; WHO prediction from Chowdhury et al., op. cit. this note, 393.

55. Helen Sewell, "Poison Threat in Bangladesh," *BBC News Online,* 6 October 1999; Sengupta, op. cit. note 53, 1; Gleick, op. cit. note 9, 165–67, 171–72; scientists debate from Nickson et al., op. cit. note 54, 338; S.K. Acharyya et al., "Arsenic Poisoning in the Ganges Delta," *Nature*, 7 October 1999, 545.

56. Figure of 2 percent from EEA, op. cit. note 23, 91; abandoned aquifers from BGS et al., op. cit. note 51, 19–20; Tushaar Shah et al., *The Global Groundwater Situation* (Colombo, Sri Lanka: International Water Management Institute, 2000), 6–7.

57. UNEP, op. cit. note 1, 18–19; Turkey and China from Shah et al., op. cit. note 56, 6.

58. WHO in BGS et al., op. cit. note 51, 86–88.

59. Table 2–5 from the following sources: Bangkok and China from BGS et al., op. cit. note 51, 19–20, 3; China from Foster, Lawrence, and Morris, op. cit. note 10, 25; Santa Monica from Richard Johnson et al., "MTBE: To What Extent Will Past Releases Contaminate Community Water Supply Wells?" *Environmental Science & Technology*, 1 May 2000, 2A. U.S. utilities from Marc O. Ribaudo and Aziz Bouzaher, *Atrazine: Environmental Characteristics and Economics of Management*, Agricultural Economic Report No. 699 (Washington, DC: U.S. Department of Agriculture, September 1994), 5.

60. National Research Council, *Alternatives for Ground Water Cleanup* (Washington, DC: National Academy Press, 1994), viii, 1–6.

61. Ibid., 1; Janet Raloff, "Hanford Tanks: Leaks Reach Groundwater," *Science News*, 20 December 1997, 410.

62. Andrew Skinner, Secretary, International Association of Hydrogeologists, London, e-mail to author, 18 November 1999.

63. T.J. Logan, "Sustainable Agriculture and Water Quality," in Clive A. Edwards et al., *Sustainable Agricultural Systems* (Ankeny, IA: Soil and Water Conservation Society, 1990),

582–608; 85–90 percent from Repetto and Baliga, op. cit. note 36; 13; FAO, "FAO: Unsafe Application of Pesticides Causes Health and Environmental Damage—Training and Standards Required," press release (Rome: 29 May 1997).

64. Jules Pretty, *The Living Land: Agriculture, Food and Community Regeneration in Rural Europe* (London: Earthscan Publications Ltd., 1998), 101.

65. L.E. Drinkwater, P. Wagoner, and M. Sarrantonio, "Legume-based Cropping Systems Have Reduced Carbon and Nitrogen Losses," *Nature*, 19 November 1999, 262–65; David Tilman, "The Greening of the Green Revolution," *Nature*, 19 November 1998, 211; Cass Petersen, Laurie E. Drinkwater, and Peggy Wagoner, *The Rodale Institute Farming Systems Trials* (Kutztown, PA: Rodale Institute, 1999).

66. Charles Darwin, *On the Origin of Species by Means of Natural Selection* (1859; New York: Modern Library, 1993); Martin S. Wolfe, "Crop Strength Through Diversity," *Nature*, 17 August 2000, 681–82; China from Youyung Zhu et al., "Genetic Diversity and Disease Control in Rice," *Nature*, 17 August 2000, 718–21.

67. Lori Ann Thrupp, ed., *New Partnerships For Sustainable Agriculture* (Washington, DC: World Resources Institute, 1996), 6–9; Gary Gardner, "IPM and the War on Pests," *World Watch*, March/April 1996, 26.

68. Indonesia from Gardner, op. cit. note 67, 26; Europe from Pretty, op. cit. note 64, 279–80; Sweden from Anders Emmerman, "Sweden's Reduced Risk Pesticide Policy," *Pesticides News*, December 1996, and from Short and Colburn, op. cit. note 36, 247; "Report Implicates Agriculture for Damage to the Environment, Recommends Eco-Taxes," *International Environment Reporter*, 17 March 1999; "Voynet Unveils Water Policy Reform With Extended 'Polluter Pays' Provisions," *International Environment Reporter*, 10 November 1999.

69. Pretty, op. cit. note 64, 283–84. Currency converted at November 2000 rate.

70. World Resources Institute et al., *World Resources 2000–01* (New York: Oxford University Press, 2000), 210–11; Watershed Agricultural Council, "Whole Farm Planning," Summary Report (New York: January 1996), 2–3.

71. World Resources Institute et al., op. cit. note 70; Al Appleton, former Director of the New York City Water and Sewer System, e-mail to Danielle Nierenberg, Worldwatch Institute, 2 August 2000; Jeffrey Gatz, EPA Region II NYC Watershed Team Leader, e-mail to Danielle Nierenberg, Worldwatch Institute, 4 August 2000; EPA from Faeth, op. cit. note 20, 16.

72. John Ehrenfeld and Nicholas Gertler, "Industrial Ecology in Practice: The Evolution of Interdependence at Kalundborg," *Journal of Industrial Ecology*, vol. 1, no. 1 (1997), 67–78.

73. Braden R. Allenby, *Industrial Ecology: Framework and Implementation* (Upper Saddle River, NJ: Prentice Hall, 1999); remanufacture statistics from Xerox Web site, <www.xerox.com/ehs/1997/sustain.htm>, viewed 18 September 1998; Xerox Corporation, *Environment, Health and Safety 2000 Progress Report* (Stamford, CT: 2000), 8; Liz Campbell, Regulatory Affairs Manager, Environment, Health and Safety Office, Xerox Corporation, Webster, NY, discussion with author, 2 November 2000.

74. Netherlands goal from <www.netherlands-embassy.org/env_nmp2.htm>, viewed 20 August 1998; pollution taxes from David Malin Roodman, *The Natural Wealth of Nations* (New York: W.W. Norton & Company, 1998), 151; Canberra from ACT government, <www.act.gov.au/nowaste/>, viewed 23 October 1998.

75. Chlorofluorocarbons from Molly O'Meara, "CFC Production Continues to Plummet," in Lester Brown, Michael Renner, and Christopher Flavin, *Vital Signs 1998* (New York: W.W. Norton & Company, 1998), 70–71; UNEP, "Regional Workshops Highlight Need for Effective Action Against Hazardous Chemicals," press release (Geneva: 9 July 1998); Anne Platt McGinn, *Why Poison Ourselves? A Precautionary Approach to Synthetic Chemicals*, Worldwatch Paper 153 (Washington, DC: Worldwatch Institute, November 2000), 7–8.

Chapter 3. Eradicating Hunger: A Growing Challenge

1. Henry Kissinger, speech delivered at the First World Food Summit Conference, Rome, 5–16 November 1974.

2. Figure of 1.1 billion hungry is a Worldwatch estimate from U.N. Administrative Committee on Coordination, Sub-Committee on Nutrition (UN ACC/SCN) in collaboration with International Food Policy Research Institute (IFPRI), *Fourth Report on the World Nutrition Situation* (Geneva: 1999), and from Rafael Flores, Research Fellow, IFPRI, Washington, DC, e-mail to Brian Halweil, 5 November 1999, and discussion with Gary Gardner, 3 February 2000; U.N. Food and Agriculture Organization (FAO), *The State of Food Insecurity in the World 1999* (Rome: 1999), 6.

3. FAO, op. cit. note 2, 4; quotation from ibid., 6.

4. UN ACC/SCN, op. cit. note 2; 1.3 billion from World Bank, *Rural Development: From Vision to Action*, Environmentally and Socially Sustainable Development Studies and Monographs Series No. 12 (Washington, DC: 1997), 1.

5. UN ACC/SCN, op. cit. note 2, 2.

6. Prevalence of undernourishment in China from FAO, *The State of Food Insecurity in the World 2000* (Rome: 2000), 27; China's grain harvest information from U.S. Department of Agriculture (USDA), *Production, Supply, and Distribution*, electronic database, Washington, DC, updated September 2000; underweight children in Latin America from UN ACC/SCN, op. cit. note 2, 10.

7. UN ACC/SCN, op. cit. note 2, 94–96.

8. World Bank, op. cit. note 4, 1.

9. Ibid., 17–19.

10. United Nations, *World Population Prospects: The 1998 Revision* (New York: December 1998).

11. R.K. Pachauri and P.V. Sridharan, eds., *Looking Back to Think Ahead* (abridged version), GREEN India 2047 Project (New Delhi: Tata Energy Research Institute, 1998), 7.

12. Population data from United Nations, op. cit. note 10; grain harvested area from USDA, op. cit. note 6.

13. Michael Morris, Nuimuddin Chowdhury, and Craig Meisner, *Wheat Production in Bangladesh: Technological, Economic, and Policy Issues* (Washington, DC: IFPRI, 1997), 10.

14. Population data from United Nations, op. cit. note 10; Sandra Postel, *Pillar of Sand* (New York: W.W. Norton & Company, 1999), 73–74.

15. Figure 3–1 and grain production data from USDA, op. cit. note 6; economic trends from International Monetary Fund (IMF), *World Economic Outlook (WEO) Database*, <www.imf.org/external/pubs/ft/weo/2000/02/data/index.htm>, September 2000.

16. USDA, op. cit. note 6.

17. Figure 3–2 from FAO, *Yearbook of Fishery Statistics: Capture Production* (Rome: various years), from FAO, *1948–1985 World Crop and Livestock Statistics* (Rome: 1987), and from FAO, *FAOSTAT Statistics Database*, <apps.fao.org>, updated 5 April 2000.

18. Conversion ratio for grain to beef based on Allen Baker, Feed Situation and Outlook staff, Economic Research Service (ERS), USDA, Washington, DC, discussion with author, 27 April 1992.

19. Fish catch data from FAO, *Yearbook of Fishery Statistics*, op. cit. note 17; grain equivalent of farmed fish from USDA, ERS, "China's Aquatic Products Economy: Production, Marketing, Consumption, and Foreign Trade," *International Agriculture and Trade Reports: China* (Washington, DC: July 1998), 45.

20. Bangladesh population from United Nations, op. cit. note 10; grain allotment data from USDA, op. cit. note 6; inundation estimate from World Bank, *World Development Report 1999/2000* (New York: Oxford University Press, 1999), 100; upper estimate on sea level rise from Tom M.L. Wigley, *The Science of Climate Change: Global and U.S. Perspectives* (Arlington, VA: Pew Center on Global Climate Change, June 1999).

21. USDA, *Grain: World Markets and Trade* (Washington, DC: October 2000), 6.

22. Ibid., 6, 11, 19.

23. Grain production and consumption data from USDA, op. cit. note 6.

24. Wheat and rice yields from ibid.

25. United Nations, op. cit. note 10.

26. Women's responsibility for food production in Africa from International Livestock Research Institute, "Women Dairy Farmers in Africa," *CGIAR News* (Consultative Group on International Agricultural Research), June 1997.

27. Percent photosynthate to seed from L.T. Evans, *Crop Evolution, Adaptation and Yield* (Cambridge, U.K.: Cambridge University Press, 1993), 242–44; theoretical upper limit from Thomas R. Sinclair, "Options for Sustaining and Increasing the Limiting Yield-Plateaus of Grain Crops," paper prepared for the 1998 Symposium on World Food Security, Kyoto, Japan (Washington, DC: USDA Agricultural Research Service, September 1998), 14.

28. Carol Kaesuk Yoon, "A 'Dead Zone' Grows in the Gulf of Mexico," *New York Times*, 20 January 1998.

29. Fertilizer usage numbers based on series from FAO, *Fertilizer Yearbook* (Rome: various years), and on K.G. Soh and K.F. Isherwood, "Short Term Prospects for World Agriculture and Fertilizer Use," International Fertilizer Industry Association Meeting, 30 November–3 December 1999.

30. Sandra Postel, "Redesigning Irrigated Agriculture," in Lester Brown et al., *State of the World 2000* (New York: W.W. Norton & Company, 2000), 40.

31. USDA, op. cit. note 6.

32. Grain yields and Figure 3–3 from ibid.

33. Double-cropping information for China from W. Hunter Colby, Frederick W. Crook, and Shwu-Eng H. Webb, *Agricultural Statistics of the People's Republic of China, 1949–90* (Washington, DC: USDA, ERS National Agricultural Statistics Service (NASS), December 1992), 48; India information from Praduman Kumar et al., "Sustainability of Rice-Wheat Based Cropping Systems in India: Socio-Economic and Policy Issues," *Economic and Political Weekly*, 26 September 1998, A-152–58; Argentina information from USDA, ERS, "Commodity Spotlight," *Agricultural Outlook*, September 2000, 6.

34. Conservation Technology Information Center (CTIC), "Conservation Tillage Survey Data: Crop Residue Management 1998," CTIC Core 4 Conservation Web site <www.ctic.purdue.edu/Core4/CT/CT.html>, West Lafayette, IN, updated 19 May 2000.

35. Sunflower and wheat double cropping from "Sunflowers Offer Double-crop Option After Wheat," *High Plains Journal: The Farmer-Rancher Paper*, <www.hpj.com/wsdocs/jul2000/0605secondsunldHTM.HTM>, posted July 2000; wheat-soy double cropping from Center for Sustainable Agriculture Systems, "Relay and Double Cropping Soybean/Wheat Production Systems," *Integrated Farm Information Sheet*, Winter 1996, and from Agricultural Communication Service at Purdue University and the Section of Communications and Technology at The Ohio State University, "Double-Cropping Gives Wheat Growers Second Income," *Ag Answers*, Electronic Information Center, posted 22 June 1999.

36. Figure 3–4 is Worldwatch estimate based on FAO, *FAOSTAT*, op. cit. note 17, and on USDA, *Agricultural Resources and Environmental Indicators* (Washington, DC: 1996–97).

37. Gershon Feder and Andrew Keck, *Increasing Competition for Land and Water Resources: A Global Perspective* (Washington, DC: World Bank, March 1995), 28–29.

38. Worldwatch calculation based on ratio of 1,000 tons of water for 1 ton of grain from FAO, *Yield Response to Water* (Rome: 1979), on global wheat prices from IMF, *International Financial Statistics* (Washington, DC: various years), and on industrial water intensity in Mark W. Rosegrant, Claudia Ringler, and Roberta V. Gerpacio, "Water and Land Resources and Global Food Supply," paper prepared for the 23rd International Conference of Agricultural Economists on Food Security, Diversification, and Resource Management: Refocusing the Role of Agriculture?, Sacramento, CA, 10–16 August 1997.

39. Postel, op. cit. note 14, 56–57, 252.

40. Sandra Postel, *Last Oasis*, rev. ed. (New York: W.W. Norton & Company, 1997), 170.

41. FAO, op. cit. note 38.

42. Figure of 70 percent calculated from Albert Nyberg and Scott Rozelle, *Accelerating China's Rural Transformation* (Washington, DC: World Bank, September 1999), 65.

43. Postel, op. cit. note 14.

44. Postel, op. cit. note 40, 102.

45. Postel, op. cit. note 14, 171–79.

46. Water efficiency of wheat and rice production from Postel, op. cit. note 14, 71.

47. Figure 3–5 from FAO, *FAOSTAT*, op. cit. note 17.

48. Conversion ratio for grain to beef based on Baker, op. cit. note 18; pork conversion data from Leland Southard, Livestock and Poultry Situation and Outlook Staff, ERS, USDA, Washington, DC, discussion with author, 27 April 1992; feed-to-poultry conversion ratio derived from data in Robert V. Bishop et al., *The World Poultry Market—Government Intervention and Multilateral Policy Reform* (Washington, DC: USDA, 1990); conversion ratio for fish USDA, op. cit. note 19, 45.

49. Oceanic fish catch growth rate from FAO, op. cit. note 17; for aquacultural output data, see FAO, *Yearbook of Fishery Statistics: Aquaculture Production 1998*, vol. 86/2 (Rome: 2000).

50. FAO, op. cit. note 49.

51. Ibid.

52. K.J. Rana, "China," in *Review of the State of World Aquaculture*, FAO Fisheries Circular No. 886 (Rome: 1997); information on rice and fish polyculture from Li Kangmin, "Rice Aquaculture Systems in China: A Case of Rice-Fish Farming from Protein Crops to Cash Crops," *Proceedings of the Internet Conference on Integrated Biosystems 1998* <www.ias.unu.edu/proceedings/icibs/li/paper.htm>, viewed 5 July 2000.

53. Information on China's carp polyculture from Rosamond L. Naylor et al., "Effect of Aquaculture on World Fish Supplies," *Nature*, 29 June 2000, 1022; polyculture in India from M.C. Nandeesha et al., "Breeding of Carps with Oviprim," in Indian Branch, Asian Fisheries Society, *India, Special Publication No. 4* (Mangalore, India: 1990), 1.

54. Krishen Rana, "Changing Scenarios in Aquaculture Development in China," *FAO Aquaculture Newsletter*, August 1999, 18.

55. Catfish feed requirements from Naylor et al., op. cit. note 53, 1019; U.S. catfish production data from USDA, NASS, *Catfish Production*

(Washington, DC: July 2000), 3.

56. For information on the role of ruminants in agriculture, see Council for Agricultural Science and Technology, "Animal Production Systems and Resource Use," *Animal Agriculture and Global Food Supply* (Ames, IA: July 1999), 25–54, and H.A. Fitzhugh et al., *The Role of Ruminants in Support of Man* (Morrilton, AR: Winrock International Livestock Research and Training Center, April 1978), 5.

57. Roughage conversion from A. Banerjee, "Dairying Systems in India," *World Animal Review*, vol. 79/2 (Rome: FAO, 1994), and from S.C. Dhall and Meena Dhall, "Dairy Industry—India's Strength is in its Livestock," *Business Line*, Internet Edition of *Financial Daily* from *The Hindu* group of publications, <www.indiaserver.com/businessline/1997/11/07/stories/03070311.htm>, 7 November 1997; Figure 3–6 from USDA, op. cit. note 6.

58. Worldwatch calculation based on data from USDA, op. cit. note 6.

59. Banerjee, op. cit. note 57.

60. China's crop residue production and use from Gao Tengyun, "Treatment and Utilization of Crop Straw and Stover in China," *Livestock Research for Rural Development*, February 2000.

61. Ibid.; China's "Beef Belt" from USDA, ERS, "China's Beef Economy: Production, Marketing, Consumption, and Foreign Trade," *International Agriculture and Trade Reports: China* (Washington, DC: July 1998), 28.

62. Lactose intolerance from Norman Kretchmer, "Lactose and Lactase," *Scientific American*, October 1972; China's milk production data from USDA, op. cit. note 6.

63. Proportion of grain harvest used for feed calculated from USDA, op. cit. note 6, and from USDA, *World Agricultural Supply and Demand Estimates* (Washington, DC: 12 October 2000).

64. Education of women as correlated to fertility from Lester R. Brown, Gary Gardner, and Brian Halweil, *Beyond Malthus* (New York: W.W. Norton & Company, 1999), 131; the role of education of women in regards to child nutrition from Lisa C. Smith and Lawrence Haddad, *Explaining Child Malnutrition in Developing Countries: A Cross-Country Analysis*, Food Consumption and Nutrition Division Discussion Paper No. 60 (Washington, DC: IFPRI, April 1999), 16–17.

65. Figure 3–7 from United Nations, op. cit. note 10.

66. Uganda's success with battling HIV from Joint United Nations Programme on HIV/AIDS, *Report on the Global HIV/AIDS Epidemic* (Geneva: June 2000), 9.

67. For information on sea level rise see Wigley, op. cit. note 20.

68. Farm size in India from Pachauri and Sridharan, op. cit. note 11, 7; information on China's economy from U.S. State Department, Bureau of Economic Policy and Trade Practices, *1999 Country Reports on Economic Policy and Trade Practices: People's Republic of China* (Washington, DC: March 2000).

69. World Bank, op. cit. note 4, 11; India's expenditures estimated from Christopher Hellman, *Military Budget Fact Sheet*, Center for Defense Information, <www.cdi.irg/issues/wme/spendersFY01.html>, and from World Bank, *World Development Indicators 2000* (Washington, DC: March 2000).

Chapter 4. Deciphering Amphibian Declines

1. Kathryn Phillips, *Tracking the Vanishing Frogs* (New York: Penguin Books USA Inc., 1994), 106–07.

2. C. J. Corben, G. J. Ingram, and M. J. Tyler, "Gastric Brooding: Unique Form of Parental Care in an Australian Frog," *Science*, December 1974, 946–47.

3. Michael J. Tyler, *Australian Frogs: A Natural History* (Ithaca, NY: Cornell University Press, 1998), 135–40; Phillips, op. cit. note 1, 111.

4. William F. Laurance, Keith R. McDonald, and Richard Speare, "Epidemic Disease and the Catastrophic Decline of Australian Rain Forest Frogs," *Conservation Biology*, April 1996, 406–13; southern dayfrog from Harry Hines, Michael Mahony, and Keith McDonald, "An Assessment of Frog Declines in Wet Subtropical Australia," in Alastair Campbell, ed., *Declines and Disappearances of Australian Frogs* (Canberra: Environment Australia, December 1999), 56.

5. Keith McDonald and Ross Alford, "A Review of Declining Frogs in Northern Queensland," in Campbell, op. cit. note 4, 14–22. Table 4–1 from the following: eastern Australia from Laurance, McDonald, and Speare, op. cit. note 4; Monteverde from J. Alan Pounds et al., "Tests of Null Models for Amphibian Declines on a Tropical Mountain," *Conservation Biology*, December 1997, 1307–22, and from J. Alan Pounds, Michael P.L. Fogden, and John H. Campbell, "Biological Response to Climate Change on a Tropical Mountain," *Nature*, 15 April 1999, 611; Las Tablas and Fortuna from Karen R. Lips, "Mass Mortality and Population Declines of Anurans at an Upland Site in Western Panama," *Conservation Biology*, February 1999, 117–24; Charles A. Drost and Gary M. Fellers, "Collapse of a Regional Frog Fauna in the Yosemite Area of the California Sierra Nevada, USA," *Conservation Biology*, April 1996, 414–24; Rafael Joglar and Patricia Burrowes, "Declining Amphibian Populations in Puerto Rico," in R. Powell and R.W. Henderson, eds., *Contributions to West Indian Herpetology: a Tribute to Albert Schwartz* (Ithaca, NY: Society for the Study of Amphibians and Reptiles, 1996), 371– 80; Enrique La Marca and Hans Peter Reinthaler, "Population Changes in *Atelopus* Species of the Cordillera de Merida, Venezuela," *Herpetological Review*, vol. 22 (1991), 125–28; P. Weygoldt, "Changes in the Composition of Mountain Stream Frog Communities in the Atlantic Mountains of Brasil: Frogs as Indicators of Environmental Deteriorations?" *Studies of Neotropical Fauna and Environment*, vol. 24 (1989), 249–55.

6. Harold G. Cogger and Richard G. Zweifel, eds., *Encyclopedia of Reptiles and Amphibians* (San Diego, CA: Academic Press, 1998), 24; Phillips, op. cit. note 1, 8.

7. Cogger and Zweifel, op. cit. note 6, 24–29.

8. Number of species and rate of description from James Hanken, "Why Are There So Many New Amphibian Species When Amphibians are Declining?" *Trends in Ecology and Evolution*, 1 January 1999, 7; geographic distribution from William E. Duellman, "Global Distribution of Amphibians: Patterns, Conservation, and Future Challenges," in William E. Duellman, ed., *Patterns and Distribution of Amphibians: A Global Perspective* (Baltimore MD: The Johns Hopkins University Press, 1999), 7.

9. Amphibians as bioindicators from Laurie J. Vitt et al., "Amphibians as Harbingers of Decay," *BioScience*, June 1990, 418; Andrew R. Blaustein and David B. Wake, "Declining Amphibian Populations: A Global Phenomenon?" *Trends in Ecology and Evolution*, July 1990, 203–04; Robert C. Stebbins and Nathan W. Cohen, *A Natural History of Amphibians* (Princeton, NJ: Princeton University Press, 1995), 238–41; Jamie K. Reaser, *Amphibian Declines: An Issue Overview* (Washington DC: Federal Task Force on Amphibian Declines and Deformities, 2000), 11.

10. Arroyo toad from U.S. Department of Interior, Fish and Wildlife Service (FWS), "Endangered and Threatened Wildlife and Plants: Determination of Endangered Status for the Arroyo Southwestern Toad" (Washington, DC: 16 December 1994).

11. Forest loss from the U.N. Food and Agriculture Organization, *State of the World's Forests 1999* (Rome: 1999), 1; estimates of recovery times of 50–70 years from James W. Petranka, Matthew E. Eldridge, and Katherine E. Haley, "Effects of Timber Harvesting on Southern

Appalachian Salamanders," *Conservation Biology*, June 1993, 363–69, and of 20–24 years from Andrew N. Ash, "Disappearance and Return of Plethodontid Salamanders to Clearcut Plots in Southern Blue Ridge Mountains," *Conservation Biology*, August 1997, 983–88.

12. Laura A. Herbeck and David R. Larsen, "Plethodontid Salamander Response to Silvicultural Practices in Missouri Ozark Forests," *Conservation Biology*, June 1999, 623–32; logging from Danna Smith, *Chipping Forests and Jobs* (Chattanooga, TN: The Dogwood Alliance, August 1997), 15.

13. Rohan Pethiyagoda and Kelum Manamendra-Arachchi, "Evaluating Sri Lanka's Amphibian Diversity," *Occasional Papers of the Wildlife Heritage Trust*, No. 2 (Colombo: Wildlife Heritage Trust, November 1998), 1–12; forest loss from Rohan Pethiyagoda and Kelum Manamendra-Arachchi, "Are Direct-Developing Frogs 'Immune' to the Amphibian Decline Syndrome?" *Froglog* (newsletter of the Declining Amphibian Populations Task Force), October 1999, 2, and from Rohan Pethiyagoda, e-mail to author, 2 April 2000.

14. Ashley Mattoon, "Paper Forests," *World Watch*, March/April 1998, 20–28; D. Bruce Means, John G. Palis, and Mary Baggett, "Effects of Slash Pine Silviculture on a Florida Population of Flatwoods Salamander," *Conservation Biology*, April 1996, 426–37; Phillip Bishop, "Where Have all the Frogs Gone?" in *Vision of Wildlife, Ecotourism and the Environment in Southern Africa*, Fourth Annual (Johannesburg: Endangered Wildlife Trust, 1995), 41.

15. California tiger salamander from Stebbins and Cohen, op. cit. note 9, 233, and from H. Bradley Shaffer, Robert N. Fisher, and Scott E. Stanley, "Status Report: The California Tiger Salamander *Ambystoma californiense*," Final Report for California Department of Fish and Game, 1994; United Kingdom from Timothy R. Halliday and W. Ronald Heyer, "The Case of the Vanishing Frogs," *Technology Review*, May/June 1997, 59; Arroyo toad from Stebbins and Cohen, op. cit. note 9, 217.

16. Andrew N. Misyura, "Amphibians Under Pollution Impact in Ukraine," *Froglog* (newsletter of the Declining Amphibian Populations Task Force), November 1996.

17. T. J. C. Beebee et al., "Decline of the Natterjack Toad *Bufo calamita* in Britain: Palaeoecological, Documentary and Experimental Evidence for Breeding Site Acidification," *Biological Conservation*, vol. 53 (1990), 1–18; Laura L. McConnell et al., "Wet Deposition of Current-use Pesticides in the Sierra Nevada Mountain Range, California, USA," *Environmental Toxicology and Chemistry*, October 1998, 1908–15; Michael Graf, "Amphibian Declines and Pesticides, Is There a Link?" *Global Pesticide Campaigner*, August 1999, 12.

18. Adolfo Marco, Consuelo Quilchano, and Andrew R. Blaustein, "Sensitivity to Nitrate and Nitrite in Pond-Breeding Amphibians from the Pacific Northwest, USA," *Environmental Toxicology and Chemistry*, December 1999, 2836–39; nitrate levels from U.S. Geological Survey (USGS), *The Quality of Our Nation's Waters—Nutrients and Pesticides* (Reston, VA: 1999).

19. UV-B radiation from Andrew R. Blaustein et al., "UV Repair and Resistance to Solar UV-B in Amphibian Eggs: A Link to Population Declines?" *Proceedings of the National Academy of Sciences*, 1 March 1994, 1791–95; Andrew R. Blaustein et al., "Ambient UV-B Radiation Causes Deformities in Amphibian Embryos," *Proceedings of the National Academy of Sciences*, 9 December 1997, 13735–37.

20. New Zealand from F. Harvey Pough et al., *Herpetology* (Upper Saddle River, NJ: Prentice-Hall, Inc., 1998), 465; Yosemite from Drost and Fellers, op. cit. note 5; Seth C. Gamradt and Lee B. Kats, "Effect of Introduced Crayfish and Mosquitofish on California Newts," *Conservation Biology*, August 1996, 1155–62; Jeff A. Goodsell and Lee B. Kats, "Effect of Introduced Mosquitofish on Pacific Treefrogs and the Role of Alternative Prey," *Conservation Biology*, August 1999, 921–24.

21. Peter B. Moyle, "Effects of Introduced Bullfrogs, *Rana catesbeiana*, on the Native Frogs of the San Joaquin Valley, California," *Copeia*, vol. 73, no. 1 (1973), 18–21; R. Bruce Bury and Roger A. Luckenbach, "Introduced Amphibians and Reptiles in California," *Biological Conservation*, vol. 10 (1976), 10; Sarah Kupferberg, "Hydrologic and Geomorphic Factors Affecting Conservation of a River-Breeding Frog (*Rana boylii*)," *Ecological Applications*, vol. 6 (1996), 1332–44; Michael Schuman, "This Riveting Tale Has South Koreans Just a Tad Anxious," *Wall Street Journal*, 10 September 1997.

22. Lips, op. cit. note 5.

23. Laurance, McDonald, and Speare, op. cit. note 4; Jean-Marc Hero and Graeme R. Gillespie, "Epidemic Disease and Amphibian Declines in Australia," *Conservation Biology*, August 1997, 1023–25; Ross A. Alford and Stephen J. Richards, "Lack of Evidence for Epidemic Disease as an Agent in the Catastrophic Decline of Australian Rain Forest Frogs," *Conservation Biology*, August 1997, 1026–28; William F. Laurance, Keith R. McDonald, and Richard Speare, "In Defense of the Epidemic Disease Hypothesis," *Conservation Biology*, August 1997, 1030–34; Lee Berger et al., "Chytridiomycosis Causes Amphibian Mortality Associated with Population Declines in the Rain Forests of Australia and Central America," *Proceedings of the National Academy of Sciences*, 21 July 1998, 9031–36.

24. S. Milius, "Fatal Skin Fungus Found in U.S. Frogs," *Science News*, 4 July 1998, 7; USGS, "Chytrid Fungus Implicated as Factor in Decline of Arizona Frogs," press release (Reston, VA: 29 March 2000); USGS, "USGS Issues Wildlife Health Alert: Chytrid Fungus Infection Associated with Deaths of Threatened Boreal Toads in Colorado," press release (Reston, VA: 13 September 1999).

25. S. Milius, "New Frog-Killing Disease May Not Be So New," *Science News*, 26 February 2000, 133; Yosemite toad and Northern leopard frog from Cynthia Carey, Nicholas Cohen, and Louise Rollins-Smith, "Amphibian Declines—An Immunological Perspective," *Developmental and Comparative Immunology*, vol. 23 (1999), 461.

26. How it may kill from Berger et al., op. cit. note 23; Peter Daszak et al., "Emerging Infectious Diseases and Amphibian Population Declines," *Emerging Infectious Diseases*, November–December 1999, 738.

27. Aquarium fish from Laurance, McDonald, and Speare, op. cit. note 4; Daszak et al., op. cit. note 26, 739; Tim Halliday, "A Declining Amphibian Conundrum," *Nature*, 30 July 1998, 418–19; Australia from Lee Berger, Rick Speare, and Alex Hyatt, "Chytrid Fungi and Amphibian Declines: Overview, Implications and Future Directions," in Campbell, op. cit. note 4, 25–26; Virginia Morell, "Are Pathogens Felling Frogs?" *Science*, 30 April 1999, 728–31.

28. Carey, Cohen, and Rollins-Smith, op. cit. note 25, 459–72; iridoviruses from Daszak et al., op. cit. note 26, 735–46; Ian Anderson, "Sick Pond Syndrome," *New Scientist*, 25 July 1998, 21.

29. J. Alan Pounds, Michael P.L. Fogden, and John H. Campbell, "Biological Response to Climate Change on a Tropical Mountain," *Nature*, 15 April 1999, 611–14; Christopher J. Still, Prudence N. Foster, and Stephen H. Schneider, "Simulating the Effects of Climate Change on Tropical Montane Cloud Forests," *Nature*, 15 April 1999, 608–10.

30. J. M. Kiesecker and Andrew R. Blaustein, "Synergism Between UV-B Radiation and a Pathogen Magnifies Amphibian Embryo Mortality in Nature," *Proceedings of the National Academy of Sciences*, 21 November 1995, 11049–52; Andrew R. Blaustein et al., "Pathogenic Fungus Contributes to Amphibian Losses in the Pacific Northwest," *Biological Conservation*, vol. 67 (1994), 251–54; Laurance, McDonald, and Speare, op. cit. note 4.

31. Vladimir Vershinin and Irina Kamkina, "Expansion of *Rana ridibunda* in the Urals—A

Danger for Native Amphibians?" *Froglog* (newsletter of the Declining Amphibian Populations Task Force), August 1999, 3; Gary M. Fellers and Charles A. Drost, "Disappearance of the Cascades Frog *Rana cascadae*, at the Southern End of its Range, California, USA," *Biological Conservation*, vol. 65 (1993), 177–181; Drost and Fellers, op. cit. note 5.

32. Pieter T. J. Johnson et al., "The Effect of Trematode Infection on Amphibian Limb Development and Survivorship," *Science*, 30 April 1999, 802–04.

33. William F. Gergits and Robert G. Jaeger, "Field Observations of the Behavior of the Red-backed Salamander (*Plethodon cinereus*): Courtship and Agonistic Interactions," *Journal of Herpetology*, vol. 24 (1990), 93; Stebbins and Cohen, op. cit. note 9, 4–6; Thomas M. Burton and Gene E. Likens, "Salamander Populations and Biomass in the Hubbard Brook Experimental Forest, New Hampshire," *Copeia*, 1975, 541–46.

34. Marsh frog from Manuel Polls Pelaz and Claude Rougier, "A Comparative Study of Buccal Volumes and the Branchial Skeleton of *Rana ridibunda* and *R. dalmantina* Tadpoles," *Copeia*, 1990, 661; Stebbins and Cohen, op. cit. note 9, 192–93; Mike Dickman, "The Effect of Grazing by Tadpoles on the Structure of a Periphyton Community," *Ecology*, Autumn 1968, 1188–90.

35. Ecuador from Stebbins and Cohen, op. cit. note 9, 236; Karen Lips, *Talk of the Nation*, National Public Radio, 18 February 2000.

36. Bruce K. Johnson and James L. Christiansen, "The Food and Food Habits of Blanchard's Cricket Frog, *Acris crepitans blanchardi* (Amphibia, Anura, Hylidae), in Iowa," *Journal of Herpetology*, vol. 10, no. 2 (1976), 74.

37. A. H. Kirkland, "Usefulness of the American Toad," *Farmer's Bulletin*, No. 196 (U.S. Department of Agriculture), March 1915, 2–16.

38. S.S. Jeevan, "Frogs are Falling Silent," *Down to Earth*, 31 May 1999, 32–33; Bangladesh from Sohrab Uddin Sarker and Noor Jahan Sarker, "Amphibian Population Status of Bangladesh," unpublished draft report (Bangladesh: Department of Zoology, Dhaka University, 1999); pesticides from "U.K. Report Finds 'Balanced' Viewpoint on Compatibility of Trade, Environment," *International Environment Reporter*, 10 July 1996; smuggling from Indraneil Das, Associate Professor, Institute of Biodiversity and Environmental Conservation, University Malaysia Sarawak, e-mail to author, 10 May 2000.

39. Reaser, op. cit. note 9, 16; W. Bryan Jennings, David F. Bradford, and Dale F. Johnson, "Dependence of the Garter Snake *Thamnophis elegans* on Amphibians in the Sierra Nevada of California," *Journal of Herpetology*, vol. 26, no. 4 (1992), 503–05; Stebbins and Cohen, op. cit. note 9, 248–49.

40. "Conveyor belt" from Stebbins and Cohen, op. cit. note 9, 5; Thomas M. Burton and Gene E. Likens, "Energy Flow and Nutrient Cycling in Salamander Populations in the Hubbard Brook Experimental Forest, New Hampshire," *Ecology*, late summer 1975, 1068–80; nutrient bridge from Richard J. Wassersug, "The Adaptive Significance of the Tadpole Stage with Comments on the Maintenance of Complex Life Cycles in Anurans," *American Zoologist*, vol. 15 (1975), 405–17.

41. Thirty liters from Gustavo Martinelli, "The Bromeliads of the Atlantic Forest," *Scientific American*, March 2000, 86; Luiz Carlos Serramo Lopez, Pablo J.F. Pena Rodrigues, and Ricardo Iglesias Rios, "Frogs and Snakes as Phoretic Dispersal Agents of Bromeliad Ostracods (Limnocytheridae: *Elpidium*) and Annelids (Naididae: *Dero*)," *Biotropica*, vol. 31, no. 4 (1999), 705–08.

42. Stebbins and Cohen, op. cit. note 9, 207–09.

43. Richard Cannell, "Spawning a Painkiller," *Financial Times*, 17 April 1998; John W. Daly, Chief, Section on Pharmacodynamics, Laboratory of Bioorganic Chemistry, National Institutes of Health, Bethesda, MD, discussion with author, 18 October 2000; John A. Kinch, "RX: One Tricolor Frog," *Nature Conservancy*, July/August 1998, 6.

44. John W. Daly et al., "Frog Secretions and Hunting Magic in the Upper Amazon: Identification of a Peptide that Interacts With an Adenosine Receptor," *Proceedings of the National Academy of Sciences*, November 1992, 10960–63; Stebbins and Cohen, op. cit. note 9, 208–09; Gordy Slack, "Magic Frog Leads Anthropologist to New Peptide," *Pacific Discovery*, spring 1993, 4.

45. Duellman, "Global Distribution," op. cit. note 8, 24; South America from William E. Duellman, "Distribution Patterns of Amphibians in South America," in Duellman, *Patterns and Distribution*, op. cit. note 8, 288; parts of the world containing the largest number of undiscovered species from Duellman, "Global Distribution," op. cit. note 8, 9; Michael J. Tyler, "Distribution Patterns of Amphibians in the Australo-Papuan Region," in ibid., 541.

46. Geographical mismatch from David B. Wake, "Action on Amphibians," *Trends in Ecology and Evolution*, 10 October 1998, 379–80.

47. Dr. Sohrab Sarker, Department of Zoology, Dhaka University, Bangladesh, e-mail to author, 20 May 2000.

48. Santiago Ron, Museum of Zoology, Universidad Catolica del Ecuador, e-mail to author, 12 May 2000.

49. Jeff E. Houlahan et al., "Quantitative Evidence for Global Amphibian Population Declines," *Nature*, 13 April 2000, 752–58.

50. DAPTF, <www.open.ac.uk/daptf>; Tim Halliday, International Director, Declining Amphibian Task Force, World Conservation Union–IUCN, Gland, Switzerland, e-mail to author, 13 October 2000.

51. Karen Lips et al., "Amphibian Declines in Latin America: Workshops to Design a Monitoring Protocol and Database," *Froglog* (newsletter of the Declining Amphibian Populations Task Force), February 2000, 1.

52. Amphibiaweb, <www.amphibiaweb.org>; Rex Dalton, "www Project Aims to Address Worldwide Decline in Amphibians," *Nature*, 3 February 2000.

53. James Collins, "Complex Mystery of Amphibian Decline Calls for a New Kind of Science," press release (Tempe, AZ: Arizona State University, 16 February 2000).

54. Linda Weir, North American Amphibian Monitoring Program (NAAMP) Coordinator, USGS, discussion with author, 18 September 2000; Minnesota from "Meeting Summary, 4th Conference of the North American Amphibian Monitoring Program," Penn State University, State College, PA, 27–28 June 1999, <www.im.mbs.gov/amphip/naamp4/summary.html#sally>, viewed 21 August 2000.

55. Cheryl Perusse Daigle, "The Creatures Beneath Our Feet: Amphibian Monitors Take to the Road," *Orion Afield*, winter 1998/99, 11; Eddie Nickens, "Salamander Sanctuary," *Inside Audubon*, March–April 1998, 143.

56. Mayan Forest Anuran Monitoring Project, <fwie.fw.vt.edu/mayanmon/maya_home.html>, viewed 19 August 2000; John R. Meyer, "The Maya Forest Anuran Monitoring Project: A Cooperative Tri-National Effort," *Froglog* (newsletter of the Declining Amphibian Populations Task Force), October 1999, 2–3.

57. Duellman, "Global Distribution," op. cit. note 8, 22; "Bucket Patrol Saves Toads," (Devon) *Western Morning News*, 15 February 2000.

58. FWS, "U.S. Listed Vertebrate Animal Species Report by Taxonomic Group as of 8/23/2000," <ecos.fws.gov/webpage>, viewed 23 August 2000.

59. Permanent listing from FWS, "Endangered

and Threatened Wildlife and Plants: Final Rule to List the Santa Barbara County Distinct Population of the California Tiger Salamander as Endangered" (Washington, DC: 21 September 2000); Allen Salzberg, "Tiger Salamander Receives Emergency Listing," *Herpdigest*, online journal, 19 February 2000.

60. Leo J. Borkin, "Distribution of Amphibians in North Africa, Europe, Western Asia, and the Former Soviet Union," in Duellman, *Patterns and Distribution*, op. cit. note 8, 394.

61. Number protected from <www.cites.org/CITES/eng/index.shtml>, 11 October 2000.

Chapter 5. Decarbonizing the Energy Economy

1. Jesse H. Ausubel, "Where Is Energy Going?" *The Industrial Physicist*, February 2000, 16–20.

2. Ibid.

3. Ibid.

4. Ibid.

5. Nebojsa Nakićenović, "Freeing Energy From Carbon," in Jesse H. Ausubel and H. Dale Langford, eds., *Technological Trajectories and the Human Environment* (Washington, DC: National Academy Press, 1997), 74–88.

6. Figure 5–1 based on G. Marland, T.A. Boden, and R.J. Andres, "Global, Regional, and National CO_2 Emission Estimates from Fossil Fuel Burning, Cement Production, and Gas Flaring: 1751–1997 (Revised August 2000)," Carbon Dioxide Information Analysis Center (CDIAC), Oak Ridge National Laboratory (ORNL), Oak Ridge, TN, 22 August 2000, on BP Amoco, *BP Amoco Statistical Review of World Energy 2000* (London: Group Media Publications, June 2000), 9, 25, 33, on Angus Maddison, *Monitoring the World Economy, 1820–1992* (Paris: Organisation for Economic Co-operation and Development (OECD), 1995), and on International Monetary Fund (IMF), *World Economic Outlook* (Washington, DC: May 2000).

7. J.T. Houghton et al., eds., *Climate Change 1995: The Science of Climate Change*, Contribution of Working Group I to the Second Assessment Report of the Intergovernmental Panel on Climate Change (New York: Cambridge University Press, 1996), 76–82; Worldwatch estimate and Figure 5–2 based on Mac Post, "Global Carbon Cycle (1992–1997)," CDIAC, ORNL, <cdiac.esd.ornl.gov/pns/graphics/globcarb.gif>, viewed 7 September 2000.

8. Figure 5–3 based on Marland, Boden, and Andres, op. cit. note 6, and on BP Amoco, op. cit. note 6.

9. Figure 5–4 based on J.M. Barnola et al., "Historical CO_2 Record from the Vostok Ice Core," and on J.R. Petit et al., "Historical Isotopic Temperature Record from the Vostok Ice Core," both in CDIAC, ORNL, U.S. Department of Energy, *Trends: A Compendium of Data on Global Change* (Oak Ridge, TN: 1999); Paul N. Pearson and Martin R. Palmer, "Atmospheric Carbon Dioxide Concentrations Over the Past 60 Million Years," *Nature*, 17 August 2000, 695–99; C.D. Keeling and T.P. Whorf, "Atmospheric CO_2 Concentrations (ppmv) Derived From In Situ Air Samples Collected at Mauna Loa Observatory, Hawaii," Scripps Institution of Oceanography, La Jolla, CA, 16 August 2000.

10. Houghton et al., op. cit. note 7, 76–82; Petit et al., op. cit. note 9, 429; James Hansen et al., Goddard Institute for Space Studies, National Aeronautic and Space Administration, Surface Temperature Analysis, "Global Temperature Anomalies in .01 C," <www.giss.nasa.gov/ data/update/gistemp>, viewed 15 August 2000; M.E. Mann, R.S. Bradley, and M.K. Hughes, "Northern Hemisphere Influences During the Past Millennium: Inferences, Uncertainties, and Limitations," *Geophysical Research Letters*, 15 March 1999, 759–62.

11. Thomas J. Crowley, "Causes of Climate

Change Over the Past 1000 Years," *Science*, 14 July 2000, 270–77.

12. Sydney Levitus et al., "Warming of the World Ocean," *Science*, 24 March 2000, 2225–29; Hansen cited in Richard A. Kerr, "Globe's 'Missing Warming' Found in the Ocean," *Science*, 24 March 2000, 2126–27.

13. James Hansen et al., "Global Warming in the Twenty-First Century: An Alternative Scenario," *Proceedings of the National Academy of Sciences*, 16 June 2000, 1–6.

14. Robert T. Watson, Marufu C. Zinyowera, and Richard H. Moss, *Climate Change 1995: Impacts, Adaptations and Mitigation of Climate Change: Scientific-Technical Analyses* (New York: Cambridge University Press, 1996); D.A. Rotchrock, Y. Yu, and G.A. Maykut, "Thinning of the Arctic Sea-Ice Cover," *Geophysical Research Letters*, 1 December 1999, 26, 23, 1–5; Lars H. Smedsrud and Tore Furevik, "Towards an Ice-Free Arctic?" *Cicerone*, February 2000, 1–7; W. Krabill et al., "Greenland Ice Sheet: High-Elevation Balance and Peripheral Thinning," *Science*, 21 July 2000, 428–30; Carsten Ruhlemann et al., "Warming of the Tropical Atlantic Ocean and Slowdown of Thermohaline Circulation During the Last Glaciation," *Nature*, 2 December 1999, 511–14.

15. Nebojsa Nakićenović et al., *Special Report On Emission Scenarios*, Summary for Policymakers, A Special Report of Working Group III of the Intergovernmental Panel on Climate Change (Geneva: May 2000), 1–27.

16. Tom M.L. Wigley, *The Science of Climate Change: Global and U.S. Perspectives* (Arlington, VA: Pew Center on Global Climate Change, 29 June 1999).

17. Hadley Centre for Climate Prediction and Research, UK Met Office, *Climate Change and its Impacts: Stabilisation of CO_2 in the Atmosphere* (London: October 1999), 1–3.

18. Michael Grubb with Christiaan Vrolijk and Duncan Brack, *The Kyoto Protocol: A Guide and Assessment* (London: Royal Institute of International Affairs/Earthscan, 1999), 37; Houghton et al., op. cit. note 7, 25.

19. Nebojsa Nakićenović, Arnulf Grubler, and Alan McDonald, eds., *Global Energy Perspectives* (New York: Cambridge University Press, 1998), 13; Figure 5–5 based on United Nations, *World Energy Supplies 1950–74* (New York: 1976), and on BP Amoco, op. cit. note 6, 9, 25, 33; Worldwatch estimate based on BP Amoco, op. cit. note 6, 38, and on International Energy Agency (IEA), *World Energy Outlook 1998* (Paris: IEA/Organisation for Economic Cooperation and Development (OECD), 1998), 157.

20. Nakićenović, Grubler, and McDonald, op. cit. note 19, 61; Martin I. Hoffert et al., "Energy Implications of Future Stabilization of Atmospheric CO_2 Content," *Nature*, 395, 29 October 1998, 881–84; Robert M. Margolis and Daniel M. Kammen, "Underinvestment: the Energy Technology and R&D Policy Challenge," *Science*, 30 July 1999, 690–92; Worldwatch estimates based on BP Amoco, op. cit. note 6, 33.

21. Worldwatch estimates based on BP Amoco, op. cit. note 6, 33; Jonathan E. Sinton and David G. Fridley, "What Goes Up: Recent Trends in China's Energy Consumption," *Energy Policy*, August 2000, 671–87.

22. Worldwatch estimates based on BP Amoco, op. cit. note 6, 33.

23. Ibid., 9.

24. Ibid.

25. Ibid., 25.

26. Ibid.

27. Ibid.

28. Ulrich Bartsch and Benito Muller, *Fossil Fuels in a Changing Climate* (New York: Oxford University Press, 2000), 161–213.

29. Nakićenović, Grubler, and McDonald, op. cit. note 19; Bartsch and Muller, op. cit. note 28.

30. Norman Myers and Jennifer Kent, *Perverse Subsidies: How Misused Tax Dollars Harm the Environment and Economy* (Washington, DC: Island Press, 2000); Andrea Baranzini, José Goldemberg, and Stefan Speck, "A Future for Carbon Taxes," *Ecological Economics*, March 2000, 396.

31. Nakićenović, Grubler, and McDonald, op. cit. note 19, 35–43; President's Committee of Advisors on Science and Technology (PCAST), *Federal Energy Research and Development for the Challenges of the Twenty-First Century* (Washington, DC: November 1997), 3-1–3-2.

32. Nakićenović, Grubler, and McDonald, op. cit. note 19, 35–43; IEA, *CO_2 Emissions from Fossil Fuel Combustion, 1971–1996* (Paris: IEA/OECD, 1999), xx, xxxiv; Figure 5–6 based on Marland, Boden, and Andres, op. cit. note 6, on World Bank, *World Development Indicators* (Washington, DC: 1999), and on IMF, op. cit. note 6.

33. Nakićenović, Grubler, and McDonald, op. cit. note 19, 35–43.

34. PCAST, *Powerful Partnerships: The Federal Role in International Cooperation on Energy Innovation* (Washington, DC: June 1999), 4-5; Figure 5–7 based on Marland, Boden, and Andres, op. cit. note 6, on Maddison, op. cit. note 6, and on IMF, op. cit. note 6.

35. Figure 5–8 based on Marland, Boden, and Andres, op. cit. note 6, and on World Bank, *World Development Indicators 2000* (Washington, DC: 2000); U.S. Department of Energy (DOE), Energy Information Administration (EIA), *International Energy Outlook 2000* (Washington, DC: March 2000).

36. DOE, op. cit. note 35; Nakićenović, Grubler, and McDonald, op. cit. note 19, 35–43.

37. Ernst Von Weizsacker, Amory B. Lovins, and L. Hunter Lovins, *Factor Four: Doubling Wealth, Halving Resource Use* (London: Earthscan Publications Ltd., 1997); PCAST, op. cit. note 34, 4-7–4-13.

38. PCAST, op. cit. note 34, 4-16–4-22; Alex Taylor III, "Another Way to Beat High Gas Prices," *Fortune*, 30 October 2000, 58.

39. PCAST, op. cit. note 34, 4-23–4-31.

40. Mark P. Mills and Peter Huber, "Dig More Coal: The PCs are Coming," *Forbes*, 31 May 1999, 70–72; Jay E. Haikes, EIA, DOE, Statement before the Subcommittee on National Economic Growth, Natural Resources, and Regulatory Affairs, Committee on Government Reform, U.S. House of Representatives, Hearings on Energy Implications of the Digital Economy, 2 February 2000; Joseph Romm with Arthur Rosenfeld and Susan Herrmann, *The Internet Economy and Global Warming* (Arlington, VA: Center for Energy and Climate Solutions, December 1999), 5–8.

41. Romm, Rosenfeld, and Herrmann, op. cit. note 40.

42. Curtis Moore and Jack Ihle, *Renewable Energy Policy Outside the United States*, REPP Issue Brief No. 14 (Washington, DC: Renewable Energy Policy Project, October 1999).

43. Worldwatch estimate based on BTM Consult, *Wind Energy Development: World Market Update 1999* (Ringkobing, Denmark: March 2000); Janet Ginsburg, "Green Power is Gaining Ground," *Business Week*, 9 October 2000, 44–45; Moore and Ihle, op. cit. note 42; Ginsburg, op. cit. this note.

44. Worldwatch estimate based on Paul Maycock, *PV News*, various issues; Ginsburg, op. cit. note 43; BP, "BP Solarex to Become BP Solar," press release (Baltimore, MD: 25 July 2000); Paul Maycock, "The World PV Market 2000: Shifting from Subsidy to 'Fully Economic?'" *Renewable Energy World*, July–August 2000, 59–74.

45. Worldwatch estimates based on BP Amoco, op. cit. note 6, 37, on John Lund, "World Status of Geothermal Energy Use: Past and Potential," *Renewable Energy World*, July-August 2000, 122–31, and on IEA, *Biomass Energy: Data, Analysis, and Trends* (Paris: IEA/OECD, 1998); Shell International Petroleum Company, *The Evolution of the World's Energy System 1860–2060* (London: Shell Centre, December 1995).

46. Shell, op. cit. note 45.

47. Ibid.; Kim Loughran, "Big Oil's Big Adventure," *Tomorrow*, May-June 1999, 37; Shell Renewables, *Shell Renewables, Summary of Activities* (London: 15 June 2000).

48. Nakićenović, Grubler, and McDonald, op. cit. note 19; Paul Raskin et al., *Bending the Curve: Toward Global Sustainability* (Stockholm: Stockholm Environment Institute, 1998), 49–56.

49. Raskin et al., op. cit. note 48, 56.

50. Bent Sørensen, *Long-Term Scenarios for Global Energy Demand and Supply* (Roskilde, Denmark: Roskilde University, January 1999), ii, 77–150.

51. Ibid.

52. Moore and Ihle, op. cit. note 42; BTM Consult, op. cit. note 43; Steve Nadis, "US Seeks a Big Increase in Wind Power," *Nature*, 15 July 1999, 201.

53. DOE, *Technology Opportunities to Reduce U.S. Greenhouse Gas Emissions* (Washington, DC: October 1997), 3-13–3-14.

54. Ausubel, op. cit. note 1; J.H. Ausubel, "Can Technology Spare the Earth?" *American Scientist*, March/April 1996, 166–78.

55. Ausubel, op. cit. note 1; Hydrogen Technical Advisory Panel, *Realizing a Hydrogen Future: Hydrogen Technical Advisory Panel Recommendations* (Washington, DC: DOE, August 1999).

56. "The Future of Fuel Cells," *Scientific American*, July 1999, 72–93; Tom Koppel, *Powering the Future: The Ballard Fuel Cell and the Race to Change the World* (New York: John Wiley & Sons Canada Ltd., 1999).

57. Tom Koppel and Jay Reynolds, "A Fuel Cell Primer: The Promise and the Pitfalls," <www.tomkoppel.com>, viewed 15 September 2000, 26; "DaimlerChrysler Offers First Commercial Fuel Cell Buses to Transit Agencies, Deliveries in 2002," *Hydrogen & Fuel Cell Letter*, May 2000, 1–2; "New Fuel Cell Prototypes, Concepts on Display at Frankfurt, Tokyo Auto Shows," *Hydrogen & Fuel Cell Letter*, October 1999, 1–2.

58. Robert F. Service, "Bringing Fuel Cells Down to Earth," *Science*, 30 July 1999, 682–85; California Fuel Cell Partnership, "California Fuel Cell Partnership Opens Headquarters Facility," press release (Sacramento, CA: 1 November 2000).

59. "Iceland, Shell, DaimlerChrysler, Norsk Hydro Form Company to Develop Hydrogen Economy," *Hydrogen & Fuel Cell Letter*, March 1999, 1–2; "High Oil Prices, Iceland Plans, Skepticism about Small Home PEMs Mark Annual NHA Meeting," *Hydrogen & Fuel Cell Letter*, April 2000, 1–2; Bragi Arnason and Thorsteinn Sigfusson, "Iceland: A Future Hydrogen Economy," *International Journal of Hydrogen Energy*, May 2000, 389–94.

60. Koppel and Reynolds, op. cit. note 57, 9; "Five Years in the Making, $18 Million Hydrogen Production/Fueling Station Opens in Munich," *Hydrogen & Fuel Cell Letter*, June 1999, 1–2; "Europe's First Hydrogen Gas Station Opens, Renewable H_2 to Come From Iceland," *Hydrogen & Fuel Cell Letter*, February 1999, 1–2.

61. Marc W. Jensen and Marc Ross, "The Ultimate Challenge: Developing an Infrastructure for Fuel Cell Vehicles," *Environment*, September 2000, 10–22; C.E.G. Padro and V. Putsche, *Survey of the Economics of Hydrogen Technologies* (Golden, CO: National Renewable Energy Laboratory (NREL), September 1999).

62. Jensen and Ross, op. cit. note 61; Pembina Institute for Appropriate Development, *Climate-Friendly Hydrogen Fuel: A Comparison of the Life-Cycle Greenhouse Gas Emissions for Selected Fuel Cell Vehicle Hydrogen Production Systems* (Drayton Valley, AB, Canada: Pembina Institute and David Suzuki Foundation, March 2000), 26.

63. Jensen and Ross, op. cit. note 61.

64. Ibid.; John A. Turner, "A Realizable Renewable Energy Future," *Science*, 30 July 1999, 687–89; P. Weiss, "Device Ups Hydrogen Energy From Sunlight," *Science News*, 16 September 2000, 182; Anastasios Melis et al., "Sustained Photobiological Hydrogen Gas Production Upon Reversible Inactivation of Oxygen Evolution in the Green Algae *Chlamydomonas reinhardtii*," *Plant Physiology*, January 2000, 122, 127–136.

65. Jensen and Ross, op. cit. note 61; DOE Hydrogen Technical Advisory Panel, *Fuel Choice for Fuel Cell Vehicles: An Overview* (Washington, DC: May 1999), 1–6; J. Ohi, *Blueprint for Hydrogen Fuel Infrastructure Development* (Golden, CO: NREL, January 2000), 1–3.

66. Bartsch and Muller, op. cit. note 28, 201; Peter Schwartz, Peter Leyden, and Joel Hyatt, *The Long Boom: A Vision for The Coming Age of Prosperity* (Reading, MA: Perseus Books, 1999).

67. Jae Edmonds et al., *International Emissions Trading and Global Climate Change*, Prepared for Pew Center on Global Climate Change (Arlington, VA: December 1999); Grubb, Vrolijk, and Brack, op. cit. note 18, 213–17.

68. "Summary of the Workshop on Best Practices in Policies and Measures," *Earth Negotiations Bulletin*, 15 April 2000, 1–9; John Gummer and Robert Moreland, *The European Union and Global Climate Change: A Review of Five National Programmes*, Prepared for Pew Center on Global Climate Change (Arlington, VA: June 2000), iii–vi.

69. John P. Weyant, *An Introduction to the Economics of Climate Change Policy* (Arlington, VA: Pew Center on Global Climate Change, July 2000), ii-iv.

70. "UK Says to Exceed Targeted Greenhouse Gas Cuts," *Reuters*, 10 March 2000; "Sweden Sets Out Ambitious Climate Goals," *ENDS Daily*, 13 April 2000.

71. José Goldemberg and Walter Reid, eds., *Promoting Development While Limiting Greenhouse Gas Emissions: Trends and Baselines* (New York: U.N. Development Programme and World Resources Institute, 1999), xii.

72. Ibid., 1–75.

73. Crescencia Maurer with Ruchi Bhandari, "The Climate of Export Credit Agencies," *Climate Note* (Washington, DC: World Resources Institute, May 2000); World Bank, "About the Prototype Carbon Fund," <www.prototypecarbonfund.org>, viewed 18 September 2000.

74. World Economic Forum, "Business Leaders Say Climate Change is Our Greatest Challenge," press release (Davos, Switzerland: 27 January 2000).

75. Kimberly O'Neill Packard and Forest Reinhardt, "What Every Executive Ought to Know About Global Warming," *Harvard Business Review*, July-August 2000, 129–35.

76. Christopher P. Loreti, William F. Wescott, and Michael A. Isenberg, *An Overview of Greenhouse Gas Emissions Inventory Issues*, Prepared for Pew Center on Global Climate Change (Arlington, VA: August 2000), iii–iv, ii, 41; Loughran, op. cit. note 47, 36–37.

77. Environmental Defense, "Global Corporations and Environmental Defense Partner to Reduce Greenhouse Gas Emissions," press release (New York: 17 October 2000); Loretti, Wescott, and Isenberg, op. cit. note 75, 16; World Wildlife Fund, "Top Companies Join WWF, Center for Energy and Climate Solutions in Innovative Effort to Save Energy, Lower

Greenhouse Gas Emissions," press release (Washington, DC: 1 March 2000).

78. Joseph Romm, *Cool Companies: How the Best Businesses Boost Profits and Productivity by Cutting Greenhouse Gas Emissions* (Washington, DC: Island Press, 1999); Pembina Institute, "Success Stories, Tools and Resources on How to Reduce Greenhouse Gas Emissions," <www.climatechangesolutions.com>, viewed 16 October 2000.

79. Lee Schipper, "One Man's Meat is Another's Carbon: Differences in Carbon Emissions Among Industrialised Countries," presentation to Resources for the Future, Washington, DC, 22 September 1999; Global Leaders for Tomorrow Environment Task Force, World Economic Forum, *Pilot Environmental Sustainability Index*, presented to Annual Meeting 2000, Davos, Switzerland, in collaboration with Yale Center for Environmental Law and Policy, Yale University, and with the Center for International Earth Science Information Network, Columbia University; Vincent Boland, "Global Indices to Support Sustainability," *Financial Times*, 9 September 1999.

80. Board on Sustainable Development, Policy Division, National Research Council, *Our Common Journey: A Transition Toward Sustainability* (Washington, DC: National Academy Press, 1999), 13; Worldwatch estimate based on Marland, Boden, and Andres, op. cit. note 6, on BP Amoco, op. cit. note 6, on Maddison, op. cit. note 6, and on IMF, op. cit. note 6.

Chapter 6. Making Better Transportation Choices

1. Joseph A. Schumpeter, *Capitalism, Socialism, and Democracy* (New York: HarperCollins, 1984).

2. Peter Bray and Barbara Brown, *Transport Through the Ages* (New York: Taplinger, 1971).

3. G.N. Georgano, ed., *Transportation Through the Ages* (New York: McGraw-Hill, 1972); John Woodforde, *The Story of the Bicycle* (New York: Universe Books, 1970).

4. Arnulf Grübler, *The Rise and Fall of Infrastructures: Dynamics of Evolution and Technological Change in Transport* (New York: Springer-Verlag, 1990); Arnulf Grübler and Nebojsa Nakićenović, *Evolution of Transport Systems: Past and Future* (Laxenburg, Austria: International Institute for Applied Systems Analysis (IIASA), 1991).

5. Grübler, op. cit. note 4, 127–45.

6. Arnulf Grübler, "Time for a Change: On the Patterns of Diffusion of Innovation," *Daedalus*, summer 1996, 19.

7. F.M.L. Thompson, "Nineteenth Century Horse Sense," *Economic History Review*, vol. 29 (1976), 60–81, cited in in Arnulf Grübler, *Technology and Global Change* (Cambridge, U.K.: Cambridge University Press, 1998), 320.

8. Early data on motor vehicle fleet from American Automobile Manufacturers Association, *World Motor Vehicle Data, 1996 Edition* (Detroit, MI: 1995); data for 1995–99 from Standard and Poor's DRI, *World Car Industry Forecast Report, December 1999* (London: 1999); U.S. increase in household vehicles from Patricia S. Hu and Jennifer R. Young, *Summary of Travel Trends: 1995 Nationwide Personal Transportation Survey*, prepared for the U.S. Department of Transportation, Federal Highway Administration (Washington, DC: 21 December 1999), 3; Western Europe from Stephen Perkins, "CO_2 Emissions from Transport," *Oil and Gas Journal*, January 1998, 5–6; developing and transition countries from World Bank, "Why is the Transport Sector Important?" <www.worldbank.org/html/fpd/transport?whytsimp.htm>, viewed 7 June 2000; Africa from U.S. Department of Energy (DOE), Energy Information Administration, *International Energy Outlook 2000* (Washington, DC: March 2000), 148–49.

9. Rise in air travel from data provided by Attilio Costaguta, Chief, Statistics and Economic Analysis Section, International Civil Aviation

Organization, Montreal, e-mail to Lisa Mastny, Worldwatch Institute, 2 November 1998, and from "Growth Versus Environment," *Financial Times*, 24 July 2000; growth rates and regional data from Kenneth Button, "The Usefulness of Current International Air Transport Statistics," *Journal of Transportation and Statistics*, May 1999.

10. Figure 6–1 from World Bank, *World Bank's Railways Database* (Washington, DC: Transportation, Water, and Urban Development Department, August 1998), and from data provided by World Bank and by Costaguta, op. cit. note 9.

11. World Bank railways database from World Bank, "Railways Overview," <www.worldbank.org/html/fpd/transport/rail_ss.htm>, viewed 7 June 2000.

12. World bicycle production from "World Market Report," *Interbike Directory 2000* (Laguna Beach, CA: Miller-Freeman, 2000); United Nations, *Industrial Commodity Statistics Yearbook 1997* (New York: 2000); decline in Asia from *Bicycle Retailer and Industry News Online*, <www.bicycleretailer.com>, viewed 10 October 2000, and from Edward A. Gargan, "Booming China Has Fewer Bikes on Road Ahead," *Seattle Times*, 4 October 2000.

13. South Korea form Organisation for Economic Co-operation and Development (OECD), *Environmental Performance Reviews: Korea* (Paris: 1997), 143, and from International Monetary Fund, *International Financial Statistics Yearbook* (Washington, DC: 2000), 610–11; Japan from OECD, *Environmental Performance Reviews: Japan* (Paris: 1994), 115–17; air travel from DOE, op. cit. note 8, 136.

14. Jeffrey R. Kenworthy et al., *An International Sourcebook of Automobile Dependence in Cities, 1960–1990* (Boulder, CO: University Press of Colorado, 1999), 615–28.

15. Survey from Fannie Mae Foundation, *Housing Facts and Findings*, winter 1999, 7; U.S. metropolitan areas from Wendell Cox Consultancy, "US Urbanized Areas: 1950–1990," <www.demographia.com>, viewed 7 June 2000; Czech Republic from Yaakov Garb, "Fighting Sprawl: Prague on the Edge," *Sustainable Transport*, fall 2000, 8.

16. "A Breath of Fresh Air for Indonesia," *Sustainable Transport*, fall 2000, 17.

17. Ken Alternburg et al., "Just-In-Time Logistics Support for the Automobile Industry," *Production and Inventory Management Journal*, second quarter 1999, 59–66.

18. Rise in carbon emissions from transportation from Lee Schipper and Céline Marie-Lilliu, *Transportation and CO_2 Emissions: Flexing the Link—A Path for the World Bank*, Environment Department Paper No. 69 (Washington, DC: World Bank, September 1999), 3–4; Japan and Denmark from Lynn Scholl, Lee Schipper, and Nancy Kiang, "CO_2 Emissions from Passenger Transport: A Comparison of International Trends from 1973 to 1992," *Energy Policy*, vol. 24, no. 1 (1996), 24.

19. Passenger travel from Lee Schipper, Céline Lilliu, and Michael Landwehr, "More Motion, More Speed, More Emissions: Will Increases in Carbon Emissions from Transport in IEA Countries Turn Around?" in *Energy Efficiency and CO_2 Reductions: The Dimensions of the Social Challenge*, Conference Proceedings of the European Council for an Energy Efficient Economy (ECEEE) 1999 Summer Study in Mandelieu, France (Paris: ECEEE, 1999); Figure 6–2 data from OECD, *Environmental Data Compendium* (Paris: 1997), Table 9.4, 206; travel surveys from L.J. Schipper, M.J. Figueroa, and R. Gorham, *People on the Move: A Comparison of Travel Patterns in OECD Countries*, Institute of Urban and Regional Development, University of California, Berkeley, CA, cited in Schipper and Marie-Lilliu, op. cit. note 18, 57.

20. Growth in automobile energy use from Don Pickerell and Paul Schimek, "Growth in Motor Vehicle Ownership and Use: Evidence from the Nationwide Personal Transportation Survey," *Journal of Transportation and Statistics*, May

1999, 9; decline in fuel economy from Lee J. Schipper, "Carbon Emissions from Travel in the OECD Countries," in Paul C. Stern et al., eds., Committee on the Human Dimensions of Global Change, National Research Council, *Environmentally Significant Consumption: Research Directions* (Washington, DC: National Academy Press, 1997), 52; sport utility vehicles in Western Europe from DOE, op. cit. note 8, 136.

21. Figure 6–3 population data from International Energy Agency (IEA), *CO_2 Emissions from Fuel Combustion* (Paris: OECD, 1999), and transportation energy data from DOE, op. cit. note 8; U.S. share of world transport energy for 1997 from IEA, *Energy Balances of OECD Countries* (Paris: OECD, 2000), II.299, and from IEA, *Energy Balances of Non-OECD Countries* (Paris: OECD, 1999), II.403; travel from Lee Schipper, "Life-Styles and the Environment," in *Technological Trajectories and the Human Environment* (Washington, DC: National Academy Press, 1997), 95; oil use from DOE, op. cit. note 8, 137.

22. DOE, op. cit. note 8, 241.

23. United States from Jane Holtz Kay, *Asphalt Nation* (New York: Random House, 1997); gender inequity from Deike Peters, "Breadwinners, Homemakers and Beasts of Burden: A Gender Perspective on Transport and Mobility," *Habitat Debate*, vol. 4, no. 2 (1998), 12–14; Africa from World Bank, "Transport and the Village in Sub-Saharan Africa," *Findings*, March 1997.

24. Traffic deaths from World Health Organization, *The World Health Report 1995* (Geneva: 1995), and from Christopher Willoughby, *Managing Motorization*, Discussion Paper, Transport Division (Washington, DC: World Bank, April 2000), ii; cost from World Bank, "Roads and Highways Overview," <www.worldbank.org/html/fpd/transport/roads_ss.htm>, viewed 7 June 2000.

25. Ingredients in exhaust from Jack Hollander, ed., *The Energy-Environment Connection* (Washington, DC: Island Press, 1992); nitrogen and sulfur from Gywneth Howells, *Acid Rain and Acid Waters*, 2nd ed. (London: Ellis Horwood Limited, 1995); acid rain in United States from James Dao, "Acid Rain Wall Found to Fail in Adirondacks," *New York Times*, 27 March 2000.

26. Rita Seethaler, *Health Costs Due to Road Traffic-related Air Pollution: An Impact Assessment Project of Austria, France and Switzerland*, Synthesis Report, prepared for the WHO Ministerial Conference on Environment and Health, London, June 1999, 9; Nino Künzli et al., "Public-Health Impact of Outdoor and Traffic-Related Air Pollution: A European Assessment," *Lancet*, 2 September 2000, 795–801.

27. C.S. Weaver and L.M. Chan, *Bangkok Particulate Study*, Report to the Thai Government and the World Bank, 1996, cited in Kenneth M. Gwilliam and Surhid Gautam, *Pollution from Motorcycles in Asia: Issues and Options*, World Bank Infrastructure Notes (Washington, DC: World Bank, April 2000).

28. Walter Hook, "Hurdles to Easing Congestion in Asia," *Habitat Debate* (Newsletter of the United Nations Centre for Human Settlements), vol. 4, no. 2 (1998); Simon Romero, "Rich Brazilians Rise Above Rush-Hour Jams," *New York Times*, 15 February 2000.

29. Land lost to roads from Walter Hook and Michael Replogle, "Motorization and Non-Motorized Transport in Asia," *Land Use Policy*, vol. 13, no. 1 (1996), 69–84; water problems from Chester Arnold and James Gibbons, "Impervious Surface Coverage: The Emergence of a Key Environmental Indicator," *Journal of the American Planning Association*, March 1996; effects on plants and animals from Stephen C. Trombulak and Christopher A Frissell, "Review of Ecological Effects of Roads on Terrestrial and Aquatic Communities," *Conservation Biology*, February 2000, 18–30; 20 percent estimate from Richard T. Forman, "Estimate of the Area Affected Ecologically by the Road System in the United States," *Conservation Biology*, February 200, 31–35.

30. OECD, *Environment and Transport: Syn-*

thesis of OECD Work on Environment and Transport and Survey of Related OECD, IEA and ECMT Activities (Paris: November 1999), 14.

31. Schipper, op. cit. note 20, 59; share of world carbon emissions from transportation from Schipper and Marie-Lilliu, op. cit. note 18; share from road traffic from IEA, *CO_2 Emissions*, op. cit. note 21.

32. Martin Wright with Carl Frankel and Dwight Holing, "Hot Air Space," *Tomorrow*, May–June 1997, 10–12; John H. Seinfeld, "Clouds, Contrails and Climate," *Nature*, 26 February 1998, 837–38; Olivier Boucher, "Air Traffic May Increase Cirrus Cloudiness," *Nature*, 7 January 1999, 30–31; U.S. General Accounting Office, *Aviation and the Environment: Aviation's Effects on the Global Atmosphere are Potentially Significant and Expected to Grow* (Washington, DC: February 2000); "Action on Aircraft CO_2 Emissions Waits for Global Accord," *ENDS Report*, March 2000, 44–45; Intergovernmental Panel on Climate Change, *Aviation and the Global Atmosphere*, Summary for Policymakers (Geneva: 1999).

33. James J. Murphy and Mark A. Delucchi, "A Review of the Literature on the Social Cost of Motor Vehicle Use in the United States," *Journal of Transportation and Statistics*, January 1998, 14–42; United Kingdom from David Maddison et al., *The True Costs of Road Transport* (London: Earthscan, 1996); United States from Clifford Cobb, "The Roads Aren't Free," *Challenge*, May/June 1999); other examples from Willoughby, op. cit. note 24, 4.

34. Emissions reductions from Richard O. Ackerman, Kenneth M. Gwilliam, and Louis S. Thompson, "The World Bank, Transport, and The Environment," *Japan Railway and Transport Review*, December 1998, 33; OECD, op. cit. note 30, 19; "Japan to Bring Forward Diesel Emission Regulation," *Reuters*, 23 February 2000; James N. Thurman and Todd Wilkinson, "EPA Takes Aim at the Exhaust Pipes of Trucks and Buses," *Christian Science Monitor*, 18 May 2000; Warren Brown, "Diesel Trucks, Buses Face Emission Curbs," *Washington Post*, 17 May 2000.

35. Gwilliam and Gautam, op. cit. note 27; Kseniya Lvovsky and Gordon Hughes, World Bank, "Addressing the Environmental Costs of Fuels in Developing Countries," prepared for the World Congress of Environmental and Resource Economists, Venice, Italy, 25–27 June 1998.

36. Miriam Jordan, "Electric Buses Put to Test in Nepal," *Wall Street Journal*, 31 May 2000.

37. "London Launches Eco-Friendly Taxis in Clean-up Drive," *Reuters*, 5 April 2000; Mexico from DOE, op. cit. note 8, 138; H. Josef Hebert, "Cities Turn to Natural Gas Buses," *Associated Press*, 8 August 2000; Jeffrey Ball, "Will California Pull the Plug on Electric Cars?" *Wall Street Journal*, 28 March 2000; Eric C. Evarts, "Goodbye to Gas-Burners?" *Christian Science Monitor*, 27 March 2000.

38. "Running on Hydrogen," *Down to Earth*, 15 February 2000, 6; "DaimlerChrysler Sees Fuel-Cell Vehicles in 2 Years," *Reuters*, 20 June 2000; "Automakers Drive Into Green Future Without a Roadmap," *Reuters*, 28 February 2000; Iceland from Seth Dunn, "The Hydrogen Experiment," *World Watch*, November/December 2000, 14–25.

39. Schipper and Marie-Lilliu, op. cit. note 18, 15–16; OECD, *Phasing Lead out of Gasoline: An Examination of Policy Approaches in Different Countries* (Paris: 1997); Rick Weiss, "EPA Seeks to End Use of Additive in Gasoline: Clean Air Measure Has Harmed Water," *Washington Post*, 21 March 2000; DOE, op. cit. note 8, 138; Timna Tanners, "Calif. Town Sues Big Oil for Millions in MTBE Damage," *Reuters*, 21 June 2000.

40. Lewis Mumford, *The City in History: Its Origins, Its Transformations, and Its Prospects* (San Diego: Harcourt Brace and Company, 1961), 507.

41. Metropolitan Washington Council of Governments cited in Alan Sipress, "Bad Traffic Grows Worse, Study Says," *Washington Post*, 16 December 1999; Replogle quoted in David W.

Chen, "The Cost of Urban Sprawl: Unplanned Obsolescence," *New York Times*, 30 January 2000.

42. Surface Transportation Policy Project, *An Analysis of the Relationship Between Highway Expansion and Congestion in Metropolitan Areas: Lessons from the 15-year Texas Transportation Institute Study* (Washington, DC: 1998); Alan Sipress, "Widen the Roads, Drivers Will Come," *Washington Post*, 4 January 1999.

43. European Commission, *European Sustainable Cities* (Brussels: 1996); Germany from "They Couldn't Car Less," *Gist Magazine*, 10 March 2000; Car Free Cities Network on the Internet at <www.eurocities.org/networks/netprog.html>, viewed 4 February 1999.

44. Steven Ambrus, "Bogota Takes a Breather," *EcoAmericas*, March 2000, 10.

45. Mexico City from Haynes C. Goddard, "Using Tradeable Permits to Achieve Sustainability in the World's Largest Cities," *Environmental and Resource Economics*, July 1997, 63–99, and from Gunnar S. Eskeland and Shantayanan Devarajan, *Taxing Bads by Taxing Goods: Pollution Control with Presumptive Charges* (Washington, DC: World Bank, 1996); Jimmy Weiskopf, "Bogota Among First Latin American Capitals to Implement Day Restricting Use of Cars," *International Environment Reporter*, 15 March 2000, 148–49.

46. Mexico City from Schipper and Marie-Lilliu, op. cit. note 18, 17; variable costs from Todd Litman, *The Costs of Automobile Dependency* (Victoria, BC, Canada: Victoria Transport Policy Institute, 1996); U.S. costs from American Automobile Association, *Your Driving Costs*, 1998 ed., cited in Stacy C. Davis, *Transportation Energy Data Book: Edition 19* (Oak Ridge, TN: Oak Ridge National Laboratory, September 1999), 5-14–5-15.

47. Ton Sledsens, *Sustainable Aviation: The Need for a European Environmental Aviation Charge* (Brussels: European Federation for Transport and Environment, March 1998); "Norwegian Government Backs Down on Jet Fuel Tax," *Global Environmental Change Report*, vol. 11, no. 2 (1999), 4; Dutch environmental group from Hans Buitelaar, "Fly for the Right Price," *Down to Earth*, 15 May 1999, 38; "EU Commission Steps Up Push On Jet Fuel Tax," *Reuters*, 13 March 2000; "EU Excludes Going It Alone On Jet Fuel Tax," *Reuters*, 14 March 2000.

48. H. William Batt, "Motor Vehicle Transportation and Proper Pricing: User Fees, Environmental Fees, and Value Capture," *Ecological Economics Bulletin*, first quarter 1998, 10–14; Ken Gwilliam and Zmarak Shalizi, "Road Funds, User Charges, and Taxes," *The World Bank Research Observer*, August 1999, 159–85; David Weller, "For Whom the Road Tolls: Road Pricing in Singapore," *Harvard International Review*, summer 1998, 10–11; Maryland from Michael Replogle and Lisa Swann, "A Pay Raise for Not Driving to Work," press release (Washington, DC: Environmental Defense, 11 May 2000).

49. Mary Williams Walsh, "Car Sharing Holds the Road in Germany," *Los Angeles Times*, 23 July 1998; Fred Bayles, "A Hot Import: Communal Cars for Congested Streets," *USA Today*, 21 July 2000.

50. Robert Cervero, *The Transit Metropolis: A Global Inquiry* (Washington, DC: Island Press, 1998).

51. Jonas Rabinovitch and Josef Leitman, "Urban Planning in Curitiba," *Scientific American*, March 1996, 26–33; Jonas Rabinovitch, "Innovative Land Use and Public Transport Policy," *Land Use Policy*, vol. 13, no. 1 (1996), 51–67.

52. Carlos Dora, "A Different Route to Health: Implications of Transport Policies," *British Medical Journal*, 19 June 1999, 1686–89; British Medical Association, *Road Transport and Health* (London: 1997), 20; Paloma Dallas, "Quixotic Bogota Mayor Pins Hopes on Bicycle," *Reuters*, 18 July 2000.

53. Craig Savoye, "More Americans Trade Car Keys for Bus Passes," *Christian Science Monitor*, 17 May 2000; Aravind Adiga, "Americans' Love Affair With The Car May Be Starting to Fade," *Financial Times*, 15 August 2000; "Public Transportation Ridership Continues to Soar: U.S. Transit Ridership Shows 4.8 Percent Increase in First Quarter," press release (Washington, DC: American Public Transportation Association, 17 July 2000).

54. PIARC Committee on Intelligent Transport, "Fighting Traffic with Technology," *The Futurist*, September–October 2000, 30.

55. Kevin Shafizadeh et al., *The Costs and Benefits of Telecommuting: An Evaluation of Macro-Scale Literature* (Davis, CA: Partners for Advanced Transit and Highways, July 1997).

56. Grübler, op. cit. note 4, 254–58.

57. Victoria Transport Policy Institute, "Transportation Impact Bibliographic Database," Victoria, BC, Canada, <www.vtpi.org>.

58. Peep Mardiste et al., *Blueprints for Sustainable Transportation in Central and Eastern Europe* (Brussels: CEE Bankwatch Network, May 1997).

59. Pietro S. Nivola, *Laws of the Landscape: How Policies Shape Cities in Europe and America* (Washington, DC: Brookings Institution Press, 1999).

60. Robert Puentes, *Flexible Funding for Transit: Who Uses It?* (Washington, DC: Brookings Institution Center on Urban and Metropolitan Policy, May 2000).

61. Surface Transportation Policy Project, *Changing Direction: Federal Transportation Spending in the 1990s* (Washington, DC: March 2000), 20; Don Chen, Sustainable Transportation Policy Project, Washington, DC, discussion with author, 30 August 2000.

62. Magda Stoczkiewicz, "The Conditions Attached To Western Money," *T&E Bulletin* (newsletter of the European Federation for Transport and Environment, Brussels), May 2000; Magda Stoczkiewicz, CEE Bankwatch Network, Brussels, discussion with author, 2 October 2000.

63. Walter Hook, Executive Director, Institute for Transportation and Development Policy, New York, discussion with author, 28 September 2000; Garb, op. cit. note 15, 9.

64. The Center for Responsive Politics, *The Big Picture: The Money Behind the 1998 Elections*, <www.opensecrets.org/pubs/bigpicture2000/index.htm>, viewed 3 August 2000; Greg Hitt, "Political Capital: Bush's Donors Have a Long Wish-List And Expect Results," *Wall Street Journal*, 31 July 2000.

65. Fred S. McChesney, *Money for Nothing: Politicians, Rent Extraction, and Political Extortion* (Cambridge, MA: Harvard University Press, 1997); J.L. Laws, "Boehlert Seeks Bipartisan Bloc to Fight CAFE Rider," *Energy and Environment Daily*, 13 March 2000, 11–12; Holly Bailey, "The Big Winners: A Look at Tax Breaks Won by Special Interests," *Money in Politics Alert*, Center for Responsive Politics, Washington, DC, 2 August 1999; Douglas Koplow and Aaron Martin, *Fueling Global Warming: Federal Subsidies to Oil in the United States* (Washington, DC: Greenpeace USA, June 1998).

66. Mark Murray, "King of the Roads," *National Journal*, 23 September 2000, 2954–59; quote from Major Garrett, "Now He Wants To Fix Your Airport," *U.S. News and World Report*, 5 July 1999, 20.

67. Lvovsky and Hughes, op. cit. note 35; Ralph Atkins, "German Greens Do U-Turn in the Drive Against Car," *Financial Times*, 29 May 2000.

68. Roger-Mark De Souza, *Household Transportation Use and Urban Air Pollution* (Washington, DC: Population Reference Bureau, 1999), 18; Tata Energy Research Institute, *Cleaner Air and Better Transport: Making Informed Choices* (New Delhi: TERI, 2000), 17.

69. J. Solomon, *To Drive or Vote: Young Adults' Culture and Priorities* (London: Charter Institute of Transport, 1998), cited in John Adams, "The Social Implications of Hypermobility," *Project on Environmentally Sustainable Transport: The Economic and Social Implications of Sustainable Transportation—Proceedings from the Ottawa Workshop* (Paris: OECD, 1998); psychology from Keith Bradsher, "Was Freud a Minivan or S.U.V. Kind of Guy?" *New York Times*, 17 July 2000.

70. Brad Knickerbocker, "Forget Crime—But Please Fix the Traffic," *Christian Science Monitor*, 16 February 2000.

71. Todd Nissen, "Ford to Boost Fuel Economy of SUVs by 25 Percent," *Reuters*, 28 July 2000; William Maclean, "BP Goes Greener With 'Beyond Petroleum' Rebrand," *Reuters*, 25 July 2000.

72. Chen, op. cit. note 61; Hook, op. cit. note 63.

Chapter 7. Averting Unnatural Disasters

1. U.S. Department of State, Bureau of Western Hemisphere Affairs, "Background Notes: Honduras," October 1999, <www.state.gov/www/background_notes/honduras_1099_bgn.html>, viewed 23 August 2000; gross domestic product from World Bank, *World Development Indicators 1999* (Washington, DC: 1999); Christian Aid, "In Debt to Disaster: What Happened to Honduras after Hurricane Mitch," October 1999, <www.christian-aid.org.uk/reports/indebt/indebt2.html>, viewed 22 August 2000.

2. Economic losses from Munich Reinsurance Company (Munich Re), *Topics: Annual Review of Natural Catastrophes 1999* (Munich: June 2000).

3. Death toll in 1998–99 from ibid., and from Munich Re, *Topics: Annual Review of Natural Catastrophes 1998* (Munich: March 1999); Gujarat from ibid.; Orissa from "Catastrophe!" *Down to Earth*, 30 November 1999; Indonesia from Charles Victor Barber and James Schweithelm, *Trial by Fire: Forest Fires and Forestry Policy in Indonesia's Era of Crisis and Reform* (Washington, DC: World Resources Institute, WWF-Indonesia, and Telepak Indonesia Foundation, 2000), 10; fires from Andy Rowell and Dr. Peter F. Moore, "Global Review of Forest Fires," Metis Associates for World Wide Fund for Nature and IUCN–World Conservation Union, 1999, <www.panda.org/forests4life/fires/fire_report.doc>, viewed 27 July 2000; Venezuela from International Federation of Red Cross and Red Crescent Societies (Red Cross), *Venezuela: Floods*, Situation Report No. 11 (Geneva: 12 October 2000).

4. Deforestation from World Resources Institute, *World Resources 2000–2001* (New York: Oxford University Press, 2000), 252–55.

5. Recovery cost savings from John Twigg, ed., *Developments at Risk: Natural Disasters and the Third World* (Oxford, U.K.: Oxford Centre for Disaster Studies, UK Coordinated Committee for the IDNDR, May 1998).

6. Munich Re, *World Map of Natural Hazards* (Munich: 1998), 19.

7. Figure 7–1 from Munich Re, op. cit. note 3, 19.

8. Figure 7–2 from Munich Re, *Topics 2000: Natural Catastrophes—The Current Position* (Munich: December 1999), 64–65.

9. Munich Re, op. cit. note 8; homeless from Worldwatch analysis of data from the Centre for Research on the Epidemiology of Disasters (CRED), *EM-DAT Database* (Brussels, Belgium), <www.md.ucl.ac.be/cred/emdat/intro.html>, obtained June 2000.

10. Munich Re, op. cit. note 8, 123; number affected from Red Cross, *World Disasters Report 2000* (Geneva: 2000).

11. Red Cross, *World Disasters Report 1999* (Geneva: 1999), 34.

12. Deaths and economic losses from Munich Re, op. cit. note 8; homeless from CRED, op. cit. note 9; China from Vaclav Smil and Mao Yushi, coord., *The Economic Costs of China's Environmental Degradation* (Cambridge, MA: American Academy of Arts and Sciences, 1998), 35; Rhine from "Dyke Disaster," *Down to Earth*, 15 March 1995.

13. Munich Re, op. cit. note 2, 19; Kobe from Munich Re, op. cit. note 8, 14; Seth Dunn, "Weather-Related Losses Hit New High," in Lester R. Brown, Michael Renner, and Brian Halweil, *Vital Signs 1999* (New York: W.W. Norton & Company), 74–75, based on data from Munich Re.

14. Figure 7–3 from Munich Re, op. cit. note 8, 64–65.

15. Figure 7–4 from Munich Re, op. cit. note 2, 24. "Richest countries" are defined as having a per capita annual GDP greater than $9,361, while "poorest" are defined as those with less than $760.

16. Peter B. Bayley, "Understanding Large River-Floodplain Ecosystems," *Bioscience*, March 1995, 153–58; J.V. Ward and J.A. Stanford, "Riverine Ecosystems: The Influence of Man on Catchment Dynamics and Fish Ecology," in D.P. Dodge, ed., *Proceedings of the International Large River Symposium*, Canadian Special Publication of Fisheries and Aquatic Sciences 106 (Ottawa: Department of Fisheries and Oceans, 1989), 56–64; Kunda Dixit and Inam Ahmed, "Managing the Himalayan Watershed: A Flood of Questions," *Economic and Political Weekly*, 31 October 1998, 2772.

17. "Flood Impact on Economy Limited—Expert," *China Daily*, 1 September 1998; Red Cross, "China: Floods, Appeal no 21/98 (Revised)," 9 November 1998, <www.ifrc.org/news/appeals.asp>; Yangtze deforestation from Carmen Revenga et al., *Watersheds of the World* (Washington, DC: World Resources Institute and Worldwatch Institute, 1998).

18. Erik Eckholm, "China Admits Ecological Sins Played Role in Flood Disaster," *New York Times*, 26 August 1998; John Pomfret, "Yangtze Flood Jolts China's Land Policies," *Washington Post*, 22 November 1998.

19. Eckholm, op. cit. note 18; Indira A. R. Lakshaman, "China Says Policies Worsened Floods, Beijing Alleges Corruption by Local Aides," *Boston Globe*, 17 August 1998; "China Flood: Logging Ban Announced to Check Devastating Floods," *Agence France Presse*, 23 August 1998; Erik Eckholm, "Stunned by Floods, China Hastens Logging Curbs," *New York Times*, 27 September 1998; Peichang Zhang et al., "China's Forest Policy for the 21st Century," *Science*, 23 June 2000, 2135–36.

20. William Weaver and Danny K. Hagans, "Aerial Reconnaissance Evaluation of 1996 Storm Effects on Upland Mountainous Watersheds of Oregon and Southern Washington: Wildland Response to the February 1996 Storm and Flood in the Oregon and Washington Cascades and Oregon Coast Range Mountains," paper prepared for Pacific Rivers Council, Eugene, OR (Arcata, CA: Pacific Watershed Associates, May 1996); "A Tale of Two Cities—and Their Drinking Water," in Sierra Club, *Stewardship or Stumps? National Forests at the Crossroads* (Washington, DC: June 1997); Romain Cooper, "Floods in the Forest," *Headwaters' Forest News*, spring 1997; David Bayles, "Logging and Landslides," *New York Times*, 19 February 1997; William Claiborne, "When a Verdant Forest Turns Ugly: 8 Oregon Deaths Blamed on Mud Sliding Down Clear-Cut Hillsides," *Washington Post*, 18 December 1996.

21. Brazil fires and ecology from Daniel C. Nepstad et al., "Large-Scale Impoverishment of Amazonian Forests by Logging and Fire," *Nature*, 8 April 1999, 505–08; Indonesia from BAPPENAS (National Development Planning Agency), 1999, cited in Barber and Schweithelm, op. cit. note 3; Rowell and Moore, op. cit. note 3.

22. BAPPENAS, op. cit. note 21.

23. Barber and Schweithelm, op. cit. note 3.

24. Paul Duggan, "Wildfires Engulf Thousands of Western Acres," *Washington Post*, 26 July 2000; Douglas Jehl, "Population Shift in the West Raises Wildfire Concerns: Los Alamos a Reminder," *New York Times*, 30 May 2000; fire statistics available from National Interagency Fire Center, <www.nifc.gov>; Kurt Kleiner, "Fanning the Wildfires," *New Scientist*, 19 October 1996.

25. Effects of flood control devices from Edward Goldsmith and Nicholas Hildyard, *The Social and Environmental Effects of Large Dams, Vol. One: Overview* (Camelford, Cornwall, U.K.: Wadebridge Ecological Centre, 1984); Haig Simonian, "Flood of Tears on the Rhine," *Financial Times*, 8 February 1995; historic flood data from "Dyke Disaster," op. cit. note 12.

26. Impacts from Gerald E. Galloway, "The Mississippi Basin Flood of 1993," prepared for Workshop on Reducing the Vulnerability of River Basin Energy, Agriculture and Transportation Systems to Floods, Foz do Iguacu, Brazil, 29 November 1995, from Stanley A. Changnon, ed., *The Great Flood of 1993: Causes, Impacts and Responses* (Boulder, CO: Westview Press, 1996), from James M. Wright, *The Nation's Responses to Flood Disasters: A Historical Account* (Madison, WI: Association of State Floodplain Managers, Inc. (ASFPM), April 2000), and from Donald L. Hey and Nancy S. Philippi, "Flood Reduction through Wetland Restoration: The Upper Mississippi River Basin as a Case History," *Restoration Ecology*, March 1995, 4–17; costs from Committee on Assessing the Costs of Natural Disasters, *The Impacts of Natural Disasters* (Washington, DC: National Academy Press, 1999); levee damage from Mary Fran Myers and Gilbert F. White, "The Challenge of the Mississippi Flood," *Environment*, December 1993, 6–9, 25–35.

27. Evolution of Mississippi River management from Hey and Philippi, op. cit. note 26, from Wright, op. cit. note 26, and from Myers and White, op. cit. note 26; artificial channel length from Jeff Hecht, "The Incredible Shrinking Mississippi Delta," *New Scientist*, 14 April 1990, 36–41; flood heights from L.B. Leopold, "Flood Hydrology and the Floodplain," in Gilbert F. White and Mary Fran Myers, eds., *Water Resources Update: Coping with the Flood—The Next Phase* (Carbondale, IL: University Council on Water Resources, 1994), cited in Richard E. Sparks, "Need for Ecosystem Management of Large Rivers and Their Floodplains," *Bioscience*, March 1995, 168–82; historic flood costs from William Stevens, "The High Costs of Denying Rivers Their Floodplains," *New York Times*, 20 July 1993; 1993 costs from Galloway, op. cit. note 26.

28. U.S. Fish and Wildlife Service (FWS), *Figures on Wetlands Lost in Mississippi Basin Prepared for Post Flood Recovery and the Restoration of Mississippi Basin Floodplains Including Riparian Habitat and Wetlands* (St. Louis, Mo.: Association of State Wetland Managers, 1993), cited in David S. Wilcove and Michael J. Bean, eds., *The Big Kill: Declining Biodiversity in America's Lakes and Rivers* (Washington, DC: Environmental Defense Fund, 1994); Hey and Philippi, op. cit. note 26; T.E. Dahl, *Wetland Losses in the United States 1780's to 1980's* (Washington, DC: U.S. Department of the Interior (DOI), FWS, 1990); fisheries from Calvin R. Fremling et al., "Mississippi River Fisheries: A Case History," in D.P. Dodge, ed., *Proceedings of the International Large River Symposium, Canadian Special Publication of Fisheries and Aquatic Sciences 106* (Ottawa, ON, Canada: Department of Fisheries and Oceans, 1989), 309–51, and from Joseph H. Wlosinski et al., "Habitat Changes in Upper Mississippi River Floodplain" in E.T. LaRoe et al., eds., *Our Living Resources* (Washington, DC: DOI, National Biological Service, 1995), 234–36.

29. Hecht, op. cit. note 27; National Marine Fisheries Service, *Fisheries of the United States, 1990* (Washington, DC: U.S. Government Printing Office, 1991), cited in J.M. Hefner et al., *Southeast Wetlands: Status and Trends, Mid-1970s to Mid-1980s* (Atlanta, GA: DOI, FWS, 1994).

30. Deborah Moore, *What Can We Learn From the Experience of the Mississippi?* (San Francisco, CA: Environmental Defense Fund, 7 September

1994); Wright, op. cit. note 26.

31. Missouri River Coalition, "Comments on the Missouri River Master Water Control Manual Review and Update Draft Environmental Impact Assessment," 1 March 1995.

32. Myers and White, op. cit. note 26; flood payments from ASFPM, *National Flood Programs in Review—2000* (Madison, WI: 2000), 42, citing National Wildlife Federation analysis *Higher Ground*.

33. For summary of task force recommendations, see Myers and White, op. cit. note 26; Galloway, op. cit. note 26; National Research Council (NRC), *Restoration of Aquatic Ecosystems: Science, Technology, and Public Policy* (Washington, DC: National Academy Press, 1992); ASFPM, op. cit. note 32.

34. A.Z.M. Obaidullah Khan, Bangladesh Centre for Advanced Studies, "Bangladesh Floods 1998 and Food Security," paper prepared for the Conference on Natural Disasters and Policy Response in Asia: Implications for Food Security," Harvard University, 30 April–1 May 1999; Government of Bangladesh, "Damages Caused by Flood 1998 (as of October 4, 1998)," <www.bangladeshonline.com/gob/flood98/foreign_1.htm>, viewed 1 August 2000, except death and homeless from Red Cross, op. cit. note 11, 32; Dixit and Ahmed, op. cit. note 16, 2772–74.

35. Dixit and Ahmed, op. cit. note 16; Khan, op. cit. note 34; "Bangladesh's Sundarbans Forest Threatened by Global Warming: Experts," *Agence France Presse*, 29 August 2000; Alfredo Quarto et al., "Choosing the Road to Sustainability: The Impacts of Shrimp Aquaculture and the Models for Change," prepared for the International Live Aquatics '96 Conference, Seattle, WA, 13–15 October 1996.

36. Hofer cited in Dixit and Ahmed, op. cit. note 16.

37. World coastal population estimates from Joel E. Cohen et al., "Estimates of Coastal Populations," *Science*, 14 November 1997, 1209c; U.S. coastal population from Sharon Begley and Thomas Hayden, "Floyd's Watery Wrath," *Newsweek*, 27 September 1999, 20; Florida population growth from Dennis S. Mileti, *Disasters by Design: A Reassessment of Natural Hazards in the United States* (Washington, DC: Joseph Henry Press, 1999), 42; hurricane frequency from Roger A. Pielke, Jr., "Reframing the US Hurricane Problem," in Roger Pielke, Jr., and Roger Pielke, Sr., eds., *Storms: Vol. 1* (London: Routledge, 2000), 391.

38. Molly O. Sheehan, "Urban Population Continues to Rise," in Lester R. Brown, Michael Renner, and Brian Halweil, *Vital Signs 2000* (New York: W.W. Norton & Company, 2000), 104–05; megacities from U.N. Population Division, *World Urbanization Prospects: The 1999 Revision* (New York: 1999).

39. Red Cross, op. cit. note 11, 17; Robert Hamilton, "Reducing Impacts of Natural Disasters—Are We Making Any Progress?" presentation for Society for International Development Working Group, Washington, DC, 27 July 2000; Mileti, op. cit. note 37; Kobe from Munich Re, op. cit. note 8, 123.

40. G.E. Hollis, "The Effect of Urbanization on Floods of Different Reference Interval," *Water Resources Research*, June 1975, 431–35; G.E. Hollis, "Rain, Roads, Roofs, and Runoff: Hydrology in Cities," *Geography*, vol. 73, no. 1 (1988), 9–18.

41. Red Cross, op. cit. note 11, 19.

42. Nicaraguans living below poverty line from Juanita Darling, "For Those Spared by Mitch, Safe Future Hinges on Jobs," *Los Angeles Times*, 3 April 1999.

43. James Wilson and Fiona Ortiz, "Mitch Teaches a Costly Conservation Lesson," *EcoAméricas*, December 1998, 6–8; Red Cross, op. cit. note 11, 42–54.

44. Christian Aid, op. cit. note 1; Organization of American States (OAS), "Hurricane Mitch

Affects OAS Land-Mine Removal Efforts," press release (Washington, DC: 9 November 1999).

45. Delayed development and rebuilding costs from U.S. Department of State, op. cit. note 1; debt and service from World Bank, op. cit. note 1.

46. "Bank Unveils New Support for Hurricane Mitch Recovery," *World Bank News*, 17 December 1998; Doug Rekenthaler, "The Battle of Mitch May be Over, But the War Has Just Begun," DisasterRelief.org, <www.disasterrelief.org/Disasters/990108Honduras>; Christian Aid, op. cit. note 1; "Debt Relief to Mitch Victims," *BBC Online News*, 11 December 1998.

47. Patricia L. Delaney and Elizabeth Shrader, "Gender and Post-Disaster Reconstruction: The Case of Hurricane Mitch in Honduras and Nicaragua," Decision Review Draft (Washington, DC: World Bank LCSPG/LAC Gender Team, January 2000); Ben Wisner, "Disaster Vulnerability: Scale, Power, and Daily Life," *GeoJournal*, vol. 30, no. 2 (1993), 127–40; migrant workers from Bob Edwards, East Carolina University, presentation at 25th Annual Hazards Research and Applications Workshop, Boulder, CO, July 11, 2000; Mileti, op. cit. note 37, 7.

48. Quote from Laurel Reuter, "A Flood of Fire and Ice," *Northwest Report*, December 1997, 28; Delaney and Shrader, op. cit. note 47; Elaine Enarson and Betty Hearn Morrow, eds., *The Gendered Terrain of Disaster: Through Women's Eyes* (Westport, CT: Praeger/Greenwood, 1998); Betty Hearn Morrow and Brenda Phillips, eds., *International Journal of Mass Emergencies and Disasters* (Special Issue on Women and Disasters), vol. 17, no 1 (1999); Walter Peacok et al., eds., *Hurricane Andrew: Ethnicity, Gender and the Sociology of Disasters* (London: Routledge, 1997).

49. Delaney and Shrader, op. cit. note 47; Enarson and Hearn Morrow, op. cit. note 48; Hearn Morrow and Phillips, op. cit. note 48

50. Delaney and Shrader, op. cit. note 47.

51. Wisner, op. cit. note 47; North Carolina from Edwards, op. cit. note 47.

52. Quote from Delaney and Shrader, op. cit. note 47.

53. Disaster assistance budget from Board on Natural Disasters (BOND), "Mitigation Emerges as Major Strategy for Reducing Losses Caused by Natural Disasters," *Science*, 18 June 1999, 1947.

54. Mileti, op. cit. note 37, 6.

55. World Bank/U.S. Geological Survey cited in Twigg, op. cit. note 5; German T. Velasquez et al., "A New Approach to Disaster Mitigation and Planning in Mega-cities: The Pivotal Role of Social Vulnerability in Disaster Risk Management," in Takashi Inoguchi et al., eds., *Cities and the Environment: New Approaches for Eco-Societies* (Tokyo: United Nations University Press, 1999), 162.

56. Lessons not learned from L. Comfort et al., "Reframing Disaster Policy: The Global Evolution of Vulnerable Communities," *Environmental Hazards*, vol. 1, no. 1 (1999), 39–44; Nicaraguan Government inactivity from "Mitch Worsened by Environmental Neglect," *Environmental News Network*, 19 March 1999; Las Casitas death toll from Red Cross, op. cit. note 11.

57. Death toll from Munich Re, op. cit. note 3; inaction from Uday Mahurkar, "Unheeded Death Knell," *India Today*, 22 June 1998.

58. Public health sector response from Meena Gupta, "Cyclone and After: Managing Public Health," *Economic and Political Weekly*, 13 May 2000, 1705–09; deaths and homeless from "Catastrophe!" op. cit. note 3, 6; all else from Richard Mahapatra, "A State in Chaos," *Down to Earth*, 15 December 1999, 30–33; Andhra Pradesh versus Orissa from ibid., and from David Gardner, "Orissa Death Toll Doubles as Recriminations Fly," *Financial Times*, 11 November 1999.

59. "Building amnesties" and "murderers" from Frederick Krimgold, World Institute for Disaster Risk Management, presentation at 25th Annual Hazards Research and Applications Workshop, Boulder, CO, 11 July 2000; Haldun Armagan, "Turkey's Earthquake Exposed a Landscape of Deals and Corruption," *The World Paper*, November 1999, 1–2; U.S. Geological Survey (USGS), "Implications for Earthquake Risk Reduction in the United States from the Kocaeli, Turkey Earthquake of August 17, 1999," Circular No. 1193 (Denver, CO: 2000).

60. Red Cross, op. cit. note 11; Honduras from Christian Aid, op. cit. note 1.

61. Subsidization from The H. John Heinz III Center for Science, Economics, and the Environment, *The Hidden Costs of Coastal Hazards: Implications for Risk Assessment and Mitigation* (Washington, DC: Island Press, 2000); 25 percent from BOND, op. cit. note 53, 1945; damage source from Mileti, op. cit. note 37, 128–32; model codes from ibid., 163.

62. Red Cross, op. cit. note 11.

63. Barber and Schweithelm, op. cit. note 3; David Swimbanks, "Forest Fires Cause Pollution Crisis in Asia," *Nature*, 25 September 1997, 321; "Jakarta Must Act to Stop Fires Spreading, Watchdog Warns," *Agence France Presse*, 4 August 1999; peat swamp scheme from Sander Thoenes, "In Asia's Big Haze, Man Battles Man-Made Disaster," *Christian Science Monitor*, 28 October 1997.

64. Indonesia from Barber and Schweithelm, op. cit. note 3, 3; Orissa from Bishnu N. Mohapatra, "Politics in Post-Cyclone Orissa," *Economic and Political Weekly*, 15 April 2000, 1353–55; Molly Moore, "Mexico's Flood Response Fuels Political Firestorm," *Washington Post*, 12 October 1999.

65. J.T. Houghton et al., eds., *Climate Change 1995: The Science of Climate Change* (Cambridge, U.K.: Cambridge University Press, 1996), 3–7; David R. Easterling et al., "Climate Extremes: Observations, Modeling, and Impacts," *Science*, 22 September 2000, 2068–74; Barber and Schweithelm, op. cit. note 3.

66. Sea level rise from Easterling et al., op. cit. note 65, and from R.A. Warrick et al., "Changes in Sea Level," in Houghton et al., op. cit. note 65, 363–64; British Meteorological Office from Red Cross, op. cit. note 11, 11; Susan Kim, "NYC Flooding Could Be Common in Next Century," *Disaster News Network*, 8 August 1999.

67. BOND, op. cit. note 53, 1943–97; disaster assistance costs from Heinz Center for Science, Economics and the Environment, op. cit. note 61, 13.

68. Mapping from ASFPM, op. cit. note 32; 25 percent from The H. John Heinz III Center for Science, Economics, and the Environment, "Evaluation of Erosion Hazards Summary," April 2000, <www.heinzcenter.org>.

69. China from "Forestry Cuts Down on Logging," *China Daily*, 26 May 1998; restoration from NRC, op. cit. note 33, and from Bayley, op. cit. note 16; Mississippi restoration from Hey and Philippi, op. cit. note 26, 4–17.

70. USGS, op. cit. note 59.

71. India from Krimgold, op. cit. note 59; Bangladesh from Aldo Benini, Global Land Mine Survey, discussion with author, 10 July 2000; Philip R. Berke et al., "Recovery After Disaster: Achieving Sustainable Development, Mitigation, and Equity," *Disasters*, vol. 17, no. 2 (1993), 93–109.

72. North Carolina from Sue Anne Pressley, "In North Carolina, Floyd Leaves a Toxic Legacy," *Washington Post*, 22 September 1999; Mozambique from Trygve Olfarnes, "Mozambique's Plight Moves International Donors," *Choices* (U.N. Development Programme), vol. 9, no. 3 (2000), 15; Central America from OAS, op. cit. note 44.

73. Earl J. Baker, "Hurricane Evacuation in the United States," in Pielke and Pielke, op. cit. note

37, 306; Robert L. Southern, "Tropical Cyclone Warning-Response Strategies," in ibid., 286, 297, 298.

74. "Sharing Information Worldwide," session at 25th Annual Hazards Research and Applications Workshop, Boulder, CO, 10 July 2000.

75. Mileti, op. cit. note 37, 11–14.

76. Red Cross, op. cit. note 11; Indonesia from Barber and Schweithelm, op. cit. note 3; Russia from Josh Newell, Anatoly Lebedev, and David Gordon, *Plundering Russia's Far East Taiga: Illegal Logging, Corruption and Trade* (Oakland, CA: Pacific Environment and Resources Center, Bureau for Regional Oriental Campaigns (Vladivostok, Russia), and Friends of the Earth–Japan, 2000).

77. Mileti, op. cit. note 37, 8; Munich Re, op. cit. note 8.

78. National Flood Insurance Program (NFIP) standards from French Wetmore, French and Associates, presentation at 25th Annual Hazards Research and Applications Workshop, Boulder, CO, 11 July 2000; NFIP critique and recommendations from Heinz Center for Science, Economics, and the Environment, op. cit. note 68, and from ASFPM, op. cit. note 32.

79. Red Cross, op. cit. note 11, 110; Twigg, op. cit. note 5.

80. Red Cross, op. cit. note 11, chapter 7.

81. World Bank, "Reducing 'Preventable' Costs of Natural Disasters Vital for Developing Countries," press release (Washington, DC: 2 February 2000).

82. "Tegucigalpa Declaration Jubilee 2000 Latin America and Caribbean Platform: 'Yes to Life, No to Debt,'" <www.jubilee2000uk.org/latin_america/declaration.htm>, viewed 31 March 1999; debt from World Bank, op. cit. note 1.

83. Donor pledges from Olfarnes, op. cit. note 72; debt relief from Joseph Kahn, "Wealthy Nations Propose Doubling Poor's Debt Relief," *New York Times*, 17 September 2000, and from "Mozambique Debt Payments Were Suspended by Paris Club of Government Donors," *Humanitarian Times*, 17 March 2000.

84. Christian Aid, op. cit. note 1; "Debt Relief to Mitch Victims," op. cit. note 46; "Paris Club Exposed over 'Misleading' Announcement on Hurricane Mitch Debt Deal," Jubilee 2000 Coalition, <www.jubilee2000uk.org/news/paris2412.html>, viewed 3 March 1999; quote from Oscar Andres Rodriquez, "Forgive the Debts of Poor Countries" (op ed), *Miami Herald*, 18 March 1999; Joseph Hanlon, "Nicaragua & Honduras Spend as Much on Debt Service as Reconstruction," Jubilee 2000 Coalition, <www.jubilee2000uk.org/reports/hurricane2910.html>, viewed 17 October 2000.

85. Post-hurricane outlays and International Monetary Fund limits from Hanlon, op. cit. note 84; "Washington Creditors Continue to Squeeze Debt Repayments out of Victims of Hurricane Mitch," Jubilee 2000 Coalition, <www.jubilee2000uk.org/news/hurricane2910.html>, viewed 17 October 2000.

86. Rodriquez, op. cit. note 84; Jubilee 2000 from <www.jubilee2000uk.org>; 10 percent from Joe Clancy, "Debt Relief: 'Too Little, Too Slow'," <CNN.com>, 26 September 2000; 3 percent from Christian Aid, op. cit. note 1. (In contrast, debt service is 32 percent of export revenues in Nicaragua and 21 percent in Honduras; Worldwatch calculations based on World Bank, op. cit. note 1.) Joseph Hanlon, *Debt, Default and Relief in the Past—And How We Are Demanding that the Poor Pay More This Time* (London: Jubilee 2000 Coalition, 1998), <www.jubilee2000uk.org/relief_before.html>, viewed 30 October 2000.

87. Jon Ingleton, ed., *Natural Disaster Management: A Presentation to the International Decade for Natural Disaster Reduction (IDNDR) 1999–2000* (Leicester, U.K.: Tudor Rose, 1999); Gilbert White, "A Decade of Missed Opportunities?" in ibid., 284–85.

88. Daniel Sarewitz and Roger Pielke, Jr., "Breaking the Global-Warming Gridlock," *The Atlantic Monthly*, July 2000, 55–64.

89. Dixit and Ahmed, op. cit. note 16; Mileti, op. cit. note 37.

Chapter 8. Ending the Debt Crisis

1. Harry S. Truman, inaugural address, 20 January 1949, available at the Whistlestop Project, <www.whistlestop.org>.

2. Ibid.

3. Ibid.

4. Author's observations, 9 April 2000.

5. Zambia budget shares from U.N. Development Programme (UNDP), Management Development and Governance Division, *Debt and Sustainable Human Development*, Technical Advisory Paper No. 4 (New York: 1999), 14–15; other figures from UNICEF, *Children in Jeopardy: The Challenge of Freeing Poor Nations from the Shackles of Debt* (New York: 1999), 11.

6. Aid flows—gross disbursements of official development assistance (which excludes loans made on near-commercial terms) to all developing countries—from Organisation for Economic Co-operation and Development (OECD), *Development Assistance Committee*, electronic database, 29 September 2000.

7. Karin Lissakers, *Banks, Borrowers, and the Establishment: A Revisionist Account of the International Debt Crisis* (New York: BasicBooks, 1991), 1-113.

8. Christian Suter, *Debt Cycles in the World-Economy: Foreign Loans, Financial Crises, and Debt Settlements, 1820–1990* (Boulder, CO: Westview Press, 1992), 106.

9. International Monetary Fund (IMF), *International Financial Statistics*, electronic database, June 2000; Thailand from Susan George, *A Fate Worse Than Debt* (New York: Grove Press, 1988), 61.

10. Figure 8–1 has Worldwatch updates of Angus Maddison, *Monitoring the World Economy, 1820–1992* (Paris: OECD, 1995), 21–22, using deflators and recent growth rates from IMF, *World Economic Outlook*, electronic database, Washington, DC, September 2000, and population data from U.S. Bureau of the Census, *International Data Base*, electronic database, Suitland, MD, 18 August 2000; Daniel Cohen, *Growth and External Debt: A New Perspective on the Africa and Latin American Tragedies*, Paper No. 9715 (Paris: Ecole Normale Supérieure, Centre d'Etudes Prospectives d'Economie Mathématique Appliquées à la Planification, July 1997), 11–15.

11. Mixed effects on poor from Tony Killick, "Structural Adjustment and Poverty Alleviation: An Interpretive Survey," *Development and Change*, vol. 26 (1995), 309; influence of middle class and civil servants from Arthur MacEwan, "Latin America: Why Not Repudiate?" *Monthly Review*, September 1986, 1–13; UNICEF from Catherine Caufield, *Masters of Illusion: The World Bank and the Poverty of Nations* (New York: Henry Holt and Co., 1996), 162.

12. Experience with adjustment lending suggests that even debtors that commit to placing savings from debt reduction in separate "poverty funds" can undermine this mechanism by reducing regular spending on basic social services or by more subtle means. Jubilee 2000 USA estimates that the debt crisis is killing 19,000 African children (under the age of five) each day (Jubilee 2000 USA, Washington, DC, <www.j2000usa.org>, viewed 30 October 2000). The calculation is flawed in several ways; in fact, UNICEF estimates that at most 11,500 children under five die each day in Africa (UNICEF, *State of the World's Children 2000* (New York: 2000), 84–87).

13. U.S. General Accounting Office (GAO), *Developing Countries: Debt Relief Initiative for Poor Countries Faces Challenges* (Washington,

DC: U.S. Government Printing Office, 2000), 57, 65; Shantayanan Devarajan, David Dollar, and Torgny Holmgren, *Aid and Reform in Africa: Lessons from Ten Case Studies* (Washington, DC: World Bank, forthcoming); debt burden from World Bank, *Global Development Finance 2000*, electronic database, Washington, DC, 2000.

14. Carl Jayarajah, William Branson, and Binayak Sen, *Social Dimensions of Adjustment: World Bank Experience, 1980–93* (Washington, DC: World Bank, 1996), 85–89; Tanzania and Zaire from George, op. cit. note 9, 100–05, 111.

15. Northeast from R.A. Becker and A. Lechtig, "Increasing Poverty and Infant Mortality in the North East of Brazil," *Journal of Tropical Pediatrics* 1986, cited in Giovanni Cornia, Richard Jolly, and Frances Stewart, eds., *Adjustment with a Human Face*, vol. 1 (Oxford: Clarendon Press, 1987), 43–44; southeast from Cornia, Jolly, and Stewart, op. cit. this note, 44–45.

16. Giovanni Andrea Cornia, "Economic Decline and Human Welfare in the First Half of the 1980s," in Cornia, Jolly, and Stewart, op. cit. note 15, 22–23; Manuel Pastor, Jr., "The Effects of IMF Programs in the Third World: Debate and Evidence from Latin America," *World Development*, vol. 15, no. 2 (1987), 249–62; Stanley Fischer, "Resolving the International Debt Crisis," in Jeffrey D. Sachs, ed., *Developing Country Debt and the World Economy* (Chicago: University of Chicago Press, 1989), 314; Mexico from Joan M. Nelson, "The Politics of Pro-Poor Adjustment," in Joan M. Nelson, ed., *Fragile Coalitions: The Politics of Economic Adjustment* (Oxford: Transaction Books, 1989), 111; Merih Celâsun and Dani Rodrik, "Turkish Experience with Debt: Macroeconomic Policy and Performance," in Sachs, op. cit. this note, 201–05.

17. London Environmental and Economics Centre, "Case Study for Thailand," in David Reed, ed., *Structural Adjustment and the Environment* (Boulder, CO: Westview Press, 1992), 107–08; Jamaica and Venezuela from David Reed, "Conclusions: Short-term Environmental Impacts of Structural Adjustment Programs," in David Reed, *Structural Adjustment, the Environment, and Sustainable Development* (London: Earthscan, 1996), 323.

18. Reed, "Conclusions," op. cit. note 17, 316.

19. James Kahn and Judith McDonald, "International Debt and Deforestation," in Katrina Brown and David W. Pearce, eds., *The Causes of Tropical Deforestation: The Economic and Statistical Analysis of Factors Giving Rise to the Loss of the Tropical Forests* (Vancouver, BC: UBC Press, 1994), 57–67. See also Ana Doris Capistrano, "Tropical Forest Depletion and the Changing Macroeconomy, 1968–85," in ibid., 68–85, which finds less significance for debt burden, but controls for currency devaluation, which is closely tied to debt burden and emerges as a statistically significant predictor of deforestation. Venezuela from Reed, *Structural Adjustment, the Environment, and Sustainable Development*, op. cit. note 17, 307.

20. World Bank, Debt Initiative for Heavily Indebted Poor Countries, Washington, DC, <www.worldbank.org/hipc>, viewed 24 October 2000; UNDP, *Human Development Report 2000* (New York: Oxford University Press, 2000), 159–160, 268. Tally of 29 includes Liberia and Somalia, which the UNDP does not classify, for lack of some data, but which almost certainly are of "low human development."

21. World Bank, *World Debt Tables 1992–93: External Finance for Developing Countries*, vol. 1 (Washington, DC: 1992), 75.

22. World Bank, op. cit. note 13 (calculations include transactions with the IMF); comparison to military spending uses an estimate of $165 billion for the present value of the debt and North American and West European military spending from Stockholm International Peace Research Institute, *SIPRI Yearbook 2000: Armaments, Disarmament and International Security* (Oxford: Oxford University Press, 2000), Table 5.1; 25 Leviticus 25–28 New English Bible;

"Who We Are," Jubilee 2000 Campaign, London, <www.jubilee2000uk.org/main.html>, viewed 23 August 2000.

23. World Bank, op. cit. note 20.

24. Cycles from Suter, op. cit. note 8, 61–63. Figure 8–2 is based on ibid., 70–71, and on Worldwatch estimates based on World Bank, *Global Development Finance 2000*, vol. 1 (Washington, DC: 2000), 158–71. Figure 8–2 includes consolidations (agreed periods during which no payments are made) and defaults; it assumes that the East German default of 1949 ended in 1990, that the Bulgarian default of 1932 and the Czechoslovakian default of 1960 ended in 1994, and that the Russian default of 1918 and the Chinese default of 1939 persist to this day.

25. Suter, op. cit. note 8, 34–38; boom-bust pattern from John Kenneth Galbraith, *A Short History of Financial Euphoria*, 2nd ed. (New York: Penguin Books, 1993), 12–25.

26. Peter H. Lindert and Peter J. Morton, "How Sovereign Debt Has Worked," in Sachs, op. cit. note 16, 227; Anthony Sampson, *The Money Lenders: Bankers in a World of Turmoil* (New York: Viking Press, 1982), 35, 231.

27. Juan Antonio Morales and Jeffrey D. Sachs, "Bolivia's Economic Crisis," in Sachs, op. cit. note 16, 63–65.

28. Lindert and Morton, op. cit. note 26, 229–30.

29. A.G. Kenwood and A.L. Lougheed, *The Growth of the International Economy, 1820–1990: An Introductory Text*, 3rd ed. (London: Routledge, 1992), 38–41; Susan M. Collins and Won-Am Park, "External Debt and Macroeconomic Performance in South Korea," in Sachs, op. cit. note 16, 129–30.

30. Lissakers, op. cit. note 7.

31. World Bank, op. cit. note 24, 158–71; debt total is from idem, op. cit. note 13; Figure 8–3 is based on ibid., and is missing the short-term debt total for 1970; problem-debt amount from Suter, op. cit. note 8, 69, 195–99.

32. World Bank, op. cit. note 13; population data from Census Bureau, op. cit. note 10; problem debt from Suter, op. cit. note 8, 69.

33. World Bank, op. cit. note 13.

34. Caufield, op. cit. note 11, 65.

35. Figure 8–4 has Worldwatch estimates based on World Bank, op. cit. note 13, and including transactions with the IMF. Low-income countries in a given year are those whose GNP per capita was $760 (in 1998 dollars) or less, following World Bank, op. cit. note 24, 141, and using the GNP per capita estimates computed using the "Atlas" method from idem, *World Development Indicators*, electronic database, Washington, DC, 2000, and the U.S. implicit GNP price deflator to convert to constant 1998 dollars.

36. Truman, op. cit. note 1.

37. Mozambique from George, op. cit. note 9, 87.

38. Official quoted in Caufield, op. cit. note 11, 89.

39. Ibid., 114–15.

40. Paul Collier, "Learning from Failure: The International Financial Institutions as Agencies of Restraint in Africa," in Andreas Schedler et al., eds., *The Self-Restraining State* (Boulder, CO: Lynn Rienner, 1999), 321.

41. Caufield, op. cit. note 11, 96–125.

42. Paul Mosley, Jane Harrigan, and John Toye, *Aid and Power: The World Bank and Policy-Based Lending*, vol. 1, 2nd ed. (London: Routledge, 1995), 32–36.

43. World Bank, "New World Bank Report Urges Broader Approach To Reducing Poverty," press release (Washington, DC: 12 September 2000).

44. Anthropologist from Caufield, op. cit. note 11, 229.

45. Zaire from George, op. cit. note 9, 106; capital flight from World Bank, op. cit. note 21, 43; Transparency International, "Transparency International Releases the Year 2000 Corruption Perceptions Index," press release (Berlin, 13 September 2000).

46. Guatemala from Caufield, op. cit. note 11, 207–08, and from World Bank, op. cit. note 13.

47. Tony Killick, *Aid and the Political Economy of Policy Change* (London: Routledge, 1998), 100–27, endpapers.

48. Countries adjusting without IFIs from Giovanni Andrea Cornia, "Adjustment Policies 1980–85: Effects on Child Welfare," in Cornia, Jolly, and Stewart, op. cit. note 15, 48–49.

49. Robert Wade, "Selective Industrial Policies in East Asia: Is *The East Asian Miracle* Right?" in Albert Fishlow et al., *Miracle or Design? Lessons from the East Asian Experience*, Policy Essay No. 11 (Washington, DC: Overseas Development Council (ODC), 1994), 59.

50. Dani Rodrik, *The New Global Economy and Developing Countries: Making Openness Work*, Policy Essay No. 11 (Washington, DC: ODC, 1999), 1–4, 150.

51. World Bank from Mosley, Harrigan, and Toye, op. cit. note 42, 166, for the 1980s, and from Killick, op. cit. note 47, 30, for the early 1990s; IMF from idem, *IMF Programmes in Developing Countries: Design and Impact* (London: Routledge, 1995), 61–63.

52. World Bank from Mosley, Harrigan, and Toye, op. cit. note 42, 166, and from Killick, op. cit. note 47, 30; IMF from idem, op. cit. note 51, 66. Table 8–4 from the following: Philippines from Paul Mosley, "The Philippines," in Mosley, Harrigan, and Toye, op. cit. note 42, 44–54, and from Killick, op. cit. note 47, 104; Argentina from Ramani Gunatilaka and Ana Marr, "Conditionality and Adjustment in Southeast Asia and Latin America," in Killick, op. cit. note 47, 74; Kenya from Collier, op. cit. note 40; sub-Saharan Africa from Devesh Kapur and Richard Webb, *Governance-Related Conditionalities of the IFIs*, prepared for the XII Technical Group Meeting of the Intergovernmental Group of 24 for International Monetary Affairs, Lima, 1–3 March 2000, revised May 2000; typical compliance from Mosley, Harrigan, and Toye, op. cit. 42, 134–36; Colin Kirkpatrick and Ziya Onis, "Turkey," in Mosley, Harrigan, and Toye, op. cit. note 42, 23–24, 30; Jane Harrigan, "Malawi," in ibid., 223–34; Bolivia from Gunatilaka and Marr, op. cit. this note, 74–78, and from Morales and Sachs, op. cit. note 27, 73–77; Zambia from Killick, op. cit. note 47, 112–13. On conventional wisdom see Collier, op. cit. note 40, and World Bank, *Assessing Aid: What Works, What Doesn't, and Why* (Oxford: Oxford University Press, 1998); Africa from Devarajan, Dollar, and Holmgren, op. cit. note 13.

53. Navroz K. Dubash and Colin Filer, "Papua New Guinea," in Frances J. Seymour and Navroz K. Dubash, *The Right Conditions: The World Bank, Structural Adjustment, and Forest Policy Reform* (Washington, DC: World Resources Institute, 2000), 35–46.

54. World Bank, op. cit. note 13.

55. World Bank, op. cit. note 21, 52; Lissakers, op. cit. note 7, 227–28.

56. World Bank, op. cit. note 21, 52–53.

57. Ibid.; deals from World Bank, op. cit. note 24, 158–62.

58. Melissa Moye, *Overview of Debt Conversion* (London: Debt Relief International, 2000), 14.

59. Ricardo Bayon, independent consultant, New York, unpublished tabulation, letter to author, 7 July 2000.

60. Moye, op. cit. note 58, 12–15.

61. George Santayana, *Life of Reason,* chapter 12, cited in Robert Andrews, *The Columbia Dictionary of Quotations* (New York: Columbia University Press, 1993).

62. Tanzania, Indonesia, and Nigeria statistics from World Bank, op. cit. note 35, and from idem, op. cit. note 24, 147–48; "arbitrary" from Euopean Network on Debt and Development (EURODAD), "Rethinking HIPC Debt Sustainability," EURODAD Background Paper (Brussels: July 2000).

63. Germany from Joseph Hanlon, *Debt, Default and Relief in the Past—And How We Are Demanding That the Poor Pay More This Time* (London: Jubilee 2000 Coalition, 1998), <www.jubilee2000uk.org/relief_before.html>.

64. Current debt from World Bank, op. cit. note 13; other information from GAO, op. cit. note 13, 42–56.

65. Figure of 28 percent from Daniel Cohen, *The HIPC Initiative: True and False Promises,* working paper (Paris: OECD Development Centre, forthcoming); other figures are Worldwatch estimates, based on ibid., and on debt indicators for 1998 from World Bank, op. cit. note 24, 102–04, and excluding Liberia and Somalia for lack of data, as well as Angola, Ghana, Laos, Viet Nam, and Yemen because they are not expected to participate in the HIPC initiative. All figures are based on present value of debt, not face value.

66. GAO, op. cit. note 13, 65.

67. EURODAD, *Poverty Reduction Strategies: What Have We Learned So Far?* (draft) (Brussels: 23 September 2000).

68. Ibid.; World Bank, op. cit. note 20.

69. Irûngûh Houghton, Manager, Policy Research Unit, ActionAid Kenya, Nairobi, interview with author, 15 June 2000.

70. Joan M. Nelson, Senior Associate, ODC, interview with author, 17 May 2000; on the World Bank's difficulties in enforcing governance-related conditions, see Kapur and Webb, op. cit. note 52.

71. GAO, op. cit. note 13, 57.

72. World Bank, op. cit. note 20.

73. EURODAD, op. cit. note 67.

74. Conable from Caufield, op. cit. note 11, 257.

Chapter 9. Controlling International Environmental Crime

1. Roberto Suro, "Fishing Boat Held After Chase," *Washington Post,* 11 May 2000; U.S. Coast Guard, "Cutter Chases, Stops Possible Drift Net Vessel," press release (Washington, DC: 8 May 2000); United Nations General Assembly, "Large-Scale Pelagic Drift-Net Fishing and its Impact on the Living Marine Resources of the World's Oceans and Seas," Resolution A/RES/46/215, adopted at the 79th Plenary Meeting, New York, 20 December 1991, available at <gopher://gopher.un.org:70/00/ga/recs/46/216>.

2. Suro, op. cit. note 1; total catch from U.S. Coast Guard, "Coast Guard Oversees Second Drift Net Retrieval," press release (Washington, DC: 11 May 2000).

3. International Fund for Animal Welfare (IFAW), "IFAW Releases Video of Major Wildlife Crime Ring Busted in Undercover Sting by Russian Police," press release (Yarmouth Port, MA: 13 April 2000). For a list of species restricted in trade, see CITES Secretariat, "CITES Databases," electronic database, <www.cites.org/CITES/eng/index.shtml>. In this chapter, the term "international environmental crime" is used loosely as a broad umbrella that encompasses a range of problems with compliance and enforcement that arise as international environmental treaties are implemented; the problems include situations in which

governments are not abiding by treaty commitments as well as cases in which private citizens or companies are flouting national laws passed to uphold international agreements.

4. For a generally positive view of the compliance record, see Abram Chayes and Antonia Handler Chayes, *The New Sovereignty* (Cambridge, MA: Harvard University Press, 1995), 3–9; George W. Downs, David M. Rocke, and Peter N. Barsoom, "Is the Good News About Compliance Good News About Cooperation?" *International Organization*, summer 1996, 379–406. For a general discussion of compliance and effectiveness issues, see David G. Victor, Kal Raustiala, and Eugene B. Skolnikoff, eds., *The Implementation and Effectiveness of International Environmental Commitments: Theory and Practice* (Cambridge: The MIT Press, 1998), 6–8, and Edith Brown Weiss and Harold K. Jacobsen, eds., *Engaging Countries: Strengthening Compliance With International Environmental Accords* (Cambridge: The MIT Press, 1998), 4–5.

5. Causes of growth in international environmental crime derived in part from Jonathan Krueger, *International Trade and the Basel Convention* (London: Royal Institute of International Affairs (RIIA), 1999), 87, and from U.N. Environment Programme (UNEP), "Workshop on Enforcement of and Compliance with Multilateral Environmental Agreements," Geneva, 12–14 July 1999, 5.

6. For international attention see, for example, "Communiqué," G8 Environment Ministers Meeting, Leeds Castle, U.K., 3–5 April 1998, and UNEP, op. cit. note 5.

7. Treaty numbers and Figure 9–1 based on UNEP, *Register of International Treaties and Other Agreements in the Field of the Environment 1999* (Nairobi: forthcoming). Protocols and amendments to existing accords are considered to be new agreements; this listing includes regional and international binding accords, but not bilateral agreements or nonbinding accords. Table 9–1 based on the Web sites of the various conventions listed in Table 9–2, plus the following Web sites: World Heritage from UNESCO, "Pathways to Preservation," <www.unesco.org/whc/nwhc/pages/doc/main.htm>; MARPOL from International Maritime Organization (IMO), "IMO Conventions," <www.imo.org/imo/convent/index.htm>; Law of the Sea from United Nations, Division for Ocean Affairs and the Law of the Sea (DOALS), home page, <www.un.org/Depts/los/index.htm>; Fish Stocks Agreement from UN, DOALS, "Marine Resources, <www.un.org/Depts/los/los_mr1.htm>; PIC Convention from <www.fao.org/ag/agp/agpp/pesticid/pic/pichome.htm>.

8. For links to the UNEP-managed conventions, see UNEP's Information Unit for Conventions home page, at <www.unep.ch/conventions/index.htm>; on the IMO conventions, see IMO, op. cit. note 7; functions of secretariats from David Hunter, James Salzman, and Durwood Zaelke, *International Environmental Law and Policy* (New York: Foundation Press, 1998), 409–11.

9. UNEP, *Global Environmental Outlook 2000* (London: Earthscan Publications Ltd., 2000), 200.

10. Victor, Raustiala, and Skolnikoff, op. cit. note 4; Ronald B. Mitchell, "Sources of Transparency: Information Systems in International Regimes," *International Studies Quarterly*, vol. 42 (1998), 109–30; Chayes and Chayes, op. cit. note 4.

11. René Coenen, Head, Office for the London Convention 1972, e-mail to Lisa Mastny, 25 August 2000; Bonn Convention reporting from Robert Vagg, Special Projects Officer, UNEP/CMS Secretariat, e-mail to Lisa Mastny, 25 August 200; quality problems in reports from Victor, Raustiala, and Skolnikoff, op. cit. note 4, 680–81, and from U. S. General Accounting Office (GAO), *International Environment: Literature on the Effectiveness of International Environmental Agreements* (Washington, DC: May 1999), 12–13.

12. CITES 1991 action from Edith Brown Weiss, "The Five International Treaties: A Living History," in Brown Weiss and Jacobsen, op. cit. note 4, 112; CITES reporting from Marceil Yeater, CITES Secretariat, e-mail to Lisa Mastny, 31 August 2000, and from Brown Weiss, op. cit. this note; trade restrictions from International Institute for Sustainable Development (IISD), "Summary of the 11th Conference of the Parties to the Convention on International Trade in Endangered Species of Wild Fauna and Flora: 10–20 April 2000," *Earth Negotiations Bulletin*, 24 April 2000, 4.

13. On the CITES secretariat, see Hunter, Salzman, and Zaelke, op. cit. note 8, 410–11, and Rosemary Sandford, "International Environmental Treaty Secretariats; a Case of Neglected Potential?" *Environmental Impact Assessment Review*, vol. 16, no. 1 (1996), 3–12; on the secretariat to the Montreal Protocol, see Winfried Lang, "Compliance Control in International Environmental Law: Institutional Necessities," *Heidelberg Journal of International Law*, vol. 56, no. 3 (1996), 685–95.

14. UNEP, op. cit. note 9, 205; Victor, Raustiala, and Skolnikoff, op. cit. note 4, 68–69; more than 20 inspections from Chayes and Chayes, op. cit. note 4, 186–87.

15. GAO, op. cit. note 11, 16–18; Sandford, op. cit. note 13. Table 9–2 based on the following sources: "Guide to the Climate Change Negotiation Process and the Secretariat," <www.unfccc.de/resource/process/components/institution/secret.html>, viewed 22 August 2000; Zoumana Bamba, Documentation Officer, Secretariat of the Convention on Biological Diversity, e-mail to Lisa Mastny, 24 August 2000; Yeater, op. cit. note 12; Dwight Peck, Executive Assistance for Communications, the Convention on Wetlands (Ramsar), e-mail to Lisa Mastny, 28 August 2000; Ozone Secretariat, "Report of the Eleventh Meeting of the Parties to the Montreal Protocol on Substances That Deplete the Ozone Layer" (Nairobi: 17 December 1999), 41–45; "About the Basel Convention," <www.basel.int/about.html>, viewed 23 August 2000; budget in 1998 from Basel Secretariat, "Institutional, Financial, and Procedural Arrangements," prepared for the Fifth Meeting of the Conference of the Parties to the Basel Convention on the Control of Transboundary Movements of Hazardous Wastes and Their Disposal, Basel, 6–10 December 1999, 4; migratory species convention from Vagg, op. cit. note 11; CCAMLR, "Administration," <www.ccamlr.org/English/e_commission/e_administration/e_admin_intro.htm>, viewed 24 August 2000; budget from CCAMLR, "Report of the Eighteenth Meeting of the Commission, Hobart, Australia, 25 October-5 November 1999" (Hobart: November 1999), 100 (converted on 27 September 2000 from Aus$2 million); London Convention from Coenen, op. cit. note 11. With regard to spending by domestic agencies, the U.S. Environmental Protection Agency's Clean and Safe Water program alone had a budget of $981 million in 2000, per U.S. Environmental Protection Agency, "Summary of the 2001 Budget" (Washington, DC: January 2000).

16. Hunter, Salzman, and Zaelke, op. cit. note 8, 430; TRAFFIC International, "What is TRAFFIC?" <www.traffic.org/about/what_is.html>, viewed 29 September 2000. For NGO monitoring, see Basel Action Network (BAN) Web site, <www.ban.org>; Greenpeace International Web site, <www.greenpeace.org>; Environmental Investigation Agency (EIA) Web site, <www.eia-international.org>; Ozone Action Web site, <www.ozone.org>; Web site of the International Southern Oceans Longline Fisheries Information Clearing House, <www.isofish.org/au>; and Global Forest Watch, World Resources Institute, <www.globalforestwatch.org>.

17. Molly O'Meara Sheehan, "Gaining Perspective," *World Watch*, March/April 2000, 14–24; Anthony Carpi and Jeffrey Mital, "The Expanding Use of Forensics in Environmental Science," *Environmental Science & Technology*, 1 June 2000, 262–66; whales from IFAW, "DNA Detectives Prove Japan's 'Scientific' Whaling Shelters Outlaw Trade," press release (Yarmouth Port, MA: 25 May 1999).

18. Chris Huxley, "CITES: The Vision," in Jon Hutton and Barnabas Dickson, eds., *Endangered*

Species Threatened Convention: The Past, Present and Future of CITES (London: Earthscan, 2000), 4–7; TRAFFIC International, "CITES at a Glance," <www.traffic.org/factfile/factfile_cites.html>, viewed 9 October 2000.

19. Reductions in trade from Caroline Taylor, "The Challenge of African Elephant Conservation," *Conservation Issues* (World Wildlife Fund-US (WWF-US)), April 1997, from Peter H. Sand, "Commodity or Taboo? International Regulation of Trade in Endangered Species," in Fridtjof Nansen Institute, *Green Globe Yearbook 1997* (Oxford: Oxford University Press, 1997), 26, and from Organisation for Economic Co-operation and Development (OECD), "Experience With the Use of Trade Measures in the Convention on International Trade in Endangered Species of Wild Fauna and Flora (CITES)" (Paris: 1997); $5 billion from Erik Madsen, Specialized Officer for Environmental Crime, International Criminal Police Organization (Interpol), letter to Lisa Mastny, 23 October 2000. Table 9–3 based on the following sources: tigers from IFAW, op. cit. note 3, and from TRAFFIC International, "Factsheet: Tiger," <www.traffic.org/factfile/factfile_tiger.html>, viewed 23 October 2000; panda from WWF-US, direct mailing to members, summer 2000; Ivory and Japan from "CITES Permits One-Time Sale of African Elephant Ivory," *Environment News Service*, 11 February 1999; Egypt from Nick Nuttall, "Rare Tiger is Being Poached to Extinction," (London) *Sunday Times*, 30 March 2000 (converted from £150,000 on 29 September 2000); elephant halving from Barrack Otieno, "African Nations Sustain Ban on Ivory Trade," *Environment News Service*, 17 April 2000; 20,000 from Alex Kirby, "Indian Minister's Elephant Alert," *BBC News Online*, 6 June 2000; parakeet from "Illegal Trade Threatens 10 'Most Wanted' Species," *Reuters*, 24 March 2000; musk deer from TRAFFIC International, "Factsheet: Musk Deer," <www.traffic.org/factfile/factfile_musk deer.html>, viewed 23 October 2000; ginseng from Kieran Murray, "Herbal Medicine Threatens Some Wild Plants," *Reuters*, 14 April 2000, and from Joel Bourne, "On the Trail of the 'Sang Poachers," *Audubon*, March/April 2000, 86; $2,000 and 80 percent from Patrick E. Tyler, "Poaching May Kill Fish That Lay the Golden Eggs," *New York Times*, 24 September 2000; more than half from Environmental Defense Fund, "Good Intentions, Promising Actions Sink During Year of the Ocean," press release (New York: 12 January 1999); leading importer from Joseph Roman, "Fishing for Evidence," *Audubon*, January/February 2000, 60.

20. Developing countries from Madsen, op. cit. note 19; Brazil from "Rio Rescues Black Market Monkeys, Turtles, Birds," *Reuters*, 1 February 2000; "Smugglers Decimating Colombia's Wildlife," *Reuters*, 14 June 1999; "Illegal Wildlife Trade in Central Asia Uncovered," *Environment News Service*, 4 December 1998.

21. Wealthy collectors from OECD, op. cit. note 19, 36; regions from CITES Secretariat and World Customs Organization (WCO), "The Customs—Wild Fauna and Flora," brochure, April 1998, <www.wcoomd.org/ENF/CITES/BROCHE/Mainfr.htm>, viewed 21 August 2000; "UK Fights Wildlife Crime with New Tools," *Environment News Service*, 17 February 2000.

22. Uses from CITES Secretariat and WCO, op. cit. note 21; exotic pets from Michael Grunwald, "U.S. Bags Alleged Trafficker in Reptiles," *Washington Post*, 16 September 1998.

23. Multibillion-dollar industry from "Endangered Wildlife Not the Best Chinese Medicine," *Reuters*, 29 October 1999; Andrea L. Gaski, *The Ongoing Sale of Tiger, Rhino, and Other Endangered Species Medicines in the United States* (Cambridge, U.K.: TRAFFIC International, February 1999); TRAFFIC International, "The Bear Facts: The East Asian Market for Bear Gall Bladder" (Cambridge, U.K.: July 1995).

24. Investment, risk, penalties, and parrot prices from CITES Secretariat and WCO, op. cit. note 21; price increases from James Heer, "Wildlife for Sale—Dead or Alive," transcript of Show 5 of *The Nature of Things with David Suzuki* (Toronto: Canadian Broadcasting Corporation, 1999), 1.

25. Legal consignments from "Animal Smuggling 'Earns Billions'," *BBC News Online*, 7 July 1999; other frauds from CITES Secretariat and WCO, op. cit. note 21; "Elephant Ivory Smuggler Jailed," *Environment News Service*, 9 August 2000.

26. Links to other illegal activities from CITES Secretariat and WCO, op. cit. note 21; snakes and turtles from "Environmental Crime—A Global Problem," *BBC News Online*, 5 April 1998; weapons from Tom Walker, "Illegal Ivory Sale Buys Guns for Mugabe," (London) *Sunday Times*, 9 July 2000.

27. Lack of central authority and agents from Amy E. Vulpio, "From the Forests of Asia to the Pharmacies of New York City: Searching for a Safe Haven for Rhinos and Tigers," *Georgetown International Environmental Law Review*, vol. 11 (1999), 466–67; 1993 report cited in OECD, op. cit. note 19, 33.

28. Developing-country challenges from UNEP, "Global Wildlife Treaty Celebrates 25 Years of Saving Endangered Species," press release (Geneva: 3 July 2000), and from Gordon Binder, "Does CITES Need More Teeth?" *Conservation Issues* (WWF-US), October 1994; CITES authorities and corruption from CITES Secretariat, "Review of Alleged Infractions and Other Problems of Implementation of the Convention," prepared for the 11th Conference of the Parties of CITES, Gigiri, Kenya, 10–20 April 2000, 3, 7.

29. "UK Fights Wildlife Crime with New Tools," op. cit. note 21; South Korea from TRAFFIC International, "Partners in Crime Prevention: Developments in the Enforcement of CITES," prepared for the 11th Conference of the Parties of CITES, available at <www.traffic.org/cop11/breifingroom/partnersincrimeprevention.html>, viewed 24 April 2000.

30. "Governments Meet to Review Efforts to Protect African Wildlife," *U.N. News Service*, 5 July 2000.

31. For a listing of key global fishing agreements, see U.N. Food and Agriculture Organization (FAO), "International Agreements and Initiatives," <www.fao.org/fi/agreem/agreem.asp>, viewed 23 October 2000; scope of illegal fishing from Kevin Bray, ed., "A Global Review of Illegal, Unreported and Unregulated (IUU) Fishing," paper prepared for the Expert Consultation on Illegal, Unreported and Unregulated (IUU) Fishing Organized by the Government of Australia in Cooperation with FAO, Sydney, 15–19 May 2000, 2; tuna prices from Greenpeace International, *Eradicating Pirate Fishing* (Amsterdam: May 2000), 4.

32. Fishing in foreign waters from Greenpeace International, op. cit. note 31, 3; Mike Mande, "Now French Navy to Police Tanzania Waters," *The East African* (Nairobi), 2 June 2000.

33. Movement to high seas and reporting fraud from Bray, op. cit. note 31, 5–6; specific oceans from Greenpeace International, "Greenpeace Acts to Protect Antarctic Waters While Ministers Talk," press release (Amsterdam: 22 January 1999); evasion and offloading from Greenpeace International, op. cit. note 31, 2, 8; penalties from Gavin Hayman, RIIA, London, e-mail to authors, 16 October 2000.

34. International Confederation of Free Trade Unions (ICFTU) et al., "Troubled Waters: Fishing, Pollution, and FOCs" (London: International Transport Workers' Federation (ITF), March 1999), 11, 14; flag state arrests from William Edeson, "Tools to Address IUU Fishing: The Current Legal Situation," paper prepared for the Expert Consultation on IUU Fishing Organized by the Government of Australia in Cooperation with FAO, Sydney, 15–19 May 2000, 7–8; changing flags and unfair competition from Greenpeace International, op. cit. note 31, 3, 7.

35. Estimate of 5 percent from Bray, op. cit. note 31, 2; higher estimate of 10 percent and 27 countries from ICFTU et al., op. cit. note 34, 15; 1,300 from Greenpeace International, op. cit. note 31, 7, 9; Law of the Sea requirements from United Nations, DOALS, "United Nations Convention on the Law of the Sea (UNCLOS):

Provisions of UNCLOS That Could Be Relevant for the Issue of 'Responsible Fisheries and IUU Fisheries'," prepared for the United Nations Open-Ended Informal Consultative Process on Ocean Affairs, at <www.un.org/Depts/los/Docs/UNICPO/convfishlink.htm>, updated 15 May 2000, 2; tax revenue from Rebecca Becker, "MARPOL 73/78: An Overview in International Environmental Enforcement," *Georgetown International Environmental Law Review*, vol. 10 (1998), 632.

36. Europe and Taiwan from Lloyd's Maritime Information Service 1999, cited in Greenpeace International, "What Do We Mean By Pirate Fishing Anyway?" <www.greenpeace.org/%7Eoceans/stoppiratefishing/pirate/whatpirate.html>, viewed 26 September 2000; subsidy from European Commission, "Council Regulation (EC) No. 2792/1999 of 17 December 1999: Laying Down the Detailed Rules and Arrangements Regarding Community Structural Assistance in the Fisheries Sector," <www.europa.eu.int/eur-lex/en/lif/dat/1999/en_399R2792.html>, viewed 14 August 2000, 4–5.

37. "Preliminary Draft: International Plan of Action to Prevent, Deter and Eliminate Illegal, Unreported and Unregulated Fishing," Appendix D of the "Report of the Expert Consultation," prepared during the Expert Consultation on IUU Fishing Organized by the Government of Australia in Cooperation with FAO, Sydney, 15–19 May 2000, 11–23; Compliance Agreement from FAO, "Agreement to Promote Compliance With International Conservation and Management Measures by Fishing Vessels on the High Seas," <www.fao.org/fi/agreem/com plian.complian.asp>, viewed 8 May 2000.

38. United Nations, "Agreement for the Implementation of the Provisions of the United Nations Convention on the Law of the Sea of 10 December 1982 Relating to the Conservation and Management of Straddling Fish Stocks and Highly Migratory Fish Stocks," adopted at the Sixth Session of the U.N. Conference on Straddling Fish Stocks and Highly Migratory Fish Stocks, New York, 24 July–4 August 1995, available at <gopher://gopher.un.org:70/00/LOS/CONF164/164_37.TXT>, 12–13; ratifications from United Nations, DOALS, "Status—Agreement on Fish Stocks," <www.un.org/Depts/los/los164st.htm>, viewed 9 October 2000.

39. Enforcement activities from Bray, op. cit. note 31, 3–6, and from Greenpeace International, op. cit. note 31, 21; landing and transshipment rights from Lee Kimball, consultant, Washington, DC, e-mail to authors, 12 October 2000; Allan Dowd, "Canada Patrols See Drop in Illegal Driftnet Fishing," *Reuters*, 9 June 2000.

40. International Commission for the Conservation of Atlantic Tunas (ICCAT), "Resolution: Interpretation and Application of the ICCAT Bluefin Tuna Statistical Document Program," adopted at the Ninth Special Meeting of ICCAT, Madrid, November–December 1994; other trade measures from Masayuki Komatsu, "The Importance of Taking Cooperative Action Against Specific Fishing Vessels that are Diminishing Effectiveness of Tuna Conservation and Management Measures," prepared for the Expert Consultation on IUU Fishing Organized by the Government of Australia in Cooperation with FAO, Sydney, 15–19 May 2000, 3–8.

41. For critique of the convention, see Kal Raustiala and David G. Victor, "Biodiversity Since Rio: The Future of the Convention on Biological Diversity," *Environment*, May 1996.

42. IMO, "IMO—The First Fifty Years," <www.imo.org/imo/50ann/history3.htm>, viewed 14 August 2000; members from IMO, op. cit. note 7; conventions from IMO, "International Conventions," <www.imo.org/imo/convent/treaty.htm>, viewed 19 June 2000; incorporation from UN, DOALS home page, op. cit. note 7.

43. London Convention successes from Brown Weiss, op. cit. note 12, 133–35; MARPOL success from U.S. National Research Council, cited in IMO, "Marpol—25 Years," *Focus on IMO*, October 1998, 8.

44. MARPOL violations from United Nations,

"Oceans and the Law of the Sea: Report of the Secretary General, 2000," document prepared for the 55th Session of the General Assembly, 20 March 2000, <www.un.org/Depts/los/GA55_61.htm>, para. 182; ongoing spills from Paul Nelson, "Pollution From Ships: A Global Perspective," in Neil Gunningham, Jennifer Norberry, and Sandra McKillop, eds., *Environmental Crime: Proceedings of a Conference Held 1–3 September 1993 in Hobart, Australia* (Canberra: Australian Institute of Criminology, 1995), 1–3; 1.1 million from International Tanker Owners Pollution Federation Limited, "Past Spills," <www.itopf.com/stats.html>, viewed 21 June 2000.

45. GAO, *Marine Pollution: Progress Made to Reduce Marine Pollution by Cruise Ships, but Important Issues Remain* (Washington, DC: February 2000); "floating cities," 77 percent, and types of waste from Kira Schmidt, "Cruising for Trouble: Stemming the Tide of Cruise Ship Pollution" (San Francisco, CA: Bluewater Network, March 2000), 1; capacity from Hayman, op. cit. note 33; costs of disposal from Douglas Franz, "Gaps in Sea Laws Shield Pollution by Cruise Lines," *New York Times*, 3 January 1999; monitoring and enforcement from Becker, op. cit. note 35, 626; "Royal Caribbean Sentenced for Ocean Dumping," *Reuters*, 4 November 1999.

46. ITF, "ITF Maritime Department," information sheet, <www.itf.org.uk/SECTIONS/Mar/mar.html>, viewed 6 July 2000; reasons for re-flagging from ICFTU et al., op. cit. note 34, 11–12, and from ITF, "Malta: A Snapshot of an FOC Register," <www.itf.org.uk/SECTIONS/Mar/malta1.htm>, viewed 6 July 2000; pollution and accidents from Becker, op. cit. note 35, 634; detentions from ITF, "ITF Maritime Department," op. cit. this note.

47. Limited IMO enforcement power from IMO, "International Conventions," op. cit. note 42; legal expenses from Becker, op. cit. note 35, 632; strengthening IMO role from ICFTU et al., op. cit. note 34, 18–19; other solutions from Becker, op. cit. note 35, 638–39.

48. Shift to non-flag states from Ten Hoopen and G.H. Henk, "Compliance and Enforcement of Internationally Agreed Upon Regulations in the International Shipping Industry," in *Proceedings of the Fifth International Conference on Environmental Compliance and Enforcement, Monterey, California, November 1998*, Vol. 1 (Washington, DC: International Network for Environmental Compliance and Enforcement (INECE), 1999), 695–96; Paris Memorandum of Understanding Web site, <www.parismou.org>, viewed 17 October 2000.

49. For full text of Basel Convention, see <www.basel.int/text/text.html>; Charles W. Schmidt, "Trading Trash: Why the U.S. Won't Sign on to the Basel Convention," *Environmental Health Perspectives*, August 1999.

50. Estimates of 300–500 million tons and one tenth from UNEP; 90 percent and 20 percent from OECD; both estimates, as well as types of waste, from Jonathan Krueger, "What's to Become of Trade in Hazardous Wastes? The Basel Convention One Decade Later," *Environment*, November 1999, 12–13.

51. "Asia Seen as Dump Site for Toxic Waste," *Bangkok Post*, 12 February 2000; Greenpeace India, "India Remains a Favoured Dumping Ground for Global Toxic Wastes," press release (New Delhi: 11 September 2000).

52. Developing-country and other resource challenges from Krueger, op. cit. note 5, 87, 91: Basel secretariat survey cited in Svend Auken, "Celebration and Change," *Our Planet*, vol. 10, no. 4 (1999), 4.

53. Methods of fraud from Leo C. Blankers, "Illegal Transports of Waste: Tricks of the Trade," in *Proceedings of the Fourth International Conference on Environmental Compliance and Enforcement, Chiang Mai, Thailand, April 1996*, Vol. 2 (Washington, DC: INECE, 1997), 1–7; role of organized crime from Basel secretariat, "Illegal Traffic in Hazardous Wastes Under the Basel Convention," information sheet, <www.basel.int/pub/illegaltraffic.html>, viewed 23 February 2000.

54. Recycling loophole and Greenpeace estimate from Jennifer Clapp, "The Illicit Trade in Hazardous Wastes and CFCs: International Responses to Environmental 'Bads'," in R. Friman and P. Andreas, eds., *The Illicit Global Economy and State Power* (Lanham, MD: Rowman and Littlefield, 1999), 107–09; unsafe dumping from Krueger, op. cit. note 5, 87; Japanese dumping from "Illegal Dumping," *Mainichi Daily News*, 13 January 2000.

55. BAN, "About the Basel Ban...," <www.ban.org>, viewed 25 September 2000; Clapp, op. cit. note 54, 110; required ratifications from Basel secretariat, "Ratifications," <www.basel.int/ratif/ratif.html>, viewed 22 September 2000; opposition from BAN, "BAN Report and Analysis of the Fifth Conference of the Parties to the Basel Convention, December 6–10, 1999," <www.ban.org>, viewed 22 September 2000.

56. Krueger, op. cit. note 5, 88.

57. Basel Secretariat, "Text of Basel Protocol on Liability and Compensation for Damage Resulting from Transboundary Movements of Hazardous Wastes and Their Disposal and of the Decision Regarding the Basel Protocol," at <www.basel.int/COP5/docs/prot-e.pdf>, viewed 15 August 2000; insurance and compensation from Daniel Pruzin, "Hazardous Waste Insurers Seen Ready to Meet Demand for Cross-Border Liability Coverage," *International Environmental Reporter*, 10 December 1999; increasing costs from Krueger, op. cit. note 5, 94; loopholes from BAN and Greenpeace International, "Toxic Waste Treaty Declares Next Decade: No Time for Waste," press release (Seattle, WA: 10 December 1999), and from United Nations, op. cit. note 44, para. 205.

58. Parties from Ozone Secretariat, "Status of Ratification/Accession/Acceptance/Approval of the Agreements on the Protection of the Stratospheric Ozone Layer," <www.unep.org/ozone/ratif.htm>, viewed 17 August 2000; 95 chemicals from K. Madhava Sarma, "Of Potholes and Ozone Holes," *Our Planet*, vol. 10, no. 4 (1999), 30; requirements from Ozone Secretariat, "Summary of Control Measures Under the Montreal Protocol," information sheet, <www.unep.org/ozone/control-measures.htm>, viewed 15 May 2000.

59. UNEP and Ozone Secretariat, *Synthesis of the Reports of the Scientific, Environmental Effects, and Technology and Economic Assessment Panels of the Montreal Protocol* (Nairobi: UNEP, February 1999), 21; World Meteorological Organization, "Unprecedented Rate of Ozone Loss Measured One to Two Weeks Earlier Over and Near Antarctica," press release (Geneva: 15 September 2000); U.S. National Aeronautics and Space Administration, Ames Research Center, "Arctic Ozone May Not Recover as Early as Predicted," press release (Moffett Field, CA: 25 May 2000).

60. Production declines from UNEP, *Data Report on Production and Consumption of Ozone Depleting Substances, 1986–1998* (Nairobi: October 1999), 12; problems in countries in transition from David G. Victor, "The Operation and Effectiveness of the Montreal Protocol's Non-Compliance Procedure," in Victor, Raustiala, and Skolnikoff, op. cit. note 4, 155; UNEP, "High-Level Meeting Brings Compliance With Montreal Protocol Within Sight for Central Asian and Caucasus Region," press release (Nairobi: 3 May 2000); Russian production from UNEP, "Global Efforts to Protect Ozone Layer Get Major Boost; Donors Commit Funds to Close Russia's CFC Factories," press release (Nairobi: 8 October 1998).

61. World Bank action from World Bank, "World Bank, Russian Federation Sign Grant Agreement to Eliminate Production of Ozone Depleting Substances," press release (Washington, DC: 25 October 2000), and from Karin Shepardson, World Bank/GEF Regional Coordinator for Europe and Central Asia, discussion with Lisa Mastny, 3 November 2000; use of incentives elsewhere from UNEP, "Report of the Implementation Committee Under the Non-Compliance Procedure for the Montreal Protocol on the Work of Its Twenty-Third Meeting," 10 December 1999, prepared for the Twenty-Third Meeting of the Implementation Committee Under the Non-Compliance Procedure for the Montreal Protocol, Beijing, 27 November 1999.

62. UNEP, "UNEP Releases Analysis of Trends in Developing Country Compliance With Ozone Treaty," press release (Nairobi: 22 August 2000); UNEP, "Ozone Talks Focus on Action by Developing Countries," press release (Nairobi: 10 July 2000).

63. Jennifer Clapp, "The Illegal CFC Trade: An Unexpected Wrinkle in the Ozone Protection Regime," *International Environmental Affairs*, fall 1997, 263–65.

64. Illegal trade estimates from Duncan Brack, "Illegal Trade in Ozone-Depleting Substances as an International Environmental Crime," paper presented at the UNEP DTIE Regional Workshop on Implementation and Enforcement of National ODS Licensing Systems, Budapest, 15–17 May 2000, 8; pathways of trade from Julian Newman and Steve Trent, *A Crime Against Nature: The Worldwide Illegal Trade in Ozone-Depleting Substances* (London: EIA, November 1998); roughly half from UNEP, op. cit. note 61, 12; domestic import laws from Clapp, op. cit. note 63, 263, 265.

65. Recycling loophole from Clapp, op. cit. note 63, 265; chemical analysis from Brack, op. cit. note 64, 9; profits from re-sale from David Spurgeon, "Ozone Treaty 'Must Tackle CFC Smuggling'," *Nature*, 18 September 1997.

66. Scale and value of U.S. trade from Duncan Brack, "The Growth and Control of Illegal Trade in Ozone-Depleting Substances," paper presented at the 1997 Taipei International Conference on Ozone Layer Protection, 9–10 December 1997, 6; second most valuable from Ozone Action, *Deadly Complacency: US CFC Production, the Black Market, and Ozone Depletion* (Washington, DC: September 1995); crackdown and shift from "The Ozone Racket," *The Economist*, 8 February 1997; reported seizures from ICF Incorporated, "Report on the Supply and Demand of CFC-12 in the United States, 1999," prepared for the Stratospheric Protection Division, Office of Air and Radiation, U.S. Environmental Protection Agency, 9 June 1999, 6.

67. ICF Incorporated, op. cit. note 66, 3–5, 9–11.

68. Gavin Hayman, Julian Newman, and Steve Trent, *Chilling Facts About a Burning Issue* (London: EIA, 1997); abundant supplies from European Chemical Industry Council (CEFIC), European Fluorocarbon Technical Committee, "EU's First Major Victory in CFC Fraud—Only the Tip of the Iceberg," press release (Brussels: 30 July 1997); low prices from Jitendra Verma, "It Comes and It Goes," *Down to Earth*, 31 October 1997; three fifths an Elf Atochem estimate, cited in "China Emerging as Source of Illicit CFCs," *Chemical Business Newsbase*, 22 August 1997.

69. Poor organization and consumer perceptions from Verma, op. cit. note 68; Central and Eastern Europe from UNEP, "Conclusions and Recommendations from the Regional Workshop on Implementation and Enforcement of Ozone-Depleting Substances Licensing Systems for the Central, Eastern European and Baltic States, Budapest, 15–17 May 2000," prepared for the 20th meeting of the Open-Ended Working Group of the Parties to the Montreal Protocol on Substances that Deplete the Ozone Layer, Geneva, 11–13 July 2000, and from "Illegal ODS Trade Addressed in Central Europe," *Global Environmental Change Report*, vol. 12, no. 10 (2000), 4.

70. Rise in halons from UNEP, "Statement by Shafqat Kakakhel, UNEP Deputy Executive Director, on Behalf of the Executive Director of UNEP at the UNEP Workshop on Enforcement and Compliance with Multilateral Environmental Agreements," Geneva, 12 July 1999; uses from EIA, "New Evidence of Illegal Trade in Ozone-Depleting CFCs and Halons From China," press release (London: November 1998); 40–60 times from Newman and Trent, op. cit. note 64, 1; rise in bromine concentrations from UNEP and Ozone Secretariat, op. cit. note 59, 11; "Atmospheric Rise in Halons Could Delay Recovery of Earth's Ozone Layer, NOAA Says," *International Environmental*

Reporter, 4 March 1998.

71. Halon phaseout schedule from Ozone Secretariat, "Summary of Control Measures," op. cit. note 58; China and recycling fraud from Newman and Trent, op. cit. note 64, 8–10; Chinese production from UNEP, *Data Report, 1986–1998*, op. cit. note 60, 13.

72. Operation Cool Breeze from James Blair, "Newest Item on Black Market: Refrigerator Coolants," *Christian Science Monitor*, 21 January 1999; Frio Tejas from U.S. Customs Service, "U.S. Customs Agents Receive Award for Investigating Smuggling of Chloroflourocarbons and Protecting the Earth's Ozone Layer," press release (Washington, DC: 18 November 1997); U.S Department of Justice, Environment and Natural Resources Division, *Fiscal Year 1999: Summary of Litigation Accomplishments* (Washington, DC: November 1999).

73. European regional body from Molly O'Meara, "Europe's Illegal Trade in Ozone Altering Substances," *World Watch*, January/February 1998; lack of data and interest from Brack, op. cit. note 65; member state differences from "Methyl Bromide, HCFCs to Dominate Montreal Meeting," *Global Environmental Change Reporter*, 12 September 1997.

74. Crime ring from CEFIC, op. cit. note 68; "Regulation (EC) No. 2037/2000 of the European Parliament and of the Council of 29 June 2000 on Substances that Deplete the Ozone Layer," *Official Journal of the European Communities*, 29 September 2000.

75. UNEP, "Licensing Agreement on Trade in Ozone Depleting Substances Enters Into Force," press release (Nairobi: 23 August 1999).

76. Potential black market explosion from "CFC Smuggling Could Mar Ozone Protection Efforts," *Hindustan Times Online*, 16 September 1999; exports to developing world from Ozone Secretariat, "In Brief: ODS Being Dumped in Africa," *OzonAction*, April 1999.

77. "Communiqué," op. cit. note 6; UNEP, op. cit. note 6; UNEP, "Report of the Working Group of Experts on Compliance and Enforcement of Environmental Conventions—Preparatory Session, Geneva, 13–15 December 1999" (Nairobi: 16 December 1999).

78. Contributions from Secretariat for the Multilateral Fund for the Implementation of the Montreal Protocol, "General Information," <www.unmfs.org/general.htm>, viewed 23 October 2000; projects from UNEP, "Report of the Twentieth Meeting of the Sub-Committee on Project Review," prepared for the Thirty First Meeting of the Executive Committee of the Multilateral Fund for the Implementation of the Montreal Protocol, Geneva, 5–7 July 2000; allocations as of 2000 from Global Environment Facility (GEF), "GEF-2 Pledges and Funding Situation," prepared for the Meeting of the Third Replenishment of the GEF Trust Fund, Washington, DC, 30 October 2000.

79. World Customs Organization efforts from "Work on the Harmonized System Amendments is Underway," *WCO News*, June 1998; INTERPOL from International Criminal Police Organization–INTERPOL, "Initiatives to Combat Environmental Crime," information sheet (Lyon, France: undated), and from Søren Klem, "Environmental Crime and the Role of ICPO-INTERPOL," in *Proceedings of the Third International Conference on Environmental Compliance and Enforcement, Oaxaca, Mexico, April 1998*, Vol. 1 (Washington, DC: INECE, 1994), 335–41; cooperation with secretariats from UNEP, op. cit. note 5, 6.

80. Hunter, Salzman, and Zaelke, op. cit. note 8, 481; Glenn M. Wiser, "Compliance Systems Under Multilateral Agreements, A Survey for the Benefit of Kyoto Protocol Policy Makers," Center for International Environmental Law, Washington, DC, October 1999, 30–36.

81. Discussion of Montreal, Basel, and CITES trade provisions from UNEP, *Policy Effectiveness and Multilateral Environmental Agreements*, Environment and Trade Series No. 17 (Geneva: Economics, Trade, and Environment Unit, 1998); CITES from J.M. Hutton, "Who Knows

Best? Controversy Over Unilateral Stricter Domestic Measures," in Hutton and Dickson, op. cit. note 18, 57–66; China, Italy, and Thailand from Sand, op. cit. note 19, 21; Taiwan from "Taipei to Meet CITES Requirements," *Taipei CAN*, 21 April 1994; Greece from TRAFFIC International, "Wildlife Ban Called for Against Greece," press release (Cambridge, U.K.: 20 August 1998); success of sanctions from Harold K. Jacobsen and Edith Brown Weiss, "Assessing the Record and Designing Strategies to Engage Countries, in Brown Weiss and Jacobsen, op. cit. note 4, 528.

82. "Illegal Fishing: Trade Measures May Dominate FAO Meeting," *U.N. Wire*, 19 September 2000; ICCAT, "Resolution Concerning the Unreported and Unregulated Catches of Tuna by Large-Scale Longline Vessels in the Convention Area," adopted at the 11th Special Meeting of ICCAT, Santiago de Compostela, Spain, November 1998.

83. Steve Charnovitz, "Encouraging Environmental Cooperation Through the Pelly Amendment," *Journal of Environment and Development*, winter 1994, 3–28; "US, Italy Cooperate to End Driftnetting in Mediterranean," *Environment News Service*, 16 July 1999.

84. Equity concerns from Anil Agarwal, Sunita Narain, and Anju Sharma, *Green Politics* (New Delhi: Centre for Science and the Environment, 1999), 9–10; George Mwangi, "Halt Aid to India Tiger Programs, Report Says," *Seattle Times*, 12 April 2000.

85. On the relationship between WTO rules and environmental treaties, see Gary P. Sampson, *Trade, Environment, and The WTO: The Post-Seattle Agenda* (Washington, DC: Overseas Development Council, 2000), 81–101, and "World Trade Rules Put Eco-Treaties at Risk," *Environment News Service*, 6 October 1999.

86. For a proposal to insulate multilateral environmental agreements from challenge at the WTO, see World Wide Fund for Nature-International, Greenpeace, and Friends of the Earth, "Joint Statement on the Relationship Between the WTO and Multilateral Environmental Agreements," position statement (Gland, Amsterdam, and London: July 1996); Article 104 in Governments of Canada, Mexico, and the United States of America, "The North American Free Trade Agreement," 6 September 1992; Center for International Environmental Law, "WTO 'Supremacy Clause' in the POPs Convention," e-mail to pops-network list-serve, 7 July 1999; Aaron Cosbey and Stas Burgeil, *The Cartagena Protocol on Biosafety: An Analysis of Results* (Winnipeg, MN, Canada: IISD, 2000).

87. WTO dispute resolution procedures from Jeffrey S. Thomas and Michael A. Meyer, *The New Rules of Global Trade: A Guide to the World Trade Organization* (Scarborough, ON, Canada: Carswell Thomson Professional Publishing, 1997), 307–27; lack of binding and compulsory dispute resolution in environmental treaties from Lee Kimball, "Reflections on International Institutions for Environment and Development," background paper prepared 11 May 2000 for a LEAD International Workshop, Bellagio Study and Conference Center, Bellagio, Italy, 10–14 July 2000, 5.

88. Law of the Sea dispute resolution procedures from United Nations, DOALS, "Settlement of Disputes," <www.un.org/Depts/los/los_disp.htm>, viewed 3 August 2000, and from Chayes and Chayes, op. cit. note 4, 217–18; tribunal rulings from United Nations, International Tribunal for the Law of the Sea (ITLOS) homepage, <www.un.org/Depts/los/ITLOS/ITLOShome.htm>; *Camouco* case from ITLOS, "Tribunal Delivers Judgement in the 'Camouco' Case," press release (Hamburg: 7 February 2000).

89. "After ICC, How About a Green Court?" *Terraviva* (Inter Press Service), undated, <www.ips.org/icc/tv260603.htm>, viewed 25 February 2000; Alfred Rest, "The Indispensability of an International Court for the Environment," paper presented to Environmental Law Conference, Washington, DC, 15–17 April 1999; "EU Working Group Formed to Examine Creation of World Environment Organization,"

International Environment Reporter, 19 July 2000, 556–57; Adam Entous, "Gore Adviser Proposes Global Environmental Watchdog," *Reuters*, 15 August 2000.

90. U.N. General Assembly, "Rio Declaration on Environment and Development," Annex I of the Report of the United Nations Conference on Environment and Development (Rio de Janeiro, 3–14 June 1992), <www.un.org/documents/ga/conf151/aconf15126-1annex1.htm>, viewed 23 October 2000; Elena Petkova with Peter Veit, "Environmental Accountability Beyond the Nation-State: The Implications of the Aarhus Convention," *Environmental Governance Note* (Washington, DC: World Resources Institute, April 2000); Jan Gustav Strandenaes, "Politicians and Civil Society Meet to Discuss Environmental Transparency Progress," *Network 2002*, October 2000, 7.

91. "Brazil: New Program Fights Illegal Wildlife Trade," *U.N. Wire*, 25 August 1999; "Brazil Steps Up the Fight Against Animal Smuggling," *Reuters*, 25 October 1999; for Basel Action Network, see <www.ban.org>; for Climate Action Network, see <www.climatenetwork.org>.

92. Margot Higgins, "China Pledges Sustainable Trade in Traditional Medicine," *Environmental News Network*, 4 November 1999; Mitsubishi from Bray, op. cit. note 31, 16, 44, and from Greenpeace International, "Greenpeace Calls on Mitsubishi to Stop Buying Illegal Fish," press release (Amsterdam: 17 December 1999).

Chapter 10. Accelerating the Shift to Sustainability

1. Everett Rogers, *Diffusion of Innovations*, 4th ed. (New York: The Free Press, 1995), 7–8.

2. Ibid.

3. U.S. Agency for International Development, "Central American Program Mitch Special Objective: Improved Regional Capacity to Mitigate Transnational Effects of Disasters," <hurricane.info.usaid.gov/spoca.html>, viewed 28 October 2000.

4. Strategic thinking from Paul Ehrlich, *Human Natures* (Washington, DC: Island Press, 2000), 327.

5. Ibid.

6. Cultural change from ibid., 63.

7. Agricultural revolution from Clive Ponting, *A Green History of the World* (New York: Penguin Group, 1991), 37–67; adoption of technology from Norman Myers, "Making Sense, Making Money," *Nature*, 4 November 1999, 14.

8. Energy sources from John Goyder, *Technology and Society: A Canadian Perspective* (Peterborough, ON, Canada: Broadview Press, 1997), 15; diversity of raw materials from U.S. Congress, Office of Technology Assessment, *Green Products by Design: Choices for a Cleaner Environment* (Washington, DC: U.S. Government Printing Office, September 1992), 26; journals from Rudi Volti, *Society and Technological Change* (New York: St. Martin's Press, 1995), 258.

9. Rogers, op. cit. note 1, 22–23; computers from Matt Richtel, "Signs of Market Saturation in PC World," *New York Times*, 9 October 2000; Figure 10–1 from European Wind Energy Association (EWEA), Forum for Energy and Development, and Greenpeace International, *Wind Force 10: A Blueprint to Achieve 10% of the World's Electricity from Wind Power by 2020* (London: 1999), 42.

10. Rogers, op. cit. note 1, 263–66.

11. L.N. Plummer and E. Busenberg, "Chlorofluorocarbons," on U.S. Geological Survey Web site, <water.usgs.gov/lab/cfc/background/chapter>, viewed 27 October 2000.

12. Gerald T. Gardner and Paul C. Stern, *Environmental Problems and Human Behavior* (Boston: Allyn and Bacon, 1996), 27–29.

13. Ibid., 27; Hammurabi from Sandra Postel, *Pillar of Sand* (New York: W.W. Norton & Company, 1999), 22.

14. Mary E. Clark, *Ariadne's Thread* (New York: St. Martin's Press, 1989), 234–36.

15. Gardner and Stern, op. cit. note 12, 125–27.

16. Wind power in 1990s from Christopher Flavin, "Wind Power Booms," in Lester R. Brown, Michael Renner, and Brian Halweil, *Vital Signs 2000* (New York: W.W. Norton & Company, 2000), 57; projection for 2020 from EWEA, Forum for Energy and Development, and Greenpeace International, op. cit. note 9, 3.

17. Market and Opinion Research International, "Citizens Worldwide Want Tough Environmental Action Now," press release (London: April 1998).

18. Paul H. Ray and Sherry Ruth Anderson, *The Cultural Creatives* (New York: Harmony Books, February 2000).

19. Study cited in Doug McKenzie-Mohr and William Smith, *Fostering Sustainable Behavior* (Gabriola Island, BC, Canada: New Society Publishers, 1999), 9–11.

20. Gardner and Stern, op. cit. note 12, 239–43; effectiveness of fear from Doug McKenzie-Mohr, discussion with author, 20 September 2000.

21. McKenzie-Mohr and Smith, op. cit. note 19.

22. E. Aronson and M. Leary, "Conserving Energy and Water at University of California at Santa Cruz with the use of Norms and Prompts," summary at Fostering Sustainable Behavior, <www.cbsm.com/case>, viewed 14 July 2000.

23. McKenzie-Mohr and Smith, op. cit. note 19, 48–49; Canadian Cancer Society from ibid., 47–48.

24. Gandhi quote from <www.motivationalquotes.com/pages/volunteer-quotes.html>, viewed 1 November 2000.

25. See McKenzie-Mohr and Smith, op. cit. note 19.

26. Ibid., 101–02; John Noble Wilford, "Ages-Old Icecap at North Pole Is Now Liquid, Scientists Find," *New York Times*, 19 August 2000.

27. McKenzie-Mohr, op. cit. note 19, 104.

28. Johns Hopkins study from Lester M. Salamon and Helmut K. Anheier and Associates, *The Emerging Sector Revisited: A Summary* (Baltimore, MD: Center for Civil Society Studies, The Johns Hopkins University, 1998), 4; Europe from Curtis Runyan, "NGOs Proliferate Worldwide," in Lester R. Brown, Michael Renner, and Brian Halweil, *Vital Signs 1999* (New York: W.W. Norton & Company, 1999), 144; international NGOs from Union of International Associations, *Yearbook of International Organizations* (Munich: K.G. Saur Verlag, various years).

29. Salamon and Anheier and Associates, op. cit. note 28, 1–2.

30. Craig Warentin and Karen Mingst, "International Institutions, the State, and Global Civil Society in the Age of the World Wide Web," *Global Governance*, June 2000, 241–42.

31. Ibid., 246–47.

32. Ibid., 245–50.

33. Ibid., 244, 249–50.

34. Hartman and New Hope Industry, "The Evolving Organic Market Place" (executive summary) (Rochester, NY: fall 1997), 2; growth of organics in United States from Organic Trade Association, "Business Facts," at <www.ota.com>, viewed 28 October 2000; tuna and unwillingness to pay premiums from Forest L. Reinhardt, *Down to Earth: Applying Business Principles to Environmental Manage-*

ment (Boston: Harvard Business School Press, 2000), 33.

35. 3M Company from John Ehrenfeld, "Sustainability and Enterprise: An Inside View of the Corporation," in Bette K. Fishbein, John Ehrenfeld, and John Young, *Extended Producer Responsibility* (New York: INFORM, 2000), 216; Interface carpet from <www.interfaceinc.com/us/company/ourwar>, viewed 26 October 2000.

36. Asahi and reasons for growth of zero-waste production in Japan from "Zero-Waste Factories: A Win-Win Situation for Manufacturers," *Trends in Japan*, <jin.jcic.or.jp/trends98/honbun/ ntj990330.html>; zero-waste survey from *Nikkei Shimbun*, 24 December 1999.

37. DuPont from Forest L. Reinhardt, "Market Failure and the Environmental Policies of Firms: Economic Rationales for 'Beyond Compliance' Behavior," *Journal of Industrial Ecology*, winter 1999, 12.

38. Rechargeable Battery Recycling Corporation (RBRC), *Measures of Success: A Progress Report On Recycling Nickel Cadmium Rechargeable Batteries in the U.S. and Canada* (Atlanta, GA: 2000); inconsistent laws from Bette K. Fishbein, "EPR: What Does It Mean? Where Is It Headed?" *P2: Pollution Prevention Review*, October 1998, 49; recycling rate from Bette K. Fishbein, "The EPR Policy Challenge for the United States," in Fishbein, Ehrenfeld, and Young, op. cit. note 35, 88; Figure 10–2 from ibid. (for 1993 and 1994) and from RBRC, op. cit. this note, 4.

39. Company examples from Allen White, Mark Stoughton, and Linda Feng, *Servicing: The Quiet Transition to Extended Producer Responsibility*, a report for the U.S. Environmental Protection Agency, Office of Solid Waste (Boston: Tellus Institute, May 1999), 80, 87, 71; motivations from John R. Ehrenfeld, "Sustainability and Enterprise: An Inside View of the Corporation," in Fishbein, Ehrenfeld, and Young, op. cit. note 35, 81.

40. Forest Reinhardt, "Global Climate Change and BP Amoco," Harvard Business School reprint, 7 April 2000, 7.

41. Ibid., 8, 9.

42. Ibid., 10–11, 13.

43. Greg Bourne, BP Amoco Regional Director for Australia and New Zealand, interview with Alexandra de Blas for "Earthbeat," *Radio National*, Australian Broadcasting Corporation, 19 June 1999, from transcript at <www.abc.net.au/rn/science/earth/stories/s32089.htm>; ARCO and SOLAREX from Kenny Bruno, "Summer Greenwash Award: BP Amoco's 'Plug in the Sun' Program," Corporate Watch, <www.corpwatch.org/trac/greenwash/bp.html>, viewed 27 October 2000.

44. Interface as "restorative enterprise" from <www.interfaceinc.com/us/company/sustainability/frontpage.asp>, viewed 11 October 2000.

45. Free rider from "The Producer Pays," *E: The Environmental Magazine*, May-June 1997; battery recycling from Fishbein, "The EPR Policy Challenge," op. cit. note 38, 87.

46. AIDS infection rates in 1990 from Jon Jeter, "Death Watch: South Africa's Advances Jeopardized by AIDS," *Washington Post*, 6 July 2000; Figure 10–3 and 1999 rates of infection for adults from the Joint United Nations Program on HIV/AIDS (UNAIDS) and the World Health Organization (WHO), "Epidemiological Fact Sheet: Thailand, 2000 Update" (Geneva: UNAIDS, 2000), 3–4, with updates from Ying-Ru Lo, Thai Ministry of Public Health, e-mail to author, 8 November 2000, and South Africa from UNAIDS, <www.unaids.org/epidemic_update/report/epi_ext/sld003.htm>, viewed 8 November 2000.

47. Initial reaction from World Bank, "Overcoming Political Impediments to Effective AIDS Policy," at <www.worldbank.org/aids-econ/confront/confrontfull/chapter5/chp5sub4.html>; 44 percent from Sidney B. Westley, "Country Finally Makes Headway in HIV/AIDS Struggle," *Washington Times*, 27

August 1999; budget from World Bank, op. cit. this note.

48. Condoms from Robert Hanenberg and Wiwat Rojanapithayakorn, "Prevention as Policy: How Thailand Reduced STD and HIV Transmission," *AIDScaptions*, May 1996, 26; premarital sex from Westley, op. cit. note 47; low incidence from Jessica Berman, "Thailand Attacks AIDS with Two-Pronged Approach," *Lancet*, vol. 353, issue 9164 (1999), 1598–1600.

49. Action plan from Catherine Hankins, "HIV: Evolution of a Pandemic," *Canadian Medical Association Journal*, 1 December 1995, 1613–16; all ministries from Pakdee Pothisiri et al., "Funding Priorities for HIV/AIDS Crisis in Thailand," paper delivered at 12th World AIDS Conference, Geneva, June 1998, at <www.worldbank.org/aids-econ/thaifund.htm>, viewed 27 October 2000.

50. Tourism from Hankins, op. cit. note 49; even enforcement from Hanenberg and Rojanapithayakorn, op. cit. note 48.

51. Dr. Ying-Ru Lo, WHO Medical Officer, Thai Ministry of Public Health, e-mail to David Ruppert, Worldwatch Institute, 9 October 2000; clinics from Hanenberg and Rojanapithayakorn, op. cit. note 48.

52. Jon Jeter, "Free of Apartheid, Divided by Disease," *Washington Post*, 6 July 2000; prevalence from Lawrence K. Altman, "U.N. Warning AIDS Imperils Africa's Youth," *New York Times*, 28 June 2000.

53. Health and military from Lisa Garbus, "South Africa," HIV InSite web page, <hiv insite.ucsf.edu/international/africa/2098.410f.html>, viewed 1 November 2000.

54. Curtis Moore and Jack Ihle, *Renewable Energy Policy Outside the United States*, Issue Brief No. 14 (Washington, DC: Renewable Energy Policy Project, October 1999); current share from J.B. Petersen, director, Alpha Wind Energy, Denmark, e-mail to David Ruppert, Worldwatch Institute, 7 November 2000.

55. Danish share of global wind turbine exports and Energy 21 from Louise Guey-Lee, "Wind Energy Developments: Incentives in Selected Countries," in Energy Information Administration (EIA), U.S. Department of Energy (DOE), *Renewable Energy Issues and Trends 1998* (Washington, DC: March 1999), 84; near doubling of Denmark's energy generating capacity is based on ibid., and on American Wind Energy Association, "Global Wind Energy Market Report," <www.awea.org/faq/global99.html>, viewed 30 October 2000; EIA, DOE, <www.eia.doe.gov/cneaf/solar.renewables/rea_issues/windart.html>, viewed 11 October 2000.

56. Moore and Ihle, op. cit. note 54.

57. Ibid.

58. N. Eldridge and S.J. Gould, "Punctuated Equilibria: An Alternative to Phyletic Gradualism," in T.J.M. Schopf, ed., *Models in Paleobiology* (San Francisco, CA: Freeman, Cooper and Company, 1972); fall of communism from Paul Hollander, *Political Will and Personal Belief: The Decline and Fall of Soviet Communism* (New Haven, CT: Yale University Press, 1999).

59. Andrew C. Revkin, "Scientists Now Acknowledge Role of Humans in Climate Change," *New York Times*, 26 October 2000.

60. Ehrlich, op. cit. note 4, 327.

61. Ivor Gaber, "The Greening of the Public, Politics, and the Press: 1985–1999," in Joe Smith, ed., *The Daily Globe: Environmental Change, the Public, and the Media* (London: Earthscan, 2000), 117, 126.

62. Ed Ayres, *God's Last Offer: Negotiating for a Sustainable Future* (New York: Four Walls Eight Windows, 1999), 171.

Index